FRACTURE MECHANICS OF POLYMERS

ELLIS HORWOOD SERIES IN MECHANICAL ENGINEERING

STRENGTH OF MATERIALS
J. M. ALEXANDER, University College, Swansea
TECHNOLOGY OF ENGINEERING MANUFACTURE
J. M. ALEXANDER, University College, Swansea, G. W. ROWE and R. C. BREWER, Imperial College of Science and Technology
VIBRATION ANALYSIS AND CONTROL SYSTEM DYNAMICS
C. BEARDS, Imperial College of Science and Technology
STRUCTURAL VIBRATION ANALYSIS
C. BEARDS, Imperial College of Science and Technology
COMPUTER AIDED DESIGN AND MANUFACTURE 2nd Edition
C. B. BESANT, Imperial College of Science and Technology
BASIC LUBRICATION THEORY 3rd Edition
A. CAMERON, Imperial College of Science and Technology
SOUND AND SOURCES OF SOUND
A. P. DOWLING and J. E. FFOWCS-WILLIAMS, University of Cambridge
MECHANICAL FOUNDATIONS OF ENGINEERING SCIENCE
H. G. EDMUNDS, University of Exeter
ADVANCED MECHANICS OF MATERIALS 2nd Edition
Sir HUGH FORD, F.R.S., Imperial College of Science and Technology, and
J. M. ALEXANDER, University College of Swansea.
MECHANICAL FOUNDATIONS OF ENGINEERING SCIENCE
H. G. EDMUNDS, Professor of Engineering Science, University of Exeter
ELASTICITY AND PLASTICITY IN ENGINEERING
Sir HUGH FORD, F.R.S. and R. T. FENNER, Imperial College of Science and Technology
TECHNIQUES OF FINITE ELEMENTS
BRUCE M. IRONS, University of Calgary, and S. AHMAD, Bangladesh University, Dacca
FINITE ELEMENT PRIMER
BRUCE IRONS and N. SHRIVE, University of Calgary
CONTROL OF FLUID POWER: ANALYSIS AND DESIGN 2nd (Revised) Edition
D. McCLOY, Ulster Polytechnic, N. Ireland and H. R. MARTIN, University of Waterloo, Ontario, Canada
UNSTEADY FLUID FLOW
R. PARKER, University College, Swansea
DYNAMICS OF MECHANICAL SYSTEMS 2nd Edition
J. M. PRENTIS, University of Cambridge
ENERGY METHODS IN VIBRATION ANALYSIS
T. H. RICHARDS, University of Aston Birmingham
ENERGY METHODS IN STRESS ANALYSIS: With Intro. to Finite Element Techniques
T. H. RICHARDS, University of Aston in Birmingham
COMPUTATIONAL METHODS IN STRUCTURAL AND CONTINUUM MECHANICS
C. T. F. ROSS, Portsmouth Polytechnic
FINITE ELEMENT PROGRAMS FOR AXISYMMETRIC PROBLEMS IN ENGINEERING
C. T. F. ROSS, Portsmouth Polytechnic
ENGINEERING DESIGN FOR PERFORMANCE
K. SHERWIN, Liverpool University
ROBOTS AND TELECHIRS
M. W. THRING, Queen Mary College, University of London

FRACTURE MECHANICS OF POLYMERS

J. G. WILLIAMS, D.Sc.(Eng), Ph.D., FEng, FPRI, FIMechE
Professor of Polymer Engineering
Imperial College of Science and Technology
University of London

ELLIS HORWOOD LIMITED
Publishers · Chichester

Halsted Press: a division of
JOHN WILEY & SONS
New York · Chichester · Brisbane · Toronto

First published in 1984 by

ELLIS HORWOOD LIMITED
Market Cross House, Cooper Street, Chichester, West Sussex, PO19 1EB, England

The publisher's colophon is reproduced from James Gillison's drawing of the ancient Market Cross, Chichester.

Distributors:

Australia, New Zealand, South-east Asia:
Jacaranda-Wiley Ltd., Jacaranda Press,
JOHN WILEY & SONS INC.,
G.P.O. Box 859, Brisbane, Queensland 40001, Australia

Canada:
JOHN WILEY & SONS CANADA LIMITED
22 Worcester Road, Rexdale, Ontario, Canada.

Europe, Africa:
JOHN WILEY & SONS LIMITED
Baffins Lane, Chichester, West Sussex, England.

North and South America and the rest of the world:
Halsted Press: a division of
JOHN WILEY & SONS
605 Third Avenue, New York, N.Y. 10016, U.S.A.

British Library Cataloguing in Publication Data
Williams, J.G.
Fracture mechanics of polymers. –
(Ellis Horwood series in mechanical engineering.)
1. Polymers and polymerization – Fracture
I. Title
620.1'92426 TA455.P58

Library of Congress Card No. 83-26593

ISBN 0-85312-685-2 (Ellis Horwood Limited)
ISBN 0-470-20013-8 (Halsted Press)

Printed in Great Britain by Butler & Tanner, Frome, Somerset.

Contents

Acknowledgements

This text is based on a programme of research in the Department of Mechanical Engineering at Imperial College over the past fifteen years and as such draws heavily on the efforts of many research students, research assistants and colleagues. I am deeply indebted to all of them and I hope they recognise due acknowledgement in the references. I must, however, make particular mention of Professor C. E. Turner and Dr. D. P. Isherwood, who have provided a sympathetic ear for trying out many of the arguments given here. In addition I acknowledge the very large contribution made by Dr. G. P. Marshall during his time working with us. His outstanding experimental skills contributed greatly to our early efforts to understand fracture. Much of the incentive for this work has come from the research group of Mr. J. M. Cann and Dr. M. J. Cawood at BP Chemicals. Their willingness to share their experience and provide support has been invaluable. The monumental task of typing the original manuscript was performed with great skill by Mrs. E. A. Hall. She has been of inestimable value in trying to eliminate errors in the final version. Some are inevitable in a text of this length and I apologise for them in advance.

Professor J. G. Williams
London, March 1983

List of main symbols

The following list is for those symbols used throughout the text to define the major parameters. Not all the specific uses are listed since their definition will be apparent from the text. Some symbols are used for more than one purpose and the main cases are listed. Such usage is undesirable but it has been considered to be preferable to adhere to common usage wherever possible (e.g. ρ for radius of curvature and density). The usage should be clear from the context and definitions given in the text. Symbols used infrequently or for special purposes are not listed but will be defined in the text. Symbols whose definition is of major importance are given with text references.

a	crack length, major ellipse semi-axis.
A	crack area.
b	minor ellipse semi-axis.
B	thickness of sheet.
B_c	thickness of crack.
c	line zone length, wave velocity.
C	compliance, wave velocity, specific heat.
d	dilatations component as a subscript.
D	width or depth of specimen.
e	strain, coefficient of restitution.
E	Young's Modulus, complete elliptic integrals.
G	energy release rate per unit crack area [23].
G_c	value of G at fracture [26].
J	rate of change of energy with crack area [96].
J_c	value of J at fracture [96].
k	stiffness, thermal conductivity.
K	stress intensity factor [44].
$K_{\mathrm{I,II,III}}$	K for crack opening modes I, II and III respectively.
K_c	value of K at fracture.
K_{c1}	plane strain value of K_c [98].
K_{c2}	plane stress value of K_c [99].

K_B	apparent K for a blunt crack.
l_0	critical characteristic length [51, 63].
m	mass, power law index.
M	mass, modulus, bending moment, constraint factor.
n	power law index, normal when a subscript.
p	stress.
P	load.
r	zone size, radial coordinate.
R	fracture resistance, stress ratio in fatigue.
s	tangential or shear as a subscript.
S	contour length.
t	time.
T	temperature.
u	displacement, displacement in direction x.
$\dot{u}$	velocity in direction x.
U	energy [21].
U_1	external work.
U_2	irreversible work.
U_3	stored (potential) energy.
U_4	kinetic energy.
v	crack face displacement, displacement in direction y.
$\dot{v}$	velocity in direction y.
V	volume, velocity.
w_p	energy dissipated per unit volume [262].
W	energy per unit volume.
W_0	energy per unit volume remote from crack [37].
x, y, z	cartesian coordinates.
Y^2	calibration factor [63].

Greek Symbols

α	fatigue model stress ratio [216], dynamic crack speed factor [286].
β	dynamic factor [280].
γ	surface energy or tension.
Γ	Gamma function.
δ	crack opening displacement, ($2v$) at crack tip [74].
δ_c	value of δ at fracture [97].
ε	bi-material parameter [57].
η	viscosity.
η_p	plastic compliance factor [111].
θ	angle.

λ	extension ratio.
ν	Poisson's ratio.
ρ	density, crack tip radius of curvature.
τ	time variable.
$\tau_{0,1,2}$	characteristic times [113].
ϕ	energy calibration factor [68], potential function.
ψ	potential function.
ω	frequency, angular velocity.

1

Introduction

Polymers in various forms are used widely in engineering. Thermoplastics usually include various stabilisers, fillers and toughening agents, so that a wide range of properties are encountered. Thermosetting resins are used alone, but more frequently as the matrices of composites and rubbers, both natural and synthetic, are employed extensively in engineering in springs and damping mechanisms. In all cases, the occurrence of fracture is a major concern and this is reflected in the large number of tests which are used to simulate the various conditions which promote failure. The use of such tests as standards and quality controls is designed to avoid failures in practice and quite reasonable success is achieved in this regard. The tests are designed to be easy to perform, and yet to include enough of the circumstances of a real situation to give a realistic measure of how the material will perform in practice. The Izod test, for example, simulates impact conditions by loading a notched bar at a high rate in bending, while various environmental stress cracking tests load specimens in the presence of liquids so that cracking may be observed. Each test is designed to meet a limited set of goals and it is frequently observed that there is little correlation between good and bad performance in the different tests. The outcome of all this is that large numbers of tests are performed to give a picture of a material, each contributing a separate piece of information and this may only be drawn together by those with great experience.

This picture is not, of course, confined to polymers, and indeed many of the tests used are taken from those designed for other materials. The pursuit of an understanding of the phenomena of fracture by the exploration of many variables inevitably leads to this proliferation of testing and schemes for systematising large volumes of information. What is lacking in such an approach is a basic idea to provide a logical framework in which to view the problem. There may well be many of these but one scheme which has been developed in recent years is embodied in the term Fracture Mechanics. It is this which will be developed in this text, and it is intended to show that, while it is neither perfect nor unique, it does provide a rationale for the phenomenon of fracture, and as such simplifies and gives order to the subject.

The ideas on which the approach is based were first used by A. A. Griffith [1], who was concerned with the problem that small glass objects generally exhibited higher strengths than large ones. He proposed that every body contains a distribution of imperfections or flaws and that failure occurs at the largest of these. Larger bodies have a greater likelihood of containing bigger flaws and will thus fail at lower stresses. He further developed a relationship between the stress at failure and the flaw size by noting that when a flaw grew in a stressed body, there was a *decrease* in the potential energy of the body, and so energy was released to be available to form the new surfaces of the growing flaw. The energy necessary to do this was taken as the true surface energy of the solid, γ, and the analysis gave the result:

$$p^2 \pi a = E\, 2\gamma \tag{1.1}$$

where p is the failure stress, a the flaw size, and E the Young's modulus. This form of relationship is the basis of everything which follows in this text. It embodies the scheme for characterising a material via the term $E\,2\gamma$ by measuring p for known a and then employing a design method in which either p or a may be determined if one is known. It also embodies the very important point that most scatter in fracture stress measurements arises from the variations in a and not in γ, the material property. As with many important basic studies, however, the nature of the case studied and perhaps some misreading of the original text has led to some confusion, as well as contributing the important basic ideas. In the geometry considered, fracture and instability coincided and, although Griffith discussed the matter in great detail, subsequent work has tended to confuse the two. In addition, the material considered, inorganic glass, gave almost perfectly brittle cracks in that the energy necessary to create new crack surfaces is the surface energy. For most materials, this is not so since the stresses induced at the tip of the flaw cause large local deformations and flow, which result in much more energy being dissipated. For the perfectly brittle case, it was rigorously true that an elastic analysis could be performed to determine the energy release rate and that this could be equated to the quite separate surface work energy. When some of the dissipated energy arose from deformation, however, was this still true? The statement that such a calculation could be performed, providing that the dissipation was confined to a small zone at the flaw tip was first made independently by both Irwin [2, 3] and Orowan [4] when considering the failure of steels. This meant that equation (1.1) could still be used but with 2γ replaced now with G_c (after Griffith), where G_c can now include plastic work as well as surface work.

The question which still remains here, of course, is how small is small? This remains a fundamental problem which occurs over and over again in this text and much effort is required to establish validity criteria for the various energy balances used. The analysis used rests on the same basis as that of Griffith but with some adjustment to allow for these more difficult cases to be included. The two basic assumptions used are as follows:

1. All materials contain flaws, cracks or inhomogeneities which propagate to cause failure.
2. A material's fracture strength may be characterised as the energy per unit area required to produce new surfaces.

The justification of these assumptions lies solely in whether they lead to conclusions in accordance with observations. This text contains no discussion whatever on mechanisms which may lead to the generation of flaws, etc., in an otherwise perfect body. They may occur but it is the author's opinion that the assumption that flaws exist and that observed phenomena arise from their growth is a simpler, and thus more likely explanation; an opinion based on a profound belief in Occam's Razor (hypotheses are not to be multiplied without necessity). It should be noted that incubation periods when flaws do not grow arise, and can be predicted quite often and may lead to the conclusion that their subsequent growth, when they are more easily observed, is a consequence of spontaneous generation. There are no authenticated examples of such a process in the published literature.

Much of the text will be taken up in describing ways of computing the energy per unit area to produce new surfaces. The case, used by Griffith, when the body may be treated as elastic so that an energy release rate may be found and thus equated to the energy absorbed forms the basis of Chapters 2, 3 and 4, and is the backbone of the whole subject. The extension of the technique to include larger scale dissipation using the J-integral is given in Chapter 5, where it is apparent that the problem is intrinsically more difficult. Indeed, the guiding principle in much of that chapter is to explore methods of analysis which yield practically useful solutions with reasonable amounts of effort.

The emphasis of much of the text is clearly on the use of fracture mechanics to characterise materials and much less is said about the use of such information in design. The experimental methods used generally employ quite large flaws or cracks and it is easy to use the data to predict what will happen when similarly large defects are present (usually by accident). When the flaws present are small, it is less so and, although some examples are given of the failure of such cases, information is scarce and will require attention in the future. The analysis given for the characterising rôle is, of course, equally applicable to design and the basis for such a development of the method is thus provided.

Some comments on definitions are in order here, since confusion in fracture mechanics frequently arises from the use of words with a common usage in more precise technical contexts. Here, the principle is adopted of using words in accordance with their normal usage, but taking care to define the meaning more precisely for technical use. The following list will be of value in subsequent chapters.

(i) *Brittle*: In common usage; liable to break, fragile. Here it is used to describe fractures in which sharp cracks propagate through bodies with small amounts of local deformation around the tip. In general, this will lead to lower toughness than cases where there is larger deformation, but some brittle fractures involve much larger energies than others. For example, an inorganic glass and a high strength steel can both have brittle cracks but one is much tougher than the other. No use will be made here of the idea of a brittle material since it will be shown that brittle cracks can occur in all materials if the conditions are correct.

(ii) *Ductile*: In common usage; the opposite to brittle or, more precisely, the ability to be drawn into wire. Here we will refer to a ductile failure when there is large-scale deformation around the crack, often leading to observable crack tip blunting. In general, larger energies are involved but in no case will a 'ductile material' be described, only ductile behaviour. The transition from brittle to ductile is not a rigorous matter and must rely on some arbitrary definition.

(iii) *Toughness*: In common usage; not easily broken. Used with the prefix Fracture, here it will refer to the energy per unit area necessary to give a new crack surface. A tough material is one with a high value of Fracture Toughness in some range appropriate to the class of materials considered.

(iv) *Strength*: Used here to describe the stress carrying capacity of a material and reflecting both toughness and inherent flaw size.

(v) *Crack*: This is the object described by fracture mechanics and is a planar discontinuity which is not capable of transmitting a load normal to its faces. When it grows, new surface area is created, which is of fundamental importance in determining behaviour.

(vi) *Stability*: The stability of a crack depends on the interaction of the applied loads and the material toughness, and, in principle, any crack, be it brittle or ductile, can be rendered stable or unstable by suitable driving forces. It is the whole system which has the property of stability and not the crack itself. In general, any system in which crack growth leads to an increase in crack velocity may be taken as unstable. At initiation, where the crack speed is zero, this is a fairly obvious statement, resulting from any excess of energy released over and above the fracture toughness, giving rise to crack acceleration. For running cracks, the situation is less obvious but may be regarded as the statement that a crack at constant speed, at rest or decelerating, is stable. The topic will be discussed in detail in several places in the text.

Most of the results given in the text are derived from first principles in an attempt to provide a self-contained version of the theoretical basis of the subject. The exceptions to this are several elasticity solutions used in the derivation of stress intensity factors which are usually worked out using the complex stress function method. It was felt that the extra complication and length of text which would be incurred by deriving these was not warranted. Where such results are given, their sources are referenced. The level of mathematics is limited to differential calculus and should be accessible to any engineering or science graduate. It is believed that no rigour is sacrificed by the approach adopted and all approximations and assumptions are stated.

Chapters 6, 7 and 8 describe the application of the results of Chapters 2 to 5 to the fracture of polymers. Chapter 6 deals with fracture initiation and, by discussing various types of polymers, covers most of the major phenomena encountered in slow rate testing. Chapter 7 covers crack propagation and deals with visco-elasticity, environmental effects, fatigue and ductile tearing. In Chapter 8, behaviour at high rates is covered. In all three chapters, examples are given from the literature since they attempt to justify the basic assumptions by demonstrating that they do indeed describe observed behaviour. Although a summary, it is not a comprehensive review of the literature and draws heavily on the author's own published work. This is partly because it was the most familiar, but also because it was performed originally to achieve the goal of these chapters. It is hoped, however, that there are sufficient references to allow the reader to pursue any topic further if desired.

The examples given cover most aspects of polymer fracture and demonstrate that, in general, the concepts of fracture mechanics do provide a sound basis on which to describe fracture. In some areas, such as slow crack growth in glassy materials and in impact, there is a wealth of information which gives a very clear picture of behaviour. In other areas, this is less so. Environmental crack growth, while being described in principle, presents some apparently complex phenomena which are difficult to rationalise. The same may be said for fatigue. As might be expected, ductile failure presents the most difficulty There is not much information available and, while some basic ideas have been established, the overall picture, and particularly the full justification of the J analysis, is far from clear. This will be apparent from the text where a rather wide range of results and approaches are described but no strong support for any of them is available.

The text is offered as one person's attempt to start at the beginning and derive all that is needed to use fracture mechanics for polymers and then to apply this to a wide range of materials and problems of practical interest. It is believed that what comes out is a useful contribution to understanding fracture in polymers in that it simplifies by providing a rational framework in which to view the phenomena.

References for Chapter 1

[1] Griffith, A. A., *Phil. Trans. R. Soc.*, **A221** (1920) 163–198.

[2] Irwin, G. R., in *Fracturing of Metals*, American Society of Metals, Cleveland (1948) 147–166.

[3] Irwin, G. R., *Fracture*, Encyclopaedia of Physics, Springer, Berlin, **6** (1958) 551–590.

[4] Orowan, E., in *Fatigue and Fracture of Metals* (MIT Symposium, June 1950), (W. M. Murray, ed.), Wiley, New York (1952) 139–167.

2

Energy in elastic systems

2.1 Introduction

This chapter will be concerned with the fracture of elastic bodies. We shall assume the presence of a sharp crack in which the energy absorption occurs very locally around the crack tip and that the behaviour of the body as a whole is controlled by elastic responses. Initially, the analysis will be conducted in terms of energy and subsequently developed in Chapter 3 using local parameters. The idea of the response of a bounded system is important here, as in conventional thermodynamics, and we will be concerned with the transfer of energy across the prescribed boundary. In principle, this approach is completely general and all forms of energy may be included, for example, thermal or electrical energy. However, the vast majority of practical problems are concerned with mechanical work, or can be represented as such, and this will be the primary form considered.

2.2 Energy balance

Let us consider a closed boundary S, as shown in Fig. 2.1, into which an

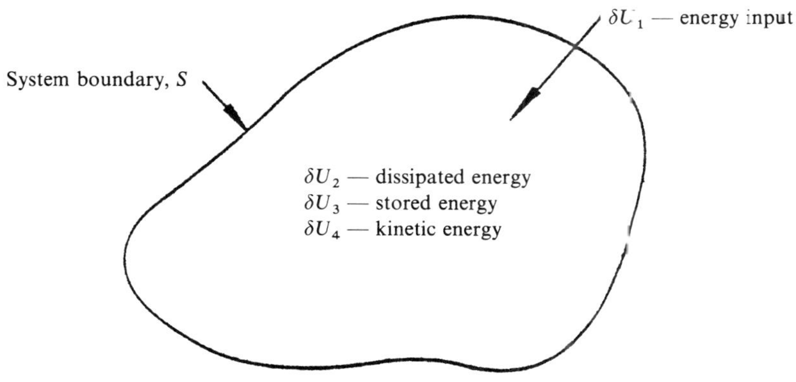

Fig. 2.1 A system bounded by S with energy changes

increment of energy δU_1 is transferred. In most fracture problems, this will consist of performing mechanical work but can include any form of energy. This energy input will now be considered to be divided into three parts:

δU_2 = the change in irreversibly dissipated energy;
δU_3 = the change in stored (potential) energy; and
δU_4 = the change in the kinetic energy of the system.

δU_2 is usually that energy which is dissipated by plastic or viscous flow and is converted to heat. In isothermal cases it is all transferred across S, while for adiabatic systems there is a temperature rise in the system. δU_3 is generally the stored elastic energy in the system but can include stored thermal energy. The inclusion of thermal effects enables situations involving thermal energy changes to be included, but unless specifically stated in this text we shall be concerned with isothermal systems in which the nett input of energy is:

$$\delta U_1 - \delta U_2$$

and the stored energy is solely elastic. Simple conservation of energy gives:

$$\delta U_1 - \delta U_2 = \delta U_3 + \delta U_4 \tag{2.1}$$

and if these increments are summed over some history, we have:

$$U_1 - U_2 = U_3 + U_4 \tag{2.2}$$

This energy balance is, of course, the basis of all mechanics analysis and it is useful to deduce some familiar results by this route in order to put fracture into perspective. If the increments of energy in equation (2.1) are taken with respect to the displacement u in some direction, say, x, then we have:

$$\frac{\delta U_1}{\delta u} - \frac{\delta U_2}{\delta u} = \frac{\delta U_3}{\delta u} + \frac{\delta U_4}{\delta u} \tag{2.3}$$

Now, an increment of energy is given by:

$$\delta U = P\,\delta u$$

where P is the force in that direction, so that equation (2.3) is simply the equilibrium of forces on the system in the direction x. In rigid body dynamics, for example, it is normally assumed that $U_3 = U_2 = 0$, so that:

$$\frac{\delta U_1}{\delta u} = \frac{\delta U_4}{\delta u}$$

If we consider a body of mass M moving at a velocity $\dot{u}$, then:

$$U_4 = \tfrac{1}{2} M\,\dot{u}^2$$

and taken in infinitesimal increments, we have:

$$\frac{dU_4}{du} = M\ddot{u}$$

Since dU_1/du is the external force P, we have:

$$P = M\ddot{u}$$

i.e. Newton's second law of motion. If the mass M is attached to a spring of stiffness k, we have:

$$U_3 = \frac{1}{2}k\,u^2 \text{ and } \frac{dU_3}{du} = k\,u$$

If there is a viscous damping element of viscosity η in parallel, then:

$$\frac{dU_2}{du} = \eta\,\dot{u}$$

Note that U_2 cannot be written unambiguously here (as can U_3 and U_4), since the evaluation of $\int \dot{u}\,du$ requires a knowledge of $\dot{u}(t)$, thus making the term history-dependent; a common feature of dissipative processes (see Section 5.7). Returning to equation (2.3), we have:

$$P - \eta\,\dot{u} = k\,u + M\ddot{u}$$

i.e.

$$M\ddot{u} + \eta\,\dot{u} + k\,u = P$$

the equation of motion for forced vibration of a viscously damped system.

2.3 Definition of G and R [1]

Let us now return to fracture where we suppose that a loaded body is in an energy state as given in equation (2.2) when a fracture increment occurs such that the area of the fracture increases by δA. Equation (2.1) now becomes:

$$\frac{\delta U_1}{\delta A} - \frac{\delta U_2}{\delta A} = \frac{\delta U_3}{\delta A} + \frac{\delta U_4}{\delta A} \tag{2.4}$$

It should be noted that the energy changes occur not because of a displacement directly, but because of displacements arising from the fracture area change δA. The term $\delta U_2/\delta A$ now has a very specific meaning in that it is the energy dissipated in propagating a fracture over an increment of area δA and is defined as the fracture resistance R. It may be a function of other variables of the system, but may be defined for a given system under a given set of conditions. i.e.

$$R = \frac{dU_2}{dA} \tag{2.5}$$

in the limit.

It is convenient now to consider the energy change in the system due to the fracture growth δA arising from the changes in external work and internal energy. This is defined as G, the energy release rate given by:

$$G=\frac{dU_1}{dA}-\frac{dU_3}{dA} \tag{2.6}$$

It must be remembered that although an overall energy balance must be maintained, there is not, as in normal deformation processes, a simple balance of external work and internal energy. This is because the usual compatibility condition for deformation is violated by fracture and results in an energy change. The occurrence of an energy release rate is not unique to fracture and can arise in any system involving discontinuous changes of boundary conditions. A simple illustration of this is given in Fig. 2.2, where a simple linearly elastic cantilever of length L and second moment of area I has an end load of P.

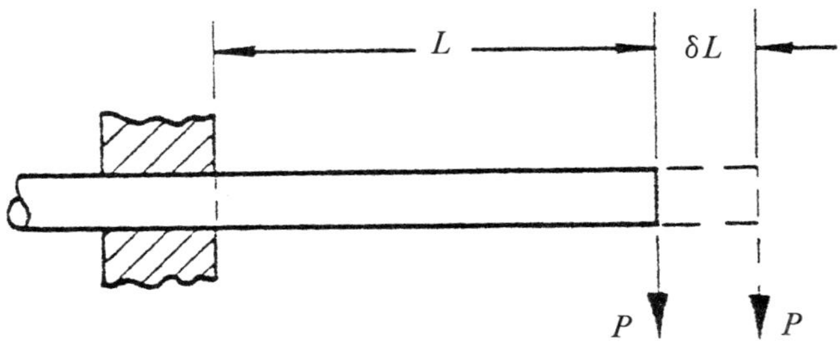

Fig. 2.2 A cantilever with a fixed load P in which the length increases

If we suppose that the length L increases by δL by simply sliding through the support, then we may compute the initial displacement:

$$u=\frac{P\,L^3}{3\,E\,I}$$

and since L increases at fixed P, the change in u is:

$$\delta u=\frac{P\,L^2\,\delta L}{E\,I}$$

Thus, the change in external work is:

$$\delta U_1=P\,\delta u=\frac{P^2\,L^2\,\delta L}{E\,I}$$

and since the stored energy is:

$$U_3=\frac{1}{2}P\,u=\frac{P^2\,L^3}{6\,E\,I}$$

then:

$$\delta U_3 = \frac{P^2\, L^2\, \delta L}{2\, E\, I}$$

and there is a change in energy, or energy release rate, given by:

$$G \equiv \frac{\delta U}{\delta L} = \frac{\delta U_1}{\delta L} - \frac{\delta U_3}{\delta L} = \frac{P^2\, L^2}{2\, E\, I}$$

The growth of the beam δL is analogous to a fracture since it is change in the boundary conditions of the problem and leads to an energy change, which, in the absence of dissipation, must be balanced by an increase in kinetic energy.

It is convenient now to rearrange equation (2.4) to give:

$$G = \frac{dU_1}{dA} - \frac{dU_3}{dA} = R + \frac{dU_4}{dA} \tag{2.7}$$

Thus, the energy release rate is balanced by R and the change in kinetic energy dU_4/dA. For fracture initiation, the body is stationary, $U_4 = 0$ and $dU_4/dA > 0$, so that for fracture to occur:

$$G - R \geqslant 0 \tag{2.8}$$

If $G > R$, then the system is unstable since dU_4/dA is positive leading to an increase in fracture velocity. If we consider a case in which G monotonically increases, then at fracture:

$$G = R$$

and the stability of the subsequent behaviour depends on G and R at $A - \delta A$, i.e. the fracture is unstable if:

$$G + \frac{dG}{dA}\delta A > R + \frac{dR}{dA}\delta A$$

and since $G = R$, we have the condition for instability of:

$$\frac{dG}{dA} > \frac{dR}{dA} \tag{2.9}$$

Note that for moving cracks where $U_4 > 0$, δU_4 can be negative and hence help drive the crack by decelerating so that the criteria become:

$$G > R + \frac{dU_4}{dA}$$

(2.10)

with:
$$\frac{dG}{dA} > \frac{dR}{dA} + \frac{d^2 U_4}{dA^2}$$

(see Section 8.7.1). It should be noted that equations (2.8) and (2.9) are the Lagrangian conditions for stable equilibrium of a minimum in potential energy.

The static criteria are usually represented on a diagram, as shown in Fig. 2.3 [2]. If we have a crack of original area A_0, the energy requirements for propagation may be represented by R as a function of A, as shown. The steep initial rise represents the resistance of the material to failure for small loads, while the tendency to a limiting value at high A is a representation of a constant energy rate requirement for large growths. The curves for G are usually drawn for constant loads as a function of A and these are shown as lines (1), (2) and (3) for increasing P in Fig. 2.3. On line (1) for a low load, $G=R$ (at B) in the very steep part of the R curve and so δA is zero. On line (2), $G=R$ at B' which corresponds to a crack length of A', thus giving some crack growth. For $A>A'$, we can see that $G<R$, i.e. $dG/dA<dR/dA$, so that the crack growth is stable. This is true up to the tangency point, after which $dG/dA>dR/dA$ and the fracture is thereafter unstable. Thus, for this system, in which the load is constant, only that part of the R curve up to B'' is accessible.

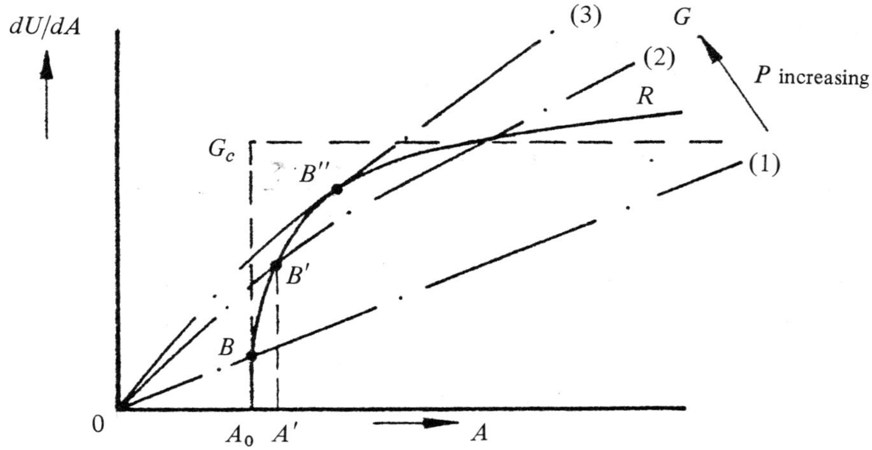

Fig. 2.3 G and R curves as functions of fracture area A

A special, and very important, case for R is shown as a broken line in Fig. 2.3, which is a limiting case of a vertical followed by a horizontal line. Here, there is no stable growth up to B'' which is, of course, the tangency point for all increasing G curves, after which there is instability so that R has a single value, usually designated G_c, for all values of A. This is a more general form of the Griffith [3] criterion for a perfectly brittle crack in which $G_c=2\gamma$, where γ is the true surface energy. Note also that in the general case, the instability point will depend on the shape of the G curve, i.e. the nature of the loading system, while G_c does not and is the same for all systems.

This is a crucially important concept since R curves define the fracture behaviour by definition and may be a function of A or system geometry. This general type of behaviour will be pursued in Chapter 5, but a greatly simplified state of affairs exists if there is a single parameter, G_c, which defines all fractures, independent of A. This implies that the energy dissipation processes are local to the crack tip and therefore unique to it. This assertion can only be justified by observation but turns out to be a very reasonable assumption for brittle cracks in elastic systems with which we are concerned here.

2.4 Evaluation of G

Irrespective of the nature of R, it is clearly important to evaluate G from the elastic characteristics of the system so that the response of that system to fracture may be deduced.

From equation (2.6), we may define G as:

$$G=\frac{dU_1}{dA}-\frac{dU_3}{dA}$$

and proceed to evaluate the two terms in more detail. Let us consider a force P on the boundary S which has a displacement u. The work done for a displacement change du arising from a fracture area change dA is given by:

$$\frac{dU_1}{dA}=P\frac{du}{dA}$$

The stored elastic energy arising from P up to the fracture point is:

$$U_3=\int_0^u P\,du$$

where we have assumed the response to be entirely elastic and that all the dissipation is a separate process embodied in R. Thus:

$$\frac{dU_3}{dA}=\frac{d}{dA}\int_0^u P\,du$$

and:

$$G=P\frac{du}{dA}-\frac{d}{dA}\int_0^u P\,du \tag{2.11}$$

In general, U_3 can be a function of u, P and A, so that:

$$\frac{dU_3}{dA}=\frac{\partial U_3}{\partial A}+\frac{\partial U_3}{\partial P}\frac{dP}{dA}+\frac{\partial U_3}{\partial u}\frac{du}{dA}$$

Now, $\partial U_3/\partial u=P$, so that equation (2.11) becomes:

$$G = -\left(\frac{\partial U_3}{\partial A} + \frac{\partial U_3}{\partial P}\frac{dP}{dA}\right) = -\frac{dU_3}{dA}\bigg|_{u\ const}$$
$$= -\frac{d}{dA}\int_0^u P\,du\bigg|_{u\ const} \tag{2.12}$$

Thus, if U_3 is evaluated, G may be determined by treating u as a constant.

Energy may also be stored independently of P because of previous history, as with residual stresses discussed in Section 2.4.4, and this must be included in U_3 when necessary.

A restriction will now be introduced in that we will consider only cracks of length a propagating in a self similar manner in a section of thickness B, as shown in Fig. 2.4, so that:

$$dA = B\,da \tag{2.13}$$

[Section 4.6 discusses situations when this is not the case]. This restricts the results to a two-dimensional problem and equation (2.12) now becomes:

$$G = \frac{1}{B}\left(P\frac{du}{da} - \frac{d}{da}\int_0^u P\,du\right) = -\frac{1}{B}\frac{dU_3}{da}\bigg|_{u\ const} \tag{2.14}$$

It is sometimes convenient to express G in terms of stresses on S and at any point we may replace P by normal and shear components $B\,dS\,p_n$ and $B\,dS\,p_s$, respectively, with displacements u_n and u_s, as shown in Fig. 2.4.

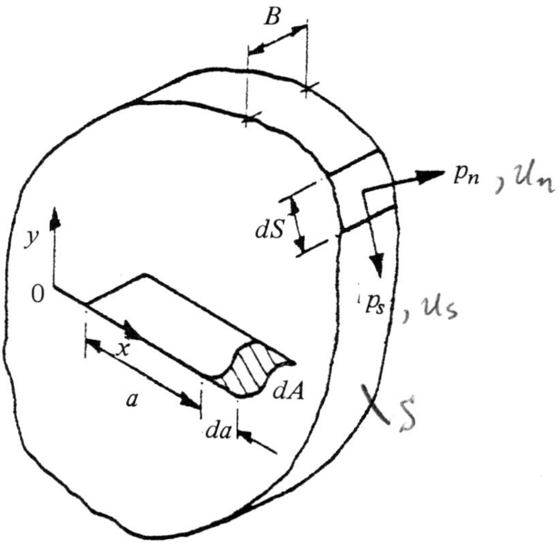

Fig. 2.4 Two-dimensional cracked body

The total G may be obtained by integrating over S to give:

$$G=\int_S\left[\left(p_n\frac{du_n}{da}+p_s\frac{du_s}{da}\right)-\frac{d}{da}\left(\int_0^{u_n}p_n\,du_n+\int_0^{u_s}p_s\,du_s\right)\right]dS$$

$$=-\frac{d}{da}\int_S\left[\int_0^{u_n}p_n\,du_n+\int_0^{u_s}p_s\,du_s\right]dS\Bigg|_{u_n,u_s\ const} \tag{2.15}$$

A further form of interest is derived by expressing U_3 in terms of the strain energy density function W within the system. Thus:

$$U_3=\int_V W\,dV$$

where V is the volume of the body. If we use the coordinate system shown in Fig. 2.4 such that x is the direction of crack propagation, then:

$$U_3=B\int W\,dx\,dy \text{ and } da=-dx$$

giving:

$$\frac{dU_3}{dA}=-\int_S W\,dy$$

$$G=\int_S W\,dy-\int_S\left(p_n\frac{du_n}{dx}+p_s\frac{du_s}{dx}\right)dS=\int_S W\,dy\Bigg|_{u_n,u_s\ const} \tag{2.16}$$

2.4.1 *Linear elasticity*

This is by far the most important practical case and results in all force-deflection curves being linear, so that:

$$U_3=\tfrac{1}{2}\,P\,u$$

and

$$\frac{dU_3}{dA}=\frac{1}{2}P\frac{du}{dA}+\frac{1}{2}u\frac{dP}{dA}$$

Since:

$$\frac{dU_1}{dA}=P\frac{du}{dA}$$

we have:

$$G=P\frac{du}{dA}-\frac{1}{2}P\frac{du}{dA}-\frac{1}{2}u\frac{dP}{dA}$$

i.e.:

$$G=\frac{1}{2}\left(P\frac{du}{dA}-u\frac{dP}{dA}\right) \tag{2.17}$$

A particular body is usually defined in terms of its compliance given by:

$$C=\frac{u}{P} \tag{2.18}$$

so that: $$\frac{du}{dA}=P\frac{dC}{dA}+C\frac{dP}{dA}$$

Equation (2.17) may now be written in terms of dC/dA and either P, u or U_3:

$$G=\frac{1}{2}P^2\frac{dC}{dA}=\frac{1}{2}\left(\frac{u}{C}\right)^2\frac{dC}{dA}=U_3\frac{1}{C}\frac{dC}{dA} \tag{2.19}$$

or, for the planar case:

$$G=\frac{P^2}{2B}\frac{dC}{da}=\frac{1}{2B}\left(\frac{u}{C}\right)^2\frac{dC}{da}=\frac{U_3}{B}\frac{1}{C}\frac{dC}{da} \tag{2.20}$$

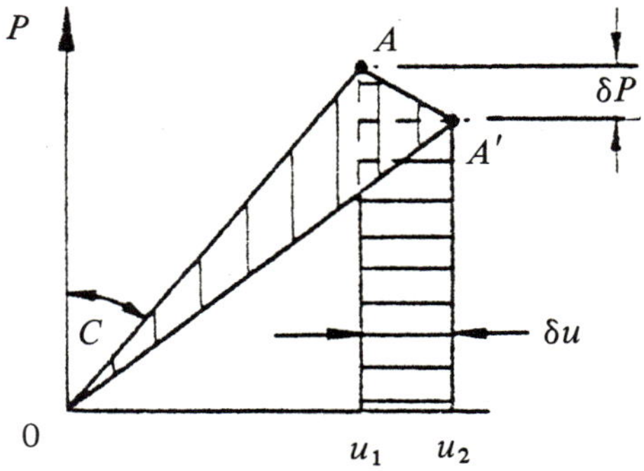

Fig. 2.5 Crack growth in a linear elastic body

The analysis is illustrated graphically in Fig. 2.5, where OA is the initial loading line. At A, the crack grows so that the load changes δP and the displacement by δu, and the unloading line would be $A'O$. The stored energy at fracture is clearly the area OAu_1 and after $OA'u_2$, so that the change is the narrow triangle with vertical shading. This is given by:

$$\frac{1}{2}P\,u-\frac{1}{2}(P+\delta P)(u+\delta u)+P\left(P+\frac{\delta P}{2}\right)=\frac{1}{2}(P\,du-u\,dP)$$

as in equation (2.17). It should be noted that G is independent of how P changes since any additional external work goes into increasing the stored energy and G arises only from compliance changes. This gives rise to the constant displacement definition given in equation (2.12) but does not imply that u must, in reality, remain constant. The basic equations may be derived by this route, since:

$$U_3=\frac{1}{2}P\,u=\frac{u^2}{2C}$$

and: $$G=\frac{1}{B}\frac{dU_3}{da}\bigg|_{u\ const}=\frac{u^2}{2B\,C^2}\frac{dC}{da}$$

as before. This result is sensible because a non-cracking body when loaded has $G=0$ since the external work is stored as a consequence of equilibrium but when a non-compatible change occurs, such as crack growth, then there is an energy imbalance due entirely to this change. Any extraneous work inputs will simply be absorbed in the system in the usual way.

The result is important in practical terms since G may be found by determining C for a cracked body either by analysis or by experiment. If $C(a)$ is found, then dC/da is known and G may be determined. It is also possible to measure G concurrently with any test by observing P, u and a [1].

In terms of stresses on the boundaries, the integrals in equation (2.15) may be expressed as $\frac{1}{2}\,p\,u$, giving a final result of:

$$G=\frac{1}{2}\int_S\left(p_n\frac{du_n}{da}-u_n\frac{dp_n}{da}\right)+\left(p_s\frac{du_s}{da}-u_s\frac{dp_s}{da}\right)ds$$

$$=-\frac{1}{2}\int_S u_n\frac{dp_n}{da}+u_s\frac{dp_s}{da}\,dS\,\bigg|_{u_n,u_s\ const} \qquad (2.21)$$

i.e. equation (2.17) integrated around S.

Local compliances $C_n(a, S)$ and $C_s(a, S)$ can also be used to give:

$$G=\frac{1}{2}\int_S p_n^{\,2}\frac{dC_n}{da}+p_s^{\,2}\frac{dC_s}{da}\,dS$$

with similar expressions for u and U.

2.4.2 *Non-linear elasticity*

The analysis here is the same as the linear case, except that the lines OA and OA' are not straight in the general case and the integral for U_3 must be evaluated. For example, a non-linear form may give rise to:

$$C_n=\frac{u^n}{P}$$

where n is a constant so that equation (2.17) now becomes:

$$G=\left[P\frac{du}{dA}-\frac{dC}{dA}\left(\frac{P\,u}{1+n}\right)\right]$$

i.e.

$$G=\frac{1}{1+n}\left[n\,P\frac{du}{dA}-u\frac{dP}{dA}\right]$$

In terms of C_n, we have:

$$G=\frac{1}{1+n}\,P^{(1+n)/n}\,C_n^{1/(1-n)}\frac{dC_n}{dA}=\frac{1}{1+n}\frac{u^{1+n}}{C_n^{\,2}}\frac{dC_n}{dA}=U_3\frac{1}{C_n}\frac{dC_n}{dA} \qquad (2.22)$$

which may be compared with equations (2.19) for $n=1$.

2.4.3 *Examples of G determination*

A series of analytical results will now be given which have some utility and also serve as examples of the application of the general methods described previously.

(i) *Double cantilever beam (DCB)*
This geometry is frequently used in practice and also yields a simple analytical solution providing zero rotation at the end of the beams is assumed. Fig. 2.6(i) shows the geometry and from simple beam theory the total deflection of the load P, for both arms, is:

$$u = \frac{8a^3}{E\,B\,D^3}\,P$$

so that:

$$C = \frac{u}{P} = \frac{8a^3}{E\,B\,D^3} \text{ and } \frac{dC}{da} = \frac{24a^2}{E\,B\,D^3}$$

Substituting in equation (2.20), we have:

$$G = \frac{12\,P^2\,a^2}{E\,B^2\,D^3} = \frac{3}{16}\,\frac{E\,D^3\,u^2}{a^4} = \frac{U_3}{B}\left[\frac{3}{a}\right] \tag{2.23}$$

For this geometry, G increases with crack length at fixed load and decreases with fixed displacement.

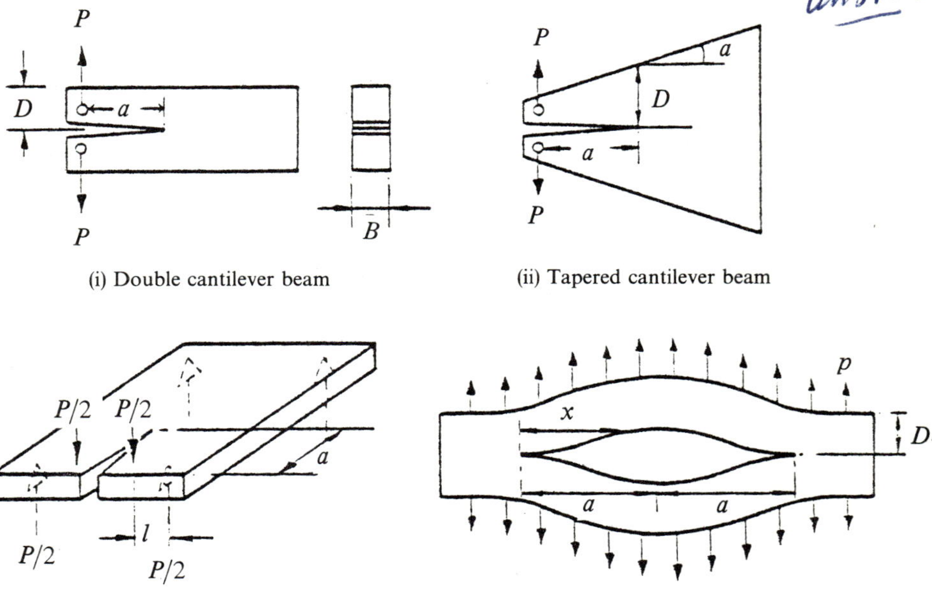

(i) Double cantilever beam

(ii) Tapered cantilever beam

(iii) Double torsion

(iv) Uniformly loaded beam test

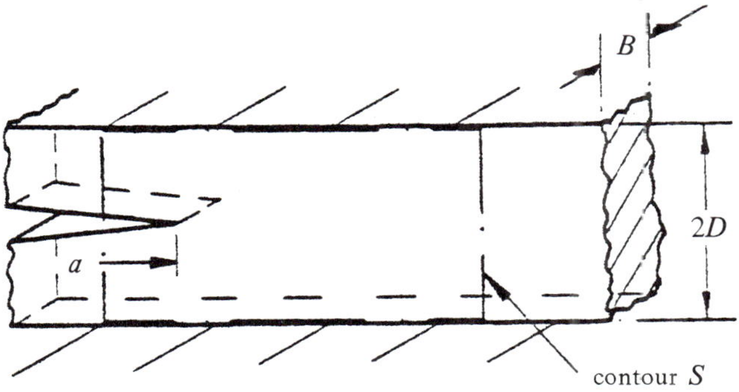

(v) Parallel strip

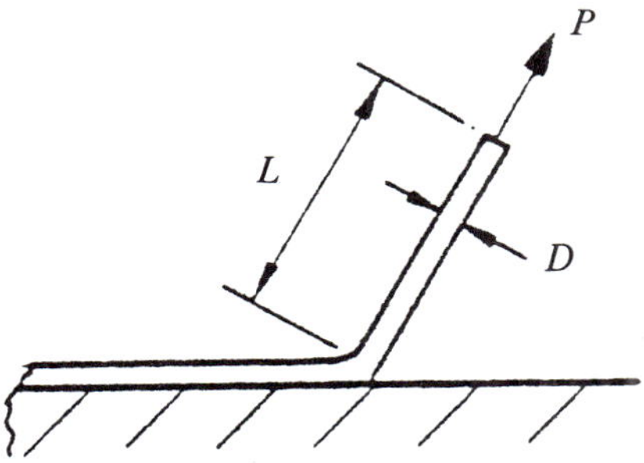

(vi) Adhesion peel test

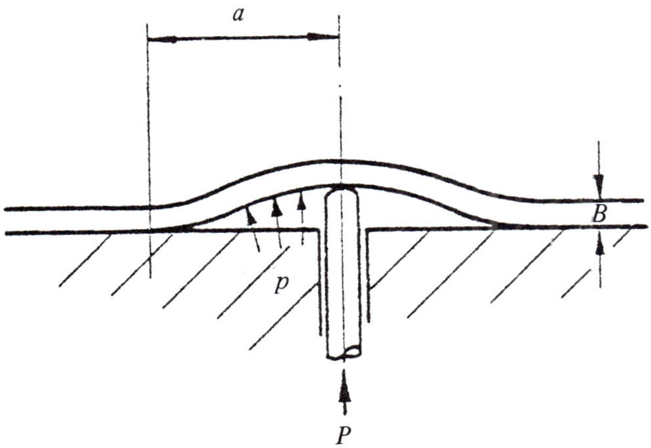

(vii) Blister test

Fig. 2.6 Geometries for G determinations

A constant G condition is provided by a loading in the form of a moment M applied to each arm, for which:

$$G = \frac{12M^2}{E\,B^2\,D^3} \tag{2.24}$$

(ii) *Contoured cantilever beam*

It is frequently useful to have a test geometry in which G is constant at a fixed load and this may be achieved by having $D(a)$; i.e. a varying section. If we take:

$$D = D_0\,a^n$$

where D_0 and n are constants, then we may deduce the displacement from simple beam theory with a varying second moment of area, giving the result:

$$\frac{dC}{da} = \frac{24}{E\,B}\,\frac{a^{2-3n}}{D_0{}^3}$$

If a profile with $n = 2/3$ is used, dC/da is constant and thus G is constant with respect to a. A rather simpler method which yields a similar result is shown in Fig. 2.6(ii), in which a taper is used which may be described as:

$$D = D_0 + \tan\alpha\,a$$

which gives:

$$\frac{dC}{da} = \frac{24}{E\,B\,D_0}\,\frac{(a/D_0)^2}{[1 + (a/D_0)\tan\alpha]^3} \tag{2.25}$$

This function has a rather flat maximum as a/D_0 increases, so that by a suitable choice of α, a sensibly constant dC/da may be achieved. For example, with $\alpha \simeq 11°$, $\tan\alpha \simeq 0.2$ and the curve is a maximum around $a/D_0 = 10$ with:

$$\frac{dC}{da} \simeq \frac{90}{E\,B\,D_0}$$

(iii) *Double torsion*

Another geometry which achieves constant G is shown in Fig. 2.6(iii), which is the double cantilever beam loaded sideways so that the arms are in torsion. In this case:

$$C = \frac{3(1+\nu)\,l^2\,a}{E\,B\,D^3} \tag{2.26}$$

so that dC/da is a constant.

The factor 3 applies only for large D/B ratios and a correction can be made for other values as listed below:

D/B	∞	10	5	2.5	1
Factor	3	3.20	3.44	4.02	7.1

(Taken from reference [4])

It is frequently necessary to employ side grooves to control the crack direction, so that the final expression for G becomes:

$$G=\frac{3(1+\nu)\, l^2\, P^2}{2\, E\, D\, B^3\, B_c} \tag{2.27}$$

where B_c is the crack thickness.

(iv) *Uniformly loaded beam*

An example involving a distributed load is shown in Fig. 2.6(iv). Here, the load is uniform and the integration in equation (2.21) is taken over a for each quarter of the specimen and since $p_s=u_s=0$, we have:

$$G=4\left[\frac{p^2}{2}\int_0^a\left(\frac{dC_n}{da}\right)dx\right]$$

The deflection at any point is given by:

$$u=\frac{p}{2\, E\, D^3}(2\, a^2\, x^2-x^4);\ \text{i.e. } C_n=\frac{2\, a^2\, x^2-x^4}{2\, E\, D^3}$$

and:

$$G=2p^2\int_0^a\frac{2\, a\, x^2}{E\, D^3}\, dx=\frac{4\, p^2\, a^4}{3\, E\, D^3} \tag{2.28}$$

(v) *Non-linear elastic DCB*

Using a power-law form of non-linearity such that:

$$p=E_0\, e^n$$

and retaining the simple beam theory assumptions, we can deduce the expression for C_n in equation (2.22) as:

$$C_n=\frac{u^n}{P}=\left(\frac{4n}{1+2n}\right)^n\frac{2(n+2)\, a^{2n+1}}{D^{2+n}\, E_0\, B}$$

and on using equation (2.23), we have:

$$G=\frac{P^{1/n+1}}{B}\left(\frac{4n}{1+n}\right)\left[\frac{2(n+2)\, a}{D^2\, B\, E_0}\right]^{1/n}\frac{a}{D}=u^{1+n}\frac{3n}{2(2+n)\,(1+n)}\left[\frac{(1+2n)\, D}{4n\, a^3}\right]^n\frac{D\, E_0}{a}$$

$$=\frac{U_3}{B\, a}(2n+1) \tag{2.29}$$

which reduces to equation (2.23) for $n=1$.

(vi) *Parallel strip*

A wide strip, as shown in Fig. 2.6(v), with clamped edges may be conveniently solved via equation (2.16) since $u_n=u_s=$const., and:

$$G=\int_S W\, dy$$

If we take the contour S, as shown by the broken line in Fig. 2.6(v), then on the cracked section remote from the tip $W=0$, and we need only evaluate W along the vertical line in front of, and remote from, the crack tip where W will be constant, so that:

$$G=2D\,W \tag{2.30}$$

where W is the energy density in the uncracked strip. This is a particularly useful system for non-linear materials where W can be found. For the power-law case:

$$W=\int_0^e p\,de=\frac{E_0\,e^{1+n}}{1+n}$$

i.e.

$$G=\frac{2\,D\,E_0}{(1+n)}\,e^{1+n} \tag{2.31}$$

and for linear elasticity ($n=1$) with an overall displacement of u_0, we have:

$$G=\frac{E\,{u_0}^2}{4D} \tag{2.32}$$

(vii) *Adhesion peel tests*

G for the fracture of adhesive joints can be measured using cantilever beam systems as in (i) but peel tests using flexible strips, as shown in Fig. 2.6(vi) are frequently employed. If the strip unpeels a length dL, then the change in external work is:

$$dU_1=P\,dL+e\,P\,dL$$

where e is the strain given by P/BDE.

The change in strain energy is:

$$dU_3=\frac{P^2}{2E\,D\,B}\,dL$$

so that we have:

$$G=\frac{1}{B}\left(\frac{dU_1}{dL}-\frac{dU_3}{dL}\right)=\frac{P}{B}(1+e) \tag{2.33}$$

For $e \ll 1$, G is determined by the external work and measured directly from P. The same result is obtained for the trouser leg tear specimen used on flexible materials.

(viii) *Blister tests*

This test configuration, shown in Fig. 2.6(vii), is sometimes used in adhesion testing. The adhered layer may be treated as a circular plate with built-in edges, for which the deflection solution is:

$$u=\frac{3}{4}\frac{(1-\nu^2)}{E\,B^3}\left\{\frac{p}{4}(a^2-r^2)^2+\frac{P}{\pi}\left[(a^2-r^2)+r^2\ln\left(\frac{r}{a}\right)^2\right]\right\}$$

where p is a constant distributed pressure, and P is a central point load. In this case, the area change on fracture is $dA = 2\pi\, a\, da$, and for the pressure term we have:

$$G = \frac{p^2}{2a} \int_0^a r \left(\frac{dC}{da}\right) dr$$

where C is the local compliance given by:

$$C = \frac{u}{p} = \frac{3}{16} \frac{(1-v^2)}{E\, B^3} (a^2 - r^2)^2$$

giving a final result of:

$$G = \frac{3}{32} \frac{p^2\, a^4\, (1-v^2)}{E\, B^3} \qquad (2.34)$$

For the constant load:

$$G = \frac{P^2}{4\pi\, a} \frac{dC}{da} \text{ with } C = \frac{3}{4\pi} \frac{(1-v^2)}{E\, B^3} a^2$$

so that:

$$G = \frac{3}{8\pi^2} \frac{P^2\, (1-v^2)}{E\, B^3} \qquad (2.35)$$

i.e. independent of a.

(ix) *Cracks in infinite bodies*

If we have an infinite body loaded with a uniform stress p remote from a crack of length a, then G in terms of W (equation (2.16)) may be non-dimensionalised to give:

$$G = W_0\, a \int_{-\infty}^{+\infty} \left(\frac{W}{W_0}\right) \left(\frac{dy}{a}\right)$$

where W_0 is the remote, uniform strain energy density $p^2/2E$. The integral must yield a constant so that the form of the expression for G is:

$$G = k\, W_0\, a = \frac{k}{2} \frac{p^2\, a}{E} \qquad (2.36)$$

where k is a constant, for infinite bodies. In Section 4.2 (equation (4.7)), it is shown that $k = 2\pi$ in linear materials but a function of strain for large strains (see Section 6.13).

2.4.4 *Residual stresses*

Residual stress systems may exist in a body when no external loads are applied. They usually arise from the restriction of thermal contraction on solidification under non-uniform temperatures and are, of course, in equilibrium internally since they exist without applied loads. A G can result if the growth of the crack gives rise to a change in this state of equilibrium. The most convenient method for computing G is to define W for the stress system and then use equation (2.16).

A simple example is provided by a DCB specimen in which a residual stress system could be produced by fabricating it from a central portion of width D and two outer sections of width $D/2$. If the three parts are then joined with the centre under tension, we will be left with a specimen with a uniform tensile stress of, say, p_0 in the centre and $-p_0$ on the outer sections. When the central crack grows, there is a nett bending moment in each arm given by $(p_0\,B\,D^2)/4$ which may be used with equation (2.24) to give G. i.e.

$$G=\frac{3}{4}\,p_0^{\,2}\,D$$

The same result may be deduced from equation (2.16) by taking a contour in front of the crack for which $W=p_0/2E$ and one behind where the stresses will have changed because of the removal of the bending moment. The stress systems are shown in Fig. 2.7(a) and we note that in following the contour, there is a sign change from one to the other giving the energy difference. We may now write G as:

$$G=2\left[\int_0^D\left(\frac{p_0^{\,2}}{2E}\right)dy-2\int_0^{D/2}\left(\frac{p_0^{\,2}}{2E}\right)\left(\frac{1}{2}-\frac{3y}{D}\right)^2 dy\right]$$

which gives the same result.

A problem of some practical interest is the pressurised pipe shown in Fig. 2.7(b), where there is often a residual stress system due to the cooling system used in manufacture. The stress system in front of the crack is that due to the internal pressure $p=PD/2B$ and an approximately parabolic form [5] of residual stress given by:

$$p_R=p_0\left(1-\frac{3z^2}{B^2}\right)$$

where p_0 is the bore value at $z=0$. After the crack has passed, the pressure stress goes to zero and the residual pattern is changed by the release of the nett moment, giving:

$$p_R'=p_0\left(-\frac{1}{2}-\frac{3z^2}{B^2}+\frac{3z}{B}\right)$$

The resultant G value is given by:

$$G=\pi\,D\left[\int_0^1\frac{(p+p_R)^2}{2E}\frac{dz}{B}-\int_0^1\frac{p_R'^{\,2}}{2E}\frac{dz}{B}\right]$$

which may be evaluated to give:

$$G=\frac{\pi\,D}{2E}\left(p^2+\frac{3}{4}\,p_0^{\,2}\right)$$

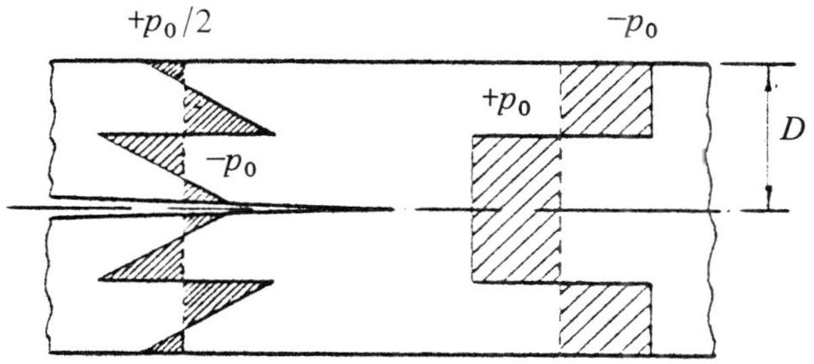

(a) The DCB specimen

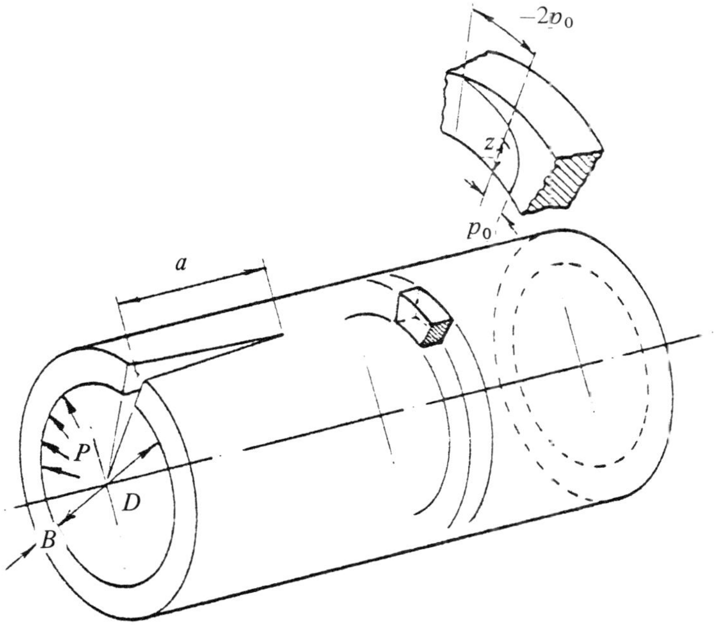

(b) Pressurised pipe

Fig. 2.7 Systems with residual stresses

References for Chapter 2

[1] Gurney, C., & Hunt, J., *Proc. R. Soc. Lond.*, **A299** (1967) 508–524.

[2] Turner, C. E., Chapter 4 in *Developments in Fracture Mechanics*, **1**, (C. G. Chell, ed.),

[3] Griffith, A. A., *Phil. Trans. R. Soc.*, **A221** (1920) 163–198.

[4] Timoshenko, S. P., & Goodier, J. W., *Theory of Elasticity*, 3rd edn., McGraw-Hill, New York (1970).

[5] Williams, J. G., *Plastics & Rubber: Processing & Applications*, **1** (1981) 369–377.

3

Elastic stresses around crack tips

3.1 Introduction

The analysis of the energy in a system in which a fracture occurs yields several useful results, as shown in Chapter 2. It provides sensible fracture criteria in G_c and R, and methods for their determination via G in several geometries. In many cases, however, the derivation, or measurement, of G is not simple since C is not accessible and the alternative route of considering the stress field associated with the crack tip is more convenient. Although perhaps less appealing than an energy analysis, the idea of analysing the local conditions at the crack tip is reasonable for a local phenomenon such as fracture, and a parameter characterising such a field could also be expected to provide a fracture criterion. This apparently alternative analysis will now be given and it will be shown that, at least in the elastic case, it is entirely equivalent to the energy method.

3.2 Stresses around a crack tip [1, 2]

Consider a crack in a body and the stress field in the vicinity of the tip defined in terms of the cylindrical polar coordinates r and θ, as shown in Fig. 3.1. We will conduct the analysis for plane stress such that the stress field is entirely described by the three components p_r, p_θ and $p_{r\theta}$ with the out of plane stress $p_z = 0$. The

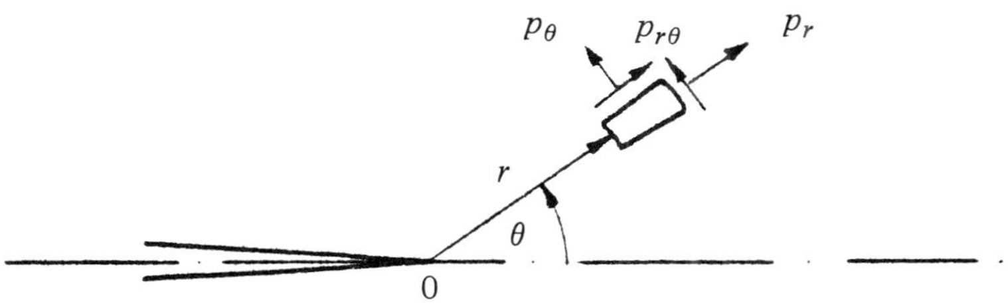

Fig. 3.1 Stresses around a crack tip

plane strain case in which $p_z = \nu(p_r + p_\theta)$ will be introduced later. Let us assume that the stresses may be described in the form:

$$p_\theta = F_\theta(\theta)\, r^{-m},\ p_r = F_r(\theta)\, r^{-m} \text{ and } p_{r\theta} = F_{r\theta}(\theta)\, r^{-m} \tag{3.1}$$

so that they become singular (i.e. $\rightarrow \infty$) as $r \rightarrow 0$ for $m > 0$. For any type of behaviour, these must satisfy the equilibrium equations:

$$p_r - p_\theta + r\frac{\partial p_r}{\partial r} + \frac{\partial p_{r\theta}}{\partial \theta} = 0$$

and:

$$2p_{r\theta} + r\frac{\partial p_{r\theta}}{\partial r} + \frac{\partial p_\theta}{\partial \theta} = 0 \tag{3.2}$$

Substituting from equations (3.1) yields the following relationships for F:

$$(1-m)(2-m)\, F_r = (2-m)\, F_\theta + \frac{d^2 F_\theta}{d\theta^2}$$

$$(2-m)\, F_{r\theta} = -\frac{dF_\theta}{d\theta} \tag{3.3}$$

The strains may be defined in terms of the displacements in the r and θ directions, u_r and u_θ:

$$e_r = \frac{\partial u_r}{\partial r},\ e_\theta = \frac{u_r}{r} + \frac{1}{r}\frac{\partial u_\theta}{\partial^\theta},\ e_{r\theta} = \frac{\partial u_\theta}{\partial r} - \frac{u_\theta}{r} + \frac{1}{r}\frac{\partial u_r}{\partial \theta} \tag{3.4}$$

If we now assume that the strains have the form:

$$e_\theta = H_\theta(\theta)\, r^{-p},\ e_r = H_r(\theta)\, r^{-p},\ e_{r\theta} = H_{r\theta}(\theta)\, r^{-p} \tag{3.5}$$

then, on substitution into equation (3.4), a compatibility condition results:

$$\left(\frac{d^2 H_r}{d\theta^2} + p\, H_r\right) = (1-p)\left(\frac{dH_{r\theta}}{d\theta} + p\, H_\theta\right) \tag{3.6}$$

Any solution to these equations requires a relationship between F and H and this is provided by the particular stress–strain relationship used to describe the material behaviour.

3.2.1 *Linear elasticity*

Using Hooke's law, we have:

$$e_r = \frac{1}{E}(p_r - \nu\, p_\theta),\ e_\theta = \frac{1}{E}(p_\theta - \nu\, p_r),\ e_{r\theta} = \frac{2(1+\nu)}{E}\, p_{r\theta} \tag{3.7}$$

which, on substitution from equations (3.1), (3.3) and (3.5), gives that $p=m$ and:

$$E\,H_r=F_r-\nu\,F_\theta=F_\theta\left(\frac{1}{1-m}-\nu\right)+\frac{d^2F_\theta/d\theta^2}{(1-m)\,(2-m)}$$

$$E\,H_\theta=F_\theta-\nu\,F_r=F_\theta\left(1-\frac{\nu}{1-m}\right)-\frac{\nu}{(1-m)\,(2-m)}\frac{d^2F_\theta}{d\theta^2}$$

and:

$$E\,H_{r\theta}=2(1+\nu)\,F_{r\theta}=-\frac{2(1+\nu)}{(2-m)}\frac{dF_\theta}{d\theta}$$

On substituting into the compatibility condition (equation (3.6)), we have:

$$\frac{d^2}{d\theta^2}\left(\frac{d^2F_\theta}{d\theta^2}+m^2F_\theta\right)+(2-m)^2\left(\frac{d^2F_\theta}{d\theta^2}+m^2\,F_\theta\right)=0 \tag{3.8}$$

which may be solved to give the result:

$$F_\theta=[A_m\cos m\,\theta+A_{2m}\cos(2-m)\,\theta]+[B_m\sin m\,\theta+B_{2m}\sin(2-m)\,\theta] \tag{3.9}$$

It can be seen that the first term is symmetrical in θ and represents loadings leading to crack opening, termed mode I, while the second is skew symmetric and represents the sliding mode II. The boundary conditions for a crack without surface tractions are:

$$p_\theta=p_{r\theta}=0 \text{ at } \theta=\pm\pi,\text{ i.e. } F_{\theta\pm\pi}=\left.\frac{dF_\theta}{d\theta}\right|_{\pm\pi}=0$$

giving, for example, for p_θ at $\theta=\pm\pi$:

$$0=A_m\cos m\,\pi+A_{2m}\cos(2-m)\,\pi+B_m\sin m\,\pi+B_{2m}\sin(2-m)\,\pi$$

i.e.

$$0=(A_m+A_{2m})\cos m\,\pi+(B_m-B_{2m})\sin m\,\pi \qquad \text{(i)}$$

and for $p_{r\theta}$:

$$0=-[m\,A_m-(2-m)\,A_{2m}]\sin m\,\pi+[m\,B_m+(2-m)\,B_{2m}]\cos m\,\pi \qquad \text{(ii)}$$

(3.10)

At $\theta=-\pi$, we have:

$$0=(A_m+A_{2m})\cos m\,\pi-(B_m-B_{2m})\sin m\,\pi \qquad \text{(iii)}$$

and:

$$0=[m\,A_m-(2-m)\,A_{2m}]\sin m\,\pi+[m\,B_m+(2-m)\,B_{2m}]\cos m\,\pi \qquad \text{(iv)}$$

These are eigenvalue equations and by subtracting and summing, we have the following conditions:

e.g. (i)+(iii) and (ii)+(iv) gives:

$$0=2(A_m+A_{2m})\cos m\pi$$

$$0=2[m\,B_m+(2-m)\,B_{2m}]\cos m\pi$$

and (i)−(iii) and (iv)−(ii) results in:

$$0=2(B_m-B_{2m})\sin m\pi$$

$$0=2[m\,A_m-(2-m)\,A_{2m}]\sin m\pi$$

Two sets of solutions are now possible:

$$\cos m\pi=0,\ m=\pm\frac{1}{2},\ \pm\frac{3}{2},\ldots \text{ with } A_m\,m=A_{2m}\,(2-m),\ B_m=B_{2m}$$

and:

$$\sin m\pi=0,\ m=0,\ \pm 1,\ \pm 2,\ldots \text{ with } A_m=-A_{2m},\ B_m\,m=-B_{2m}\,(2-m)$$

It is necessary to assume finite crack tip displacements and since from equations (3.4) and (3.5):

$$u\propto r^{1-m} \text{ then } m<1$$

For $m<0$, all stresses and strains tend to zero at $r=0$ so that the stress field in the vicinity of the tip is dominated by the singular term for $m=\frac{1}{2}$. The expressions for p_θ, for example, now become:

$$p_\theta=r^{-1/2}A_{1/2}\left(\cos\frac{\theta}{2}+\frac{1}{3}\cos\frac{3\theta}{2}\right)+r^{-1/2}B_{1/2}\left(\sin\frac{\theta}{2}+\sin\frac{3\theta}{2}\right)$$

It is usual to replace $A_{1/2}$ and $B_{1/2}$ by the stress intensity factors K_I and K_{II}, respectively, which are defined as:

$$K_I=[p_\theta(0)\sqrt{2\pi r}]_{r\to 0}=\tfrac{4}{3}\sqrt{2\pi}\,A_{1/2}$$

and:

$$K_{II}=[p_{r\theta}(0)\sqrt{2\pi r}]_{r\to 0}=-\tfrac{4}{3}\sqrt{2\pi}\,B_{1/2}$$

On recouching the angular terms, we have the final results:

$$p_\theta=\frac{K_I}{\sqrt{2\pi r}}\frac{1}{2}\cos\frac{\theta}{2}(1+\cos\theta)-\frac{K_{II}}{\sqrt{2\pi r}}\frac{3}{2}\sin\frac{\theta}{2}(1+\cos\theta)$$

$$p_r=\frac{K_I}{\sqrt{2\pi r}}\frac{1}{2}\cos\frac{\theta}{2}(3-\cos\theta)-\frac{K_{II}}{\sqrt{2\pi r}}\frac{1}{2}\sin\frac{\theta}{2}(1-3\cos\theta) \qquad (3.11)$$

$$p_{r\theta}=\frac{K_I}{\sqrt{2\pi r}}\frac{1}{2}\sin\frac{\theta}{2}(1+\cos\theta)-\frac{K_{II}}{\sqrt{2\pi r}}\frac{1}{2}\cos\frac{\theta}{2}(1-3\cos\theta)$$

and the associated displacements:

$$u_\theta = \frac{\sqrt{2\pi r}}{2\pi E}\left\{K_I \sin\frac{\theta}{2}\left[(1+\nu)(1+\cos\theta)-4\right] + K_{II}\cos\frac{\theta}{2}\left[(1+\nu)(1-3\cos\theta)+4\right]\right\}$$

$$u_r = \frac{\sqrt{2\pi r}}{2\pi E}\left\{K_I \cos\frac{\theta}{2}\left[4-(1+\nu)(1+\cos\theta)\right] - K_{II}\sin\frac{\theta}{2}\left[3(1+\nu)(1+\cos\theta)-4\right]\right\} \tag{3.12}$$

Fig. 3.2 shows these displacements plotted for each mode and the distortion of the circle illustrates clearly the symmetrical deformation and crack opening of mode I. Mode II, on the other hand, is produced by sliding the crack faces relative to each other. Any in-plane loading can thus be resolved into two components; mode I from stresses normal to the crack and mode II from shear stresses parallel to the crack. A further mode is also possible which is out-of-plane sliding, or mode III, but this is not included in this planar solution.

The solution for $m=0$ is also of interest since this term gives stresses which are

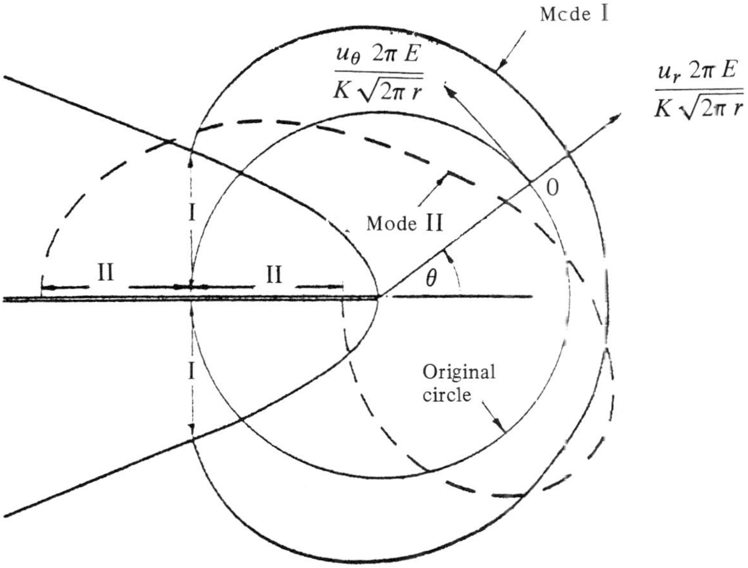

Fig. 3.2 Displacements around a crack tip — modes I and II

independent of r and the symmetrical term varies with θ in the form:

$$p_\theta = A_0\,(1-\cos 2\theta),\ p_r = A_0\,(1+\cos 2\theta),\ p_{r\theta} = -A_0 \sin 2\theta$$

(The asymmetric term is zero for $m=0$). These are the transformation relationships for a stress parallel to the crack such that $p_x = 2A_0$ which has no influence on the singular stress field, as would be expected. Mode I arises from p_y and mode II from p_{xy}.

The most important characteristic of the singular term is that the stress field local to the crack tip is identical in form (i.e. dependence on r and θ) for *all* loadings and that the loading is only reflected through the magnitude of the stresses determined by the stress intensity factor, K. Thus, the local stress field can be characterised by the value of K and a critical value would be expected to form a fracture criterion. If we confine our attention to mode I loading only at this stage, then we can postulate that fracture will occur when:

$$K_I = K_{Ic} \tag{3.13}$$

It now simply remains to determine K_I for various systems to establish a fracture prediction analysis. The local stress field method used here is not suitable for determining K_I which must be done by deducing a complete stress field and then isolating the singular term. This turns out to be very convenient for certain configurations and will be developed in Sections 4.2 and 4.3. However, we have now arrived at two separate fracture criteria, namely, G_c and K_{Ic}, and we must unify these approaches before proceeding further.

3.3 Relationship between *K* and *G*

The analysis of the previous section enabled the complete stress and displacement field around any crack tip to be determined. We may deduce an energy release rate G for this system by taking a contour around the crack tip, as shown in Fig. 3.3. For convenience, we may take a circle of radius r so that the normal stress on the contour boundary is p_r and the shear stress is $p_{r\theta}$. In addition, the normal and tangential displacements are u_r and u_θ, respectively. If the contour remains fixed and the crack tip moves da from 0 to 0′ as shown, then both θ and r change and changes in p_r, $p_{r\theta}$, u_r and u_θ result. In addition, the new directions of r and θ must be corrected to give changes in the normal and tangential directions on the contour.

G may conveniently be deduced from equation (2.21):

$$G = \frac{1}{2}\int_S \left(p_n \frac{du_n}{da} - u_n \frac{dp_n}{da}\right) + \left(p_s \frac{du_s}{da} - u_s \frac{dp_s}{da}\right) dS$$

and since $dS = r\,d\theta$, we have:

$$G = \frac{1}{2}\int_{-\pi}^{+\pi} \left(p_r \frac{du_r}{da} - u_r \frac{dp_r}{da}\right) + \left(p_{r\theta} \frac{du_\theta}{da} - u_\theta \frac{dp_{r\theta}}{da}\right) r\,d\theta$$

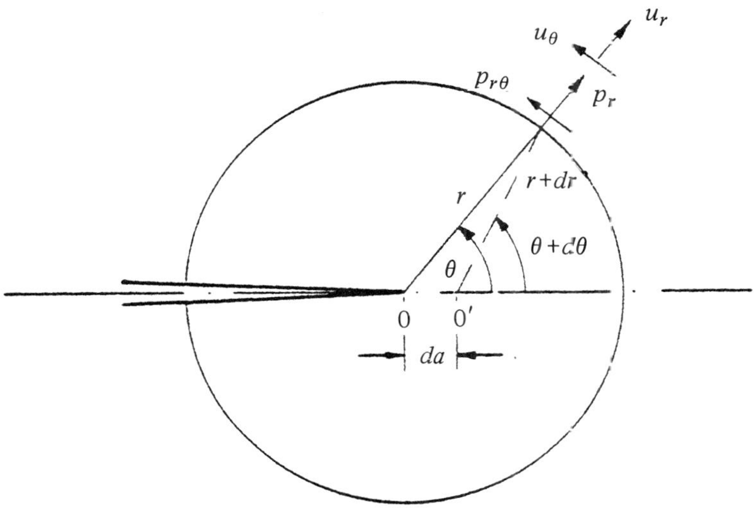

Fig. 3.3 Stresses and displacements on a circular contour

The appropriate derivatives are:

$$\frac{dp_r}{da}=\frac{\partial p_r}{\partial a}+\frac{\partial p_r}{\partial \theta}\frac{\sin\theta}{r}-\frac{\partial p_r}{\partial r}\cos\theta-2p_{r\theta}\frac{\sin\theta}{r}$$

$$\frac{dp_{r\theta}}{da}=\frac{\partial p_{r\theta}}{\partial a}+\frac{\partial p_{r\theta}}{\partial \theta}\frac{\sin\theta}{r}-\frac{\partial p_{r\theta}}{\partial r}\cos\theta+(p_r-p_\theta)\frac{\sin\theta}{r}$$

(3.14)

$$\frac{du_r}{da}=\frac{\partial u_r}{\partial a}+\frac{\partial u_r}{\partial \theta}\frac{\sin\theta}{r}-\frac{\partial u_r}{\partial r}\cos\theta-u_\theta\frac{\sin\theta}{r}$$

$$\frac{du_\theta}{da}=\frac{\partial u_\theta}{\partial a}+\frac{\partial u_\theta}{\partial \theta}\frac{\sin\theta}{r}-\frac{\partial u_\theta}{\partial r}\cos\theta+u_r\frac{\sin\theta}{r}$$

Since both p and u are proportional to the same functions of a and the expression for G contains terms of the form $p\,(du/da)-u(dp/da)$, the explicit dependence on a is eliminated and the expression for G becomes:

$$G=\frac{1}{2}\int_{-\pi}^{+\pi}\left[\left(p_r\frac{\partial u_r}{\partial r}+p_{r\theta}\frac{\partial u_\theta}{\partial r}-u_r\frac{\partial p_r}{\partial r}-u_\theta\frac{\partial p_{r\theta}}{\partial r}\right)r\cos\theta-\right.$$

$$\left.\left(p_r\frac{\partial u_r}{\partial \theta}+p_{r\theta}\frac{\partial u_\theta}{\partial \theta}-u_r\frac{\partial p_r}{\partial \theta}-u_\theta\frac{\partial p_{r\theta}}{\partial \theta}+3u_rp_r\theta-u_\theta\,(2p_r-p_\theta)\right)\sin\theta\right]d\theta$$

(3.15)

The appropriate expression for the stresses and displacements in terms of A_m, r and θ may be derived from equations (3.9) and (3.10) for all values of m and then substituted into equation (3.15). The set for $m=0, -1, -2, \ldots$ are all zero while for $m=\frac{1}{2}, -\frac{1}{2}, -\frac{3}{2}, \ldots$, we have the result:

$$G=\frac{A_m{}^2\, r^{1-2m}}{E}\,\frac{m}{(2-m)^2\,(1-m)}\left\{(3-2m)\int_{-\pi}^{+\pi}\cos(1-2m)\,\theta-\cos(3-2m)\,\theta\,d\theta\right.$$

$$\left.-\frac{(1-m)\,(1+2m)}{m}\int_{-\pi}^{+\pi}\cos(1+2m)\,\theta-\cos(1-2m)\,\theta\,d\theta\right\} \quad (3.16)$$

For all values of m, other than $\frac{1}{2}$, this expression is zero so that we have:

$$G=\frac{A_{1/2}{}^2}{E}\,\frac{4}{9}\left\{2\int_{-\pi}^{+\pi}d\theta+2\int_{-\pi}^{+\pi}d\theta\right\}=\frac{32\pi}{9}\,\frac{A_{1/2}{}^2}{E}$$

Now:

$$A_{1/2}=\frac{3}{4}\,\frac{K_I}{\sqrt{2\pi}}$$

so that:

$$G=\frac{K_I{}^2}{E} \quad (3.17)$$

A similar analysis for mode II gives the result:

$$G=\frac{K_{II}{}^2}{E}$$

and combined loading are simply described by superposition. The result does assume, however, that the crack propagates in its original direction.

Equation (3.17) is an important result since it demonstrates that the energy release rate arises entirely from the singular term and that all the other non-singular ones give no contribution. This would, of course, be expected since G arises from the breakdown of continuity which is implicit in the singular condition. It is also true that G does not depend on r and the integral may be described as path independent, again a necessary condition for a finite energy release rate and only true for $m=\frac{1}{2}$. For $m\neq\frac{1}{2}$, there is path dependence but the contour integral is zero.

It is also apparent that the fracture condition $G=G_c$ is entirely equivalent to $K_I=K_{Ic}$ and they may be interchanged using equation (3.17).

An alternative and somewhat simpler derivation may be made by taking a very thin contour a distance Δa in front of the crack tip as shown in Fig. 3.4. If the crack is now considered to grow Δa, we may return to equation (2.21) and, if we limit our attention to mode I and note that $p_s=u_s=0$ and that at some point a distance S from the contour tip $p_n=p_\theta$ $(r=\Delta a-S,\ \theta=0)$ and $du_n=2u_\theta$ $(r=S,\ \theta=\pi)$, we may write the general expressions for p_θ and u_θ as:

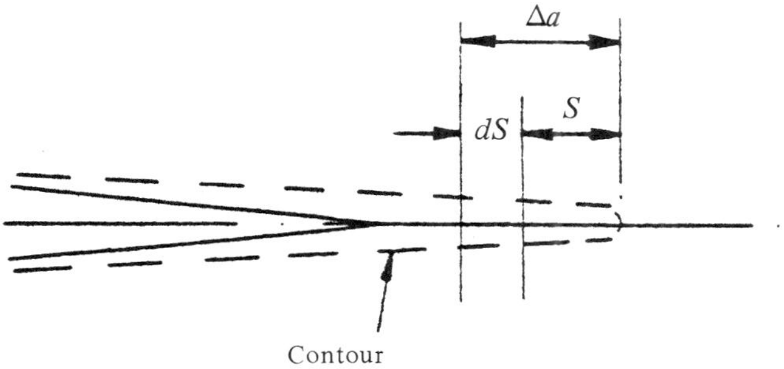

Fig. 3.4 The colinear crack contour

$$p_\theta = A_m r^{-m} f_m(\theta) \text{ and } u_\theta = \frac{A_m}{E} r^{1-m} g_m(\theta)$$

and the resulting expression for G is:

$$G = \frac{A_m{}^2}{E} f_m(0)\,[-g_m(\pi)] \left[\frac{1}{\Delta a} \int_0^{\Delta a} S^{1-m} (\Delta a - S)^{-m}\, dS \right] \tag{3.18}$$

The bracketed term may be re-expressed using the substitution $S = \Delta a \sin^2\psi$ such that:

$$I = \left(\frac{\Delta a}{2}\right)^{1-2m} \int_0^{\pi/2} (1 - \cos 2\psi)(\sin 2\psi)^{1-2m}\, d\psi$$

Remembering that $m < \frac{1}{2}$, it is easy to show that the integral is zero for all permissible values of m other than $\frac{1}{2}$ which removes the dependence on Δa (i.e. path independence) and gives:

$$I = \frac{\pi}{2}$$

and:

$$G = \frac{A_{1/2}{}^2}{E} f_{1/2}(0)\,[-g_{1/2}(\pi)]\, \frac{\pi}{2}$$

From equations (3.11) and (3.12), we have:

$$A_{1/2} = \frac{3}{4} \frac{K_I}{\sqrt{2\pi}},\; f_{1/2}(0) = \frac{4}{3}\sqrt{2\pi} \text{ and } g_{1/2}(\pi) = -\frac{16}{3}\sqrt{2\pi}$$

so that:

$$G = \frac{K_I{}^2}{E}$$

as before.

Plane strain conditions have the same expressions for p_r, p_θ and $p_{r\theta}$ as in plane stress but, in addition, we now have that $p_z = \nu(p_r + p_\theta)$. The expressions for strain are now changed and we have:

$$e_\theta = \frac{1-\nu^2}{E}\left[p_\theta - \left(\frac{\nu}{1-\nu}\right)p_r\right]$$

$$e_r = \frac{1-\nu^2}{E}\left[p_r - \left(\frac{\nu}{1-\nu}\right)p_\theta\right]$$

and:
$$e_{r\theta} = \frac{1-\nu^2}{E}\,2\left[1+\left(\frac{\nu}{1-\nu}\right)\right]p_{r\theta}$$

These are identical in form to the plane stress case used in equations (3.7), except that $(1-\nu^2)/E$ replaces $1/E$ and $\nu/(1-\nu)$ replaces ν. Thus, the plane strain results may be derived by making this change in those for plane stress (as, indeed, they can for any solution). The appropriate $G-K_I$ relationship is thus $G=(1-\nu^2)(K_I{}^2/E)$.

3.4 Path independence

The importance of this property in determining crack tip fields was illustrated in the previous section and it is useful to look at the condition in more general terms. The basis of the argument is that for any elastic system in which a crack grows, it was shown in Chapter 2 that a finite energy release rate will exist. This may be argued from the two load-deflection curves for bodies with crack lengths a and $a+da$. Since this is true, then the local contour integral must be finite and thus must be path independent. If we now write the expression for G given in equation (2.16) for a circular contour (Fig. 3.3), in which $y=r\sin\theta$ and $x=r\cos\theta$, then we have:

$$G = \int_{-\pi}^{+\pi}\left[W\,r\cos\theta + \left(p_n\frac{du_n}{d\theta} + p_s\frac{du_s}{d\theta}\right)\sin\theta\right]d\theta$$

Now, the strain energy density function, W, is derived from expressions of the form:

$$\int p\,de$$

and using the general crack tip fields of equations (3.1) and (3.5), we have that:

$$W \propto r^{-(m+p)}$$

Similarly, $p \propto r^{-m}$ and $u \propto r^{1-p}$, so that $p\,du \propto r^{1-(p+m)}$ and the expression has a dependence of the form:

$$r^{1-(m+p)}$$

For path independence, this dependence on r *must* disappear and this can only occur if:

$$m+p=1 \tag{3.19}$$

In addition, the resulting integral in θ for this condition must be non-zero.

A useful general expression for G may be derived in terms of a hoop stress distribution written in the form:

$$p_\theta = p_0\, l_0{}^m\, r^{-m} f_\theta{}^{(\theta)}$$

where $f_\theta(0)=1$, and p_0 and l_0 are, respectively, characteristic stresses and lengths of the stress distribution at the crack tip. Similarly, we may write the displacement field as:

$$u_\theta = u_0\, l_0{}^{-1+p}\, r^{1-p}\, h_\theta(\theta) \tag{3.20}$$

where $h_\theta(\pi) = -1$, and u_0 is a characteristic displacement. The relationship between the two expressions can only be determined from the particular stress–strain relationship involved but an expression for G in mode I may be written analogous to equation (3.18) and if we invoke the path independence condition to relate p and m, we have:

$$G = \frac{1}{\Delta a} \int_0^{\Delta a} p_0\, l_0{}^m\, (\Delta a - S)^{-m}\, 2u_0\, l_0{}^{-1+p}\, S^{1-p}\, dS$$

with $p+m=1$ so that:

$$G = [2\, p_0\, u_0] \int_0^1 \left(\frac{\xi}{1-\xi}\right)^m d\xi = [2\, p_0\, u_0] \left(\frac{m\,\pi}{\sin m\,\pi}\right) \tag{3.21}$$

To maintain both finite crack tip displacements and singular stresses, $m \geqslant 0$. For the linear case, $m=\frac{1}{2}$ and:

$$p_0\, l_0{}^{1/2} = \frac{K_I}{\sqrt{2\pi}} \text{ with } \frac{u_0}{l_0{}^{1/2}} = \frac{4K_I}{\sqrt{2\pi}} E$$

so that we retrieve the same expression for G as before. K_I can now be represented in terms of the characteristic length l_0 so that:

$$K_I = p_0 \sqrt{2\pi\, l_0} = \frac{\pi}{2} u_0 \frac{E}{\sqrt{2\pi\, l_0}} \tag{3.22}$$

but additional criteria for p_0 or u_0 must be introduced before l_0 can be determined (see Section 4.2). In this particular case, i.e. linear, l_0 and u_0 may be related to the distorted crack tip radius given by $1/\rho = d^2r/du^2$, i.e.

$$\rho = \frac{u_0{}^2}{2l_0} = \frac{4}{\pi} \frac{K_I{}^2}{E^2} \tag{3.23}$$

3.5 Non-linear elasticity

The crack tip stress field for a general non-linear case may be usefully modelled using a strain energy density function of the form:

$$\frac{W}{W_0}=\left[\left(\frac{p_r}{p_0}\right)^2+\left(\frac{p_\theta}{p_0}\right)^2-2\nu\frac{p_r}{p_0}\frac{p_\theta}{p_0}+2(1+\nu)\left(\frac{p_{r\theta}}{p_0}\right)^2\right]^{1+n/2n}$$

where p_0, ν and n are constants, and $W_0=p_0{}^2/2E$. This reduces to Hooke's law for $n=1$ and the Von Mises yield criterion for $n=0$ (see Section 5.1). The strains may be derived from the relationship; for example, $e_r=\partial W/\partial p_r$ which gives the results:

$$e_r=\left(\frac{1+n}{2n}\right)\left(\frac{W}{W_0}\right)^{1-n/1+n}\frac{1}{E}(p_r-\nu\, p_\theta)$$

$$e_\theta=\left(\frac{1+n}{2n}\right)\left(\frac{W}{W_0}\right)^{1-n/1+n}\frac{1}{E}(p_\theta-\nu\, p_r) \qquad (3.24)$$

and:

$$e_{r\theta}=\left(\frac{1+n}{2n}\right)\left(\frac{W}{W_0}\right)^{1-n/1+n}\frac{2(1+\nu)}{E}p_{r\theta}$$

From these equations, it is clear that $e\propto p^{1/n}$, i.e. $p\propto e^n$, as used in Section 2.4.3(v). Thus, in equation (3.5), we have:

$$p=\frac{m}{n}$$

and from the path independence condition, equation (3.19):

$$m=\frac{n}{1+n} \text{ and } p=\frac{1}{1+n} \qquad (3.25)$$

The appropriate angular distribution functions F_θ may be deduced from the equilibrium and compatibility conditions (equations (3.3) and (3.6)), as in the linear case, but the resulting fourth order differential equation is non-linear and yields only to numerical solution [3, 4].

The forms of stress–strain curve represented by various values of n are illustrated in Fig. 3.5 where simple tension curves are shown schematically. From equations (3.24), we have:

$$\frac{p}{p_0}=\left(\frac{E}{p_0}\right)^n\left(\frac{2n}{1+n}\right)^n e^n=\left(\frac{e}{e_0}\right)^n$$

and for $n=1$, we have the linear case as used previously. For $0<n\ll 1$, we have curves which, although elastic, have a form similar to that for materials exhibiting plastic yielding, and for $n=0$ are as for the perfectly plastic case. $n\gg 1$ is a condition which gives a steeply rising curve around $e=e_0$ and is a useful representation for rubber-like behaviour. The forms of crack opening profiles for the various n values may be deduced from equation (3.20) with $p+m=1$, giving:

$$\frac{u_\theta}{u_0}=\left(\frac{r}{l_0}\right)^m$$

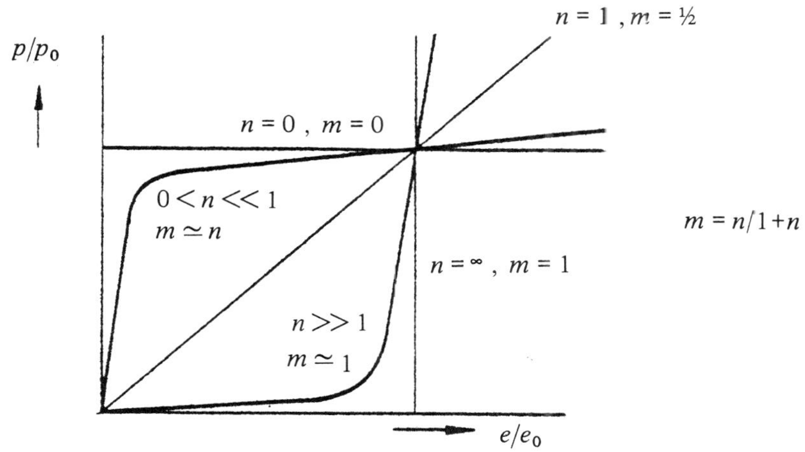

Fig. 3.5 Stress–strain curves for various n values; $n=1$ linear elasticity; $n \gg 1$ approximates to rubber elasticity; $n \ll 1$ has a form similar to plastic yielding

and these are shown in Fig. 3.6. $n=1$ ($m=\frac{1}{2}$) gives a parabola, while low values of n (and m) give blunt ended cracks with sharp corners. High n values ($m \simeq 1$) gives essentially sharp cracks which open in a wedge form.

The expression for G comes directly from equation (3.21):

$$G=[2\, p_0\, u_0]\,\frac{((n\,\pi)/1+n)}{\sin\,((n\,\pi)/1+n)} \tag{3.26}$$

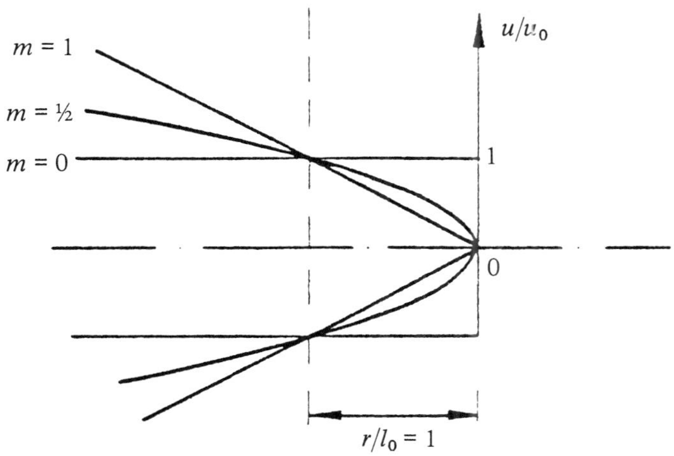

Fig. 3.6 Crack opening profiles for various m values

For small n, this is a useful result since p_0 can be taken as the yield stress, $2u_0$ is the crack opening displacement, and:

$$G \simeq 2\, p_0\, u_0 \qquad (3.27)$$

to which a correction for small work hardening of $(1-\pi^2\, n^2/6)$ may be added.

3.6 Anisotropic linear materials

If we assume that a material obeys the generalised Hooke's law, then $p \propto e$ and thus $p=m$. The crack tip singularity giving rise to the energy release rate is then determined by equation (3.19) and hence:

$$m=p=\tfrac{1}{2}$$

Thus, for *any* linear elastic body, we have an $r^{-1/2}$ singularity at the crack tip in stress (and strain) and the stresses may be written in terms of the same stress intensity factors as the isotropic case but with dimensionless functions of the elastic parameters appearing in the θ dependent terms, i.e.

$$p_\theta = \frac{K_I}{\sqrt{2\pi\, r}} f_\theta,\; p_r = \frac{K_I}{\sqrt{2\pi\, r}} f_r \text{ (elastic constants, } \theta)$$

$$\text{and } p_{r\theta} = \frac{K_I}{\sqrt{2\pi\, r}} f_{r\theta}$$

We have $f_\theta(0)=1$ to conform with the definition of K_I and, as before, f_r and $f_{r\theta}$ may be determined from the compatibility and equilibrium conditions. This may be worked out for any anisotropic condition but it is most useful for an orthotropic material with the crack in one of the principal elastic directions.

The solutions for f are difficult and best performed in terms of complex numbers. Some results for the orthotropic case [5] with elastic compliances a_{11}, a_{22}, a_{12} and a_{66}, where the crack is in direction 1, is, at $\theta=0$:

$$p_r = \frac{K_I}{\sqrt{2\pi\, r}} \left(\frac{a_{22}}{a_{11}}\right)^{1/2},\; p_\theta = \frac{K_I}{\sqrt{2\pi\, r}},\; p_{r\theta}=0 \qquad (3.27)$$

The appropriate computations for G may also be performed to give:

$$G = K_I^2 \left(\frac{a_{11}\, a_{22}}{2}\right)^{1/2} \left[\left(\frac{a_{22}}{a_{11}}\right)^{1/2} + \frac{2a_{12}+a_{66}}{2a_{11}}\right]^{1/2} \qquad (3.28)$$

For the isotropic case, $a_{11}=a_{22}=1/E$, $a_{12}=-\nu/E$ and $a_{66}=2(1+\nu)/E$, giving K_I^2/E as before.

It should be noted that the isotropic K_I value may be used to characterise fracture since any modulus ratios will appear only as scaling factors. A G_c may, of course, be measured directly by the usual compliance route or via equation (3.28) if the various parameters are known.

3.7 Cracks at interfaces [6, 7]

A situation of some practical importance in adhesion is that of a crack running along a bonded interface between dissimilar materials. Fig. 3.7 illustrates the problem in which the upper material has the elastic constants E_1 and ν_1 and the lower one E_2 and ν_2. Such a configuration would be expected to result in an energy release rate and, indeed, the DCB problem with dissimilar arms may be solved simply by adding the now different displacements so that the term E^{-1} in the expression for G (equation 2.23) may be replaced by $\frac{1}{2}(E_1^{-1}+E_2^{-1})$.

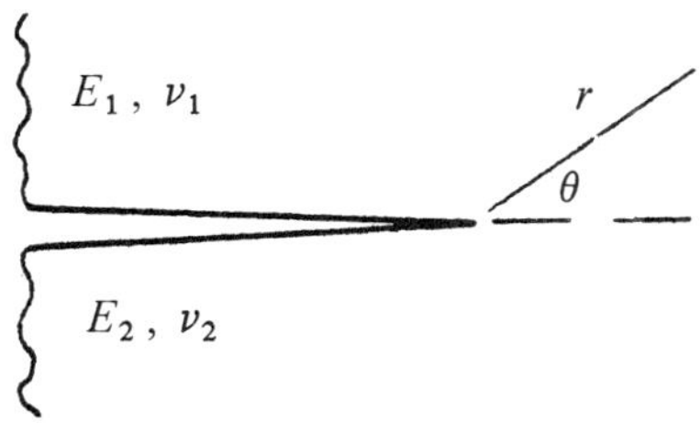

Fig. 3.7 Crack at an interface of bonded dissimilar materials

The nature of the local stress field may be found by using the results of Section 3.2.1, since, if a singularity is assumed, the function F (equation (3.9)) will apply for each half with a different set of four constants giving eight in all. The boundary conditions necessary to evaluate the constants and m are as follows:

$$p_\theta = p_{r\theta} = 0 \text{ at } \theta = \pm\pi \text{ as before}$$

and assuming perfect adhesion at the interface for which $\theta = 0$:

$$p_{\theta,1} = p_{\theta,2},\ p_{r\theta,1} = p_{r\theta,2},\ u_{\theta,1} = u_{\theta,2} \text{ and } u_{r,1} = u_{r,2}$$

giving the necessary extra four conditions. m must be the same for both halves in order to equate the stresses and displacements, and if we use the function:

$$F_{\theta,1} = A_m \cos m\,\theta + A_{2m} \cos(2-m)\,\theta + B_m \sin m\,\theta + B_{2m} \sin(2-m)\,\theta$$

for the top half, and:

$$F_{\theta,2} = C_m \cos m\,\theta + C_{2m} \cos(2-m)\,\theta + D_m \sin m\,\theta + D_{2m} \sin(2-m)\,\theta$$

for the bottom, the eight conditions result in the following equations:

$$0 = (A_m + A_{2m}) \cos m\,\pi + (B_m - B_{2m}) \sin m\,\pi \qquad \text{(i)}$$

$$0 = -[m\,A_m - (2-m)\,A_{2m}] \sin m\,\pi + [m\,B_m + (2-m)\,B_{2m}] \cos m\,\pi \qquad \text{(ii)}$$

$$0 = (C_m + C_{2m}) \cos m\,\pi - (D_m - D_{2m}) \sin m\,\pi \qquad \text{(iii)}$$

$$
\begin{aligned}
0 &= [m\, C_m - (2-m)\, C_{2m}] \sin m\pi + [m\, D_m \\
&\quad + (2-m)\, D_{2m}] \cos m\pi && \text{(iv)} \\
0 &= (A_m + A_{2m}) - (C_m + C_{2m}) && \text{(v)} \\
0 &= [m\, B_m + (2-m)\, B_{2m}] - [m\, D_m + (2-m)\, D_{2m}] && \text{(vi)} \\
0 &= \frac{1}{E_1}\left\{[m(1+\nu_1) + 2(1-\nu_1)]\frac{A_m}{2-m} - (1+\nu_1)\, A_{2m}\right\} \\
&\quad - \frac{1}{E_2}\left\{[m(1+\nu_2) + 2(1-\nu_2)]\frac{C_m}{2-m} - (1+\nu_2)\, C_{2m}\right\} && \text{(vii)} \\
0 &= \frac{1}{E_1}\left\{[m(1+\nu_1) - 4]\frac{B_m}{2-m} + (1+\nu_1)\, B_{2m}\right\} \\
&\quad - \frac{1}{E_2}\left\{[m(1+\nu_2) - 4]\frac{D_m}{2-m} + (1+\nu_2)\, D_{2m}\right\} && \text{(viii)}
\end{aligned}
\qquad (3.28)
$$

On substituting (v) into (iii) and (vi) into (v), and then evaluating (i) − (iii) and (iv) − (ii), we have:

$$0 = [(B_m - B_{2m}) + (D_m - D_{2m})] \sin m\pi$$

$$0 = \{[m\, A_m - (2-m)\, A_{2m}] + [m\, C_m - (2-m)\, C_{2m}]\} \sin m\pi$$

from which $\sin m\pi = 0$, $m = 0, \pm 1, \pm 2, \ldots$, is an eigen condition as in the homogeneous case, and:

$$0 = B_m - B_{2m} + D_m - D_{2m}$$

$$0 = m\, A_m - (2-m)\, A_{2m} + m\, C_m - (2-m)\, C_{2m}$$

may be used to evaluate the second eigen condition. Equations (iii) to (viii) now reduce to:

$$
\begin{aligned}
0 &= (A_m + A_{2m}) \cos m\pi - \left[\left(\frac{2\xi - m\,\phi}{2-m}\right) B_m - \phi\, B_{2m}\right] \sin m\pi && \text{(ix)} \\
0 &= \{[2(\xi - \phi) + \phi\, m]\, A_m - (2-m)\, \phi\, A_{2m}\} \sin m\pi \\
&\quad + [m\, B_m + (2-m)\, B_{2m}] \cos m\pi && \text{(x)}
\end{aligned}
\qquad (3.29)
$$

where the elastic constants are expressed in the parameters:

$$\xi = \frac{E_2}{E_1} \text{ and } \phi = 1 + \frac{E_2}{2}\left(\frac{1+\nu_1}{E_1} - \frac{1+\nu_2}{E_2}\right)$$

For the homogeneous case $\xi = \phi = 1$ and (ix) and (x) reduce to equations (3.10)

(iii) and (iv) as expected. Combining equations (i), (ii), (ix) and (x), we now have the results:

$$\frac{A_{2m}}{A_m}=\frac{2\alpha}{2-m}-1$$

and:

$$\frac{B_{2m}}{B_m}=-\frac{m(1+\phi)+2(\xi+1)}{(2-m)(1+\phi)}=\frac{2\alpha}{2-m}-\frac{m}{2-m}$$

where:

$$\alpha=\frac{\xi+1}{\phi+1}$$

so that the stress field may be expressed in terms of two arbitrary constants as in the homogeneous case. The eigen condition for defining m is:

$$\cot^2 m\,\pi+(\alpha-1)^2=0 \tag{3.30}$$

For $\xi=\phi$, we have $\alpha=1$ and $\cot m\,\pi=0$ as for the homogeneous case, for which:

$$m=\pm\frac{1}{2},\pm\frac{3}{2},\ldots$$

but when $\xi\neq\phi$, m is complex of the form:

$$m=\lambda+i\,\varepsilon$$

with:

$$\lambda=\pm\frac{1}{2},\pm\frac{3}{2},\ldots$$

and if we concern ourselves only with the singular term $\lambda=\frac{1}{2}$, then:

$$\varepsilon=\frac{1}{2\pi}\ln(2\alpha-1) \tag{3.31}$$

The stress distribution along the interface is thus:

$$p_{\theta,0}=\frac{2(1+e^{2\pi\varepsilon})}{9+4\varepsilon^2}\frac{A_m}{r^{1/2}}(3+2i)\,r^{-i\varepsilon} \tag{3.32}$$

Introducing an additional length constant h from A_m, the real part of this expression becomes:

$$p_{\theta,0}=\frac{A'_m}{r^{1/2}}\frac{2(1+e^{2\pi\varepsilon})}{9+4\varepsilon^2}\left[3\cos\left(\varepsilon\ln\frac{r}{h}\right)+2\varepsilon\sin\left(\varepsilon\ln\frac{r}{h}\right)\right]$$

These stresses are oscillatory for both $r\ll h$ and $r\gg h$ and have the highly improbable property of changing signs with increasing frequency as $r\to 0$. In addition, the crack face displacements also oscillate and near the crack tip interfere, giving a physically inadmissible solution. A more satisfactory solution in this regard is obtained by putting $m=\frac{1}{2}$, for which the relationships between all the constants remain unchanged but the stress-free crack face conditions of equations (i) to (iv) are violated. For symmetrical loading, the bond line stresses are given by:

$$p_{\theta,0} = \frac{K_I}{\sqrt{2\pi r}}, \; p_{r\theta,0} = 0$$

with a shear stress induced along the crack faces given by:

$$p_{r\theta,\pm\pi} = -\left(\frac{1-\alpha}{\alpha}\right)\frac{K_I}{\sqrt{2\pi r}}$$

Similarly, for mode II loading, we have:

$$p_{\theta,0} = 0, \; p_{r\theta,0} = -\frac{K_{II}}{\sqrt{2\pi r}} \text{ and } p_{\theta,\pm\pi} = \left(\frac{1-\alpha}{\alpha}\right)\frac{K_{II}}{\sqrt{2\pi r}}$$

These singular interfacial stresses represent contact between the crack faces induced during loading.

The strain energy release rate may now be computed in the usual way with the displacements now being the sum of those on the upper and lower faces. For mode I loading:

$$u_{r,\pm\pi} = 0, \; u_{\theta,+\pi} = -\frac{2}{\pi}K_I\frac{\sqrt{2\pi r}}{\alpha}\left[\frac{1}{E_1} - \left(\frac{1-\alpha}{2}\right)\left(\frac{1+\nu_1}{E_1}\right)\right]$$

$$\text{and } u_{\theta,-\pi} = +\frac{2}{\pi}K_I\frac{\sqrt{2\pi r}}{\alpha}\left[\frac{2\alpha-1}{E_2} + \left(\frac{1-\alpha}{2}\right)\left(\frac{1+\nu_2}{E_2}\right)\right]$$

Performing the necessary integration, we have:

$$G = \frac{K_I^2}{2}\left(\frac{1}{E_1} + \frac{1}{E_2}\right)\left(\frac{2\alpha-1}{\alpha^2}\right) \tag{3.34}$$

A similar analysis may be performed for a crack normal to an interface, in which case m is real but not equal to $\frac{1}{2}$. G calculations are not possible here since moving the crack tip from the interface invalidates the solution and path independence cannot be stipulated. These two examples illustrate some of the problems of exploring the crack tip stress fields when singularities are involved and clearly point to the need for some model to remove the singularity, a concept to be discussed in Chapter 5. G effectively does this by not defining the local behaviour in any precise way and it is obviously a preferable approach. Stress intensity factors are useful in the linear case, however, as a device for obtaining solutions where compliance determinations are inconvenient. This aspect will be developed in Chapter 4.

References for Chapter 3

[1] Williams, M. L., *J. Appl. Mech.*, **24**(1) (1957).

[2] Williams, J. G., *Stress Analysis of Polymers*, 2nd edn, Ellis Horwood, Chichester (1981).

[3] Rice, J. R. & Rosengren, G. F., *J. Mech. Phys. Solids*, **16** (1968) 1–12.

[4] Hutchinson, J. W., *J. Mech. Phys. Solids*, **16** (1968) 13–31.

[5] Sih, G. C., Paris, P. C. & Irwin, G. R., *Int. J. Fract. Mech.*, 1(3) (1965) 189–203.

[6] Malgslev, B. M., & Salganik, R. L., *Int. J. Fract. Mech.*, **1**(1965) 114.

[7] Williams, M. L., *Bull. Seism. Soc. Am.*, **49** (1959) 199.

4

Stress intensity factor solutions

4.1 Introduction

There are many loading configurations and geometries where compliance changes are difficult to compute so that the determination of fracture criteria via G is not convenient. An analysis of the stresses within such bodies, however, allows the stress field around the crack tip to be determined in terms of the loading and geometry, and the singular term may then be isolated and thus K found. In this chapter, some basic analytical results for K will be given and from these several important solutions derived. In addition, numerical results for specific, practically important, geometries will be given.

4.2 Stresses around an elliptical hole

Many useful results may be deduced from the solution for the stresses around an elliptical hole subjected to uniform, remote tension p normal to its major axis, as shown in Fig. 4.1. A complete solution may be effected most conveniently in terms of a complex stress function analysis [1], but for our purposes here we need only consider the stresses p_y and p_x along the line of the major axis ($y=0$). These may be expressed in terms of two functions:

$$p_x - p_y = f_1$$

and:

$$p_x + p_y = f_2 \tag{4.1}$$

where:

$$f_1 = -p\left\{1+\left(\frac{q^2-1}{\alpha^2-q}\right)\left[1+\left(\frac{q-1}{\alpha^2-q}\right)\left(\frac{3\alpha^2-q}{\alpha^2-q}\right)\right]\right\}$$

and:

$$f_2 = +p\left\{1+\frac{2(1+q)}{(\alpha^2-q)}\right\} \tag{4.2}$$

with:

$$q=\frac{a-b}{a+b} \text{ and } \alpha=\left(\frac{x}{a+b}\right)+\sqrt{\left(\frac{x}{a+b}\right)^2-q}$$

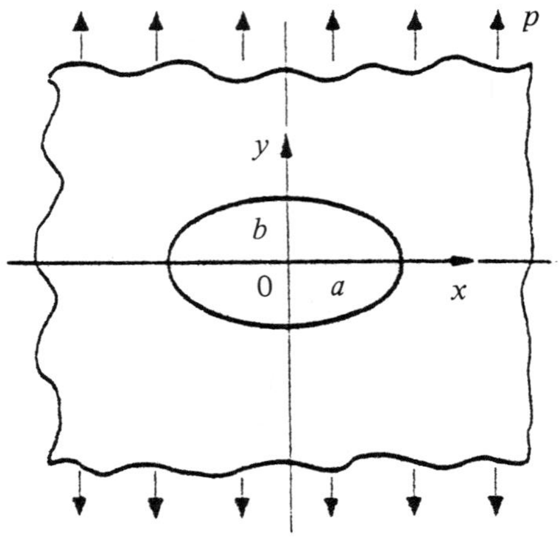

Fig. 4.1 An elliptical hole in a plate

A solution of interest which may be derived from this result is the stress at the end of the major axis at which the largest stress occurs, i.e. $x=a$. For this case, $\alpha=1$ and $f_1-f_2=-p(3+q/1-q)$ to give a final result of:

$$p_x=0 \text{ and } p_y=p\left(\frac{3+q}{1-q}\right)=p\left(1+\frac{2a}{b}\right)$$

The radius of curvature of the hole at this point $\rho=b^2/a$, so we may write the final result as:

$$p_y=p(1+2\sqrt{a/\rho}) \tag{4.3}$$

This expression is often useful for estimating stress concentration factors for notches for which a and ρ can be approximated by the ellipse. For a circular hole, $\rho=a$, giving the well known result of 3, while a crack may be modelled by $b=0$ giving, as expected, an infinite stress.

The result for a crack is of particular importance since it enables the singular term, and hence K_I, to be determined. For a crack $b=0$, so that $q=1$ and:

$$\alpha=\frac{x}{a}+\sqrt{\frac{x^2-a^2}{a^2}}$$

giving the final result:

$$p_y=\frac{p\,x}{\sqrt{x^2-a^2}}=p_x+p \tag{4.4}$$

The singular term may be deduced by rewriting the expression in terms of $r=x-a$ and noting that $p_y\equiv p_{\theta,\theta=0}$ and $p_x\equiv p_{r,\theta=0}$. Thus:

$$p_{\theta,\theta=0} = p_{r,\theta=0} + p = p\left(\frac{a}{2r}\right)^{1/2} + \frac{3}{8}p\left(\frac{2r}{a}\right)^{1/2} - \ldots \qquad (4.5)$$

and on comparing the singular term with equation (3.11), we have:

$$K_I = p\sqrt{\pi a} \qquad (4.6)$$

and a similar analysis for shear loading gives an identical expression for K_{II} with p replaced by the remote shear stress. The expression for G for a remotely loaded crack of length $2a$ in an infinite plate may now be deduced, i.e.

$$G = \frac{K_I^2}{E} = \frac{p^2 \pi a}{E} \qquad (4.7)$$

One further result of interest is that for a crack with a finite tip radius $\rho \ll a$. We now have $b = \sqrt{a\rho}$ and $q = 1 - 2\sqrt{\rho/a}$, from which:

$$\alpha = 1 - \sqrt{\rho/a} + \sqrt{(\rho + 2r)/a}$$

where $r = x - a$, as before. Thus, the stress at r may be deduced as:

$$p_y = p\left[1 + \frac{(a)^{1/2}\,(2\rho + 2r)}{(\rho + 2r)^{3/2}}\right] \qquad (4.8)$$

For $r = 0$, we recover equation (4.3) and for $\rho = 0$ the usual singular term. For $\rho \neq 0$, there is, of course, no true stress intensity factor since the stresses are not singular but a useful result may be derived by expressing via a characteristic stress and distance as in equation (3.22). Thus:

$$K_I = p_0\sqrt{2\pi\, l_0}$$

and if we now assume that an equivalent stress state occurs in the blunt notch for p_0 acting at a distance l_0, then:

$$p\sqrt{\pi a} = K_I\,\frac{(1 + \rho/2l_0)^{3/2}}{(1 + \rho/l_0)} \qquad (4.9)$$

(since $p_0 \gg p$ for $\rho \ll a$). It is important to note that the crack tip stress state may be defined in a single parameter K_I for $\rho \ll a_0$, i.e. a sharp crack, but that when this is not so, *two* parameters are needed since the relationship of ρ to l_0 must be defined. For $\rho \gg l_0$, ρ becomes the controlling length factor since:

$$p\sqrt{\pi a} \rightarrow p_0\,\frac{\sqrt{2\pi\rho}}{2\sqrt{2}}$$

4.3 Solutions for K_I and Y^2 factors

The most important result is that given in the previous section (4.2) for a uniformly loaded infinite plate containing a crack of length $2a$ for which:

$$K_I = p\sqrt{\pi a}$$

The result illustrates the value of the stress intensity factor approach in that it is deduced from the singular term derived from the stress analysis which turns out to be much more convenient than the compliance approach for certain geometries. From dimensional grounds, K_I must be a product of a stress and a square root of length and, in this case, since there is but one of each, the form $K_I \propto p\sqrt{a}$ is inevitable and the analysis serves to provide the necessary constant. For systems of finite dimensions, other length factors appear and the constant term may now be a function of dimensionless length ratios. If we define a width factor as D, then we have:

$$K_I^2 = Y^2(a/D)\, p^2\, a \tag{4.10}$$

A useful analytical result may be obtained for a periodic array of colinear cracks, as shown in Fig. 4.2 [2], and the expression for K_I is:

$$K_I = p\sqrt{\pi a}\left[\left(\frac{2D}{\pi a}\right)\tan\left(\frac{\pi a}{2D}\right)\right]^{1/2} \tag{4.11}$$

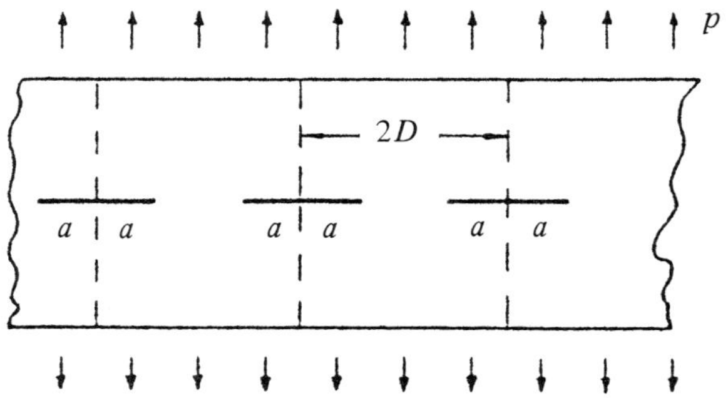

Fig. 4.2 Periodic array of colinear cracks

This gives a useful approximation to Y^2 for three practically important geometries: the centre notch plate (CN), the double-edge notch plate (DEN), and the single-edge notch plate (SEN). For the latter two, in which the crack is in a free edge, a better approximation is produced by a correction factor of $(1.12)^2$ to account for additional deformation at the free surface. Thus, we have:

For CN: $$Y^2 \simeq \pi\left(\frac{2}{\pi}\frac{D}{a}\right)\tan\left(\frac{\pi}{2}\frac{a}{D}\right)$$

(4.12)

and for DEN and SEN: $$Y^2 \simeq 3.94\left(\frac{2}{\pi}\frac{D}{a}\right)\tan\left(\frac{\pi}{2}\frac{a}{D}\right)$$

Many solutions have been obtained for Y^2 using numerical methods (see [3, 4]), and in practice it is usually most convenient to refer to these results. A selection of common testing geometries is shown in Fig. 4.3 and Table 4.1 gives numerical solutions for the various Y^2 factors. For tensile loading, the stress in equation (4.10) is taken as P/BD and for bending $6M/BD^2$. Equation (4.12) is also shown for the CN case and there is quite good agreement over most of the range (15%). The SEN case is markedly different which arises from induced bending from lack of symmetry.

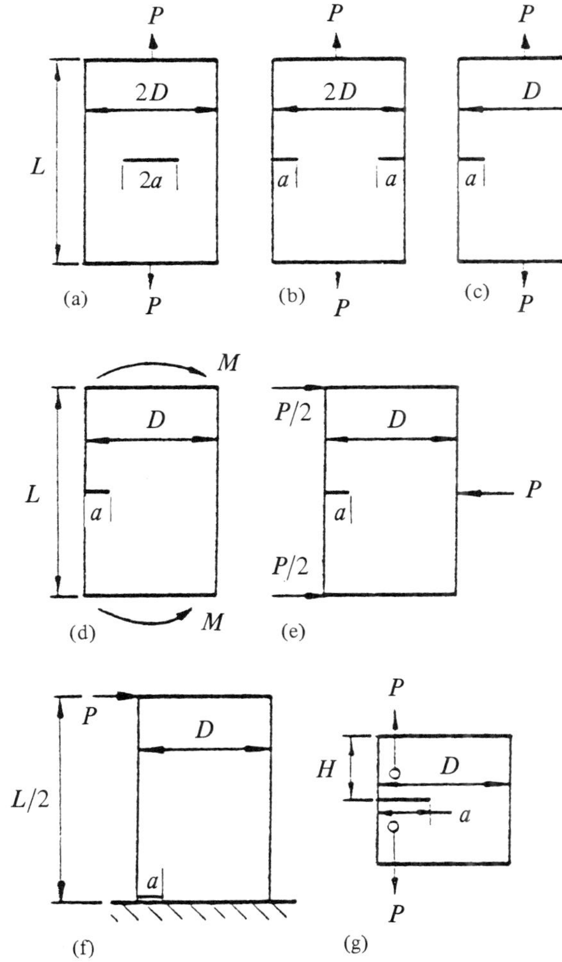

Fig. 4.3 Common testing specimen geometries; (a) centre notch plate (CN) in tension; (b) double-edge notch (DEN) in tension; (c) single-edge notch (SEN) in tension; (d) single-edge notch plate in bending; (e) three-point bending – Charpy; (f) cantilever bending – Izod; (g) compact tension (CT)

Table 4.1

Finite width correction factors, Y^2

				Tan formula	Approximate method	
				Equation (4.12)	Equation (4.13)	Equation (4.14)
a/D	CN	DEN	SEN	CN	CN, DEN	SEN
0.05	3.17	3.97	4.05			
0.1	3.20	3.99	4.42	3.18	3.32	3.94
0.15	3.25	3.99	5.05			
0.2	3.30	3.98	5.92	3.25	3.55	5.28
0.25	3.37	3.96	7.10			
0.3	3.48	3.96	8.69	3.40	3.85	7.71
0.35	3.62	3.97	10.86			
0.4	3.81	4.01	13.94	3.63	4.28	12.56
0.45	4.06	4.10	18.44			
0.5	4.38	4.23	25.16	4.00	4.91	23.76
0.55	4.80	4.43	35.36			
0.6	5.32	4.71	51.03	4.59	5.94	55.09

	CT	Pure	SEN bending	
a/D	$H/D=0.6$	bending	$L/D=4$	$L/D=8$
0.05	—	3.60	3.28	3.44
0.1	—	3.43	3.05	3.24
0.15	—	3.41	2.97	3.19
0.2	—	3.49	3.02	3.26
0.25	—	3.68	3.16	3.42
0.3	112.4	3.97	3.41	3.69
0.35	121.0	4.38	3.79	4.08
0.4	132.3	4.97	4.33	4.65
0.45	153.8	5.81	5.11	5.46
0.5	182.3	7.03	6.26	6.65
0.55	246.5	8.82	7.95	8.40
0.6	302.8	11.47	10.49	11.00

Y^2 may be computed from the polynomials given in reference [3] from:

$$Y = \sum_0^4 A_n \left(\frac{a}{D}\right)^n$$

where A_n has the values:

		A_0	A_1	A_2	A_3	A_4
Tension	CN	1.77	0.227	−0.51	2.70	—
	DEN	1.98	0.360	−2.12	3.42	—
	SEN	1.99	−0.410	18.70	−38.48	53.85
SEN	Pure bending	1.99	−2.47	12.97	−23.17	24.80
	Three-point, $L/D=8$	1.96	−2.75	13.66	−23.98	25.22
	Three-point, $L/D=4$	1.93	−3.07	14.53	−25.11	25.80

An approximate method for estimating Y^2 may be used which can incorporate bending effects [5] and gives some insight into the origins of the increase in Y^2. Here, we write:

$$K_I = p' \sqrt{\pi a}$$

where p' is now the effective stress augmented by finite width effects. This is made up of the mean stress over the cracked section (with no crack present), p_m, and p_n, the nominal stress crack tip ignoring stress concentration effects. The mean stress at the crack tip is thus $(p_m + p_n)/2$, while at the crack centre (or edge) it is p_m giving an overall average of:

$$\frac{1}{2}\left(p_m + \frac{p_m + p_n}{2}\right) = \frac{3p_m + p_n}{4} = p'$$

This may be regarded as a weighted mean of the average stress and the nominal. Both CN and DEN have the same result since $p_m = p$ and $p_n = p/(1-a/D)$, giving:

$$Y^2 = \pi \left(\frac{1 - \frac{3}{4}a/D}{1 - a/D}\right)^2 \tag{4.13}$$

while SEN has a bending effect in p_n so that $p_m = p$ and:

$$p_n = p\left[\frac{1}{1-a/D} + \frac{\frac{3}{4}a/D}{(1-a/D)^2}\right]$$

giving:

$$Y^2 = \pi \left[\frac{1 - a/D + \frac{3}{4}(a/D)^2}{(1-a/D)^2}\right]^2 \tag{4.14}$$

These results are also shown in Table 4.1 and are in quite good agreement ($\simeq 10\%$) with the computed values over the whole range. The method is useful for complicated cases as, for example, with a double-edge notched sheet of width $2D$, notch lengths a_1 and a_2 ($a_1 > a_2$), and with a load a distance e from the centre line. Here, we have:

$$Y^2 = \pi \left\{ \frac{3}{4} \left[1 + \frac{3e}{D} \left(1 - \frac{a_1}{2D} \right) \right] + \frac{1}{4} \left[\frac{1 + 3e/D + a_1/D - 2a_2/D}{(1 - (a_1 + a_2)/2D)^2} \right] \right\}^2 \quad (4.15)$$

for K_I at the crack tip a_1.

4.4 Solutions for energy calibration factor ϕ

As shown in Section 2.4.1, G may be determined from the elastic energy stored, U_3, in a body, since:

$$G = \frac{U_3}{B\,D\,C} \frac{dC}{d(a/D)}$$

(equation (2.20)) and the calibration factor is given as:

$$\phi = \frac{C}{dC/d(a/D)}$$

where C is the compliance of the body. Clearly, C can be deduced from Y^2 via the expression for G and if we take the simple case of a plate in tension with a small edge notch a, then we may write:

$$G = \frac{P^2}{2B} \frac{dC}{da} = \frac{K_I{}^2}{E} = \frac{\pi\, p^2\, a}{E}$$

and $P = p\,B\,D$. Thus:

$$\frac{dC}{da} = \frac{2\pi\, a}{B\,D^2\,E}$$

and, on integrating:

$$C = \frac{\pi\, a^2}{B\,D^2\,E} + C_0$$

where C_0 is the compliance for $a = 0$ and is L/BDE, where L is the specimen length. Finally, we may deduce an expression for ϕ:

$$\phi = \frac{C}{dc/d(a/D)} = \frac{1}{2}\left(\frac{a}{D}\right) + \frac{1}{2\pi}\left(\frac{L}{D}\right)\left(\frac{D}{a}\right) \quad (4.16)$$

and for $a/D \ll 1$, $\phi \rightarrow (1/2\pi)\,(L/a)$, i.e. inversely proportional to crack length.

If we now turn to the more practical case of three-point bending as used in the Charpy impact test (Fig. 4.3(e)), then we may write:

$$K_I{}^2 = Y^2 \left[\frac{6M}{B\,D^2} \right]^2 a = E\,\frac{P^2}{2B}\frac{dC}{da} \text{ and } M = \frac{P\,L}{4}$$

so that: $$\frac{dC}{dx}=\frac{9}{2}\frac{L^2}{B\,D^2\,E}\,Y^2(x)\,x \text{ where } x=\frac{a}{D}$$

If we now note that $C_0=L^3/4EBD^3$ in this case, we have a final result:

$$\phi=\frac{\int Y^2(x)\,x\,dx}{Y^2(x)\,x}+\frac{L}{18D}\frac{1}{Y^2(x)\,x} \tag{4.17}$$

Thus, ϕ may be computed from Y^2 functions and Table 4.2 gives values for the Charpy, Izod and SEN tension geometries. Those for Charpy and tension are deduced from the Y^2 values in Table 4.1, but the Izod values are from experimental compliance measurements [6] since the theoretical solution which would assume perfect clamping is found to be significantly in error. Fig. 4.4 shows the Charpy and Izod values compared for $L/D=6$, together with the infinite plate result for which $Y^2=\pi$ is substituted into equation (4.17).

For very deep notches, Y^2 is large and the result is dominated by the first term. As may be deduced from the approximate solution given in Section 4.3, as $x\to 1$, $Y^2\propto(1-x)^{-n}$ when $n=2$ for tension and $n=4$ for bending. The asymptotic solution of equation (4.17) at $x=1$ then gives the result:

$$\phi\to\left.\frac{(1-x)}{n-1}\right|_{x\to 1}$$

Table 4.2

Charpy energy calibration factor

	ϕ				
a/D	$L/D=4$	$L/D=6$*	$L/D=8$	$L/D=10$*	$L/D=12$*
0.05	1.383	1.996	2.612	3.231	3.845
0.1	0.782	1.101	1.424	1.748	2.070
0.15	0.576	0.789	1.005	1.224	1.433
0.2	0.469	0.624	0.781	0.942	1.100
0.25	0.402	0.520	0.639	0.761	0.880
0.3	0.354	0.446	0.538	0.633	0.724
0.35	0.318	0.389	0.461	0.535	0.605
0.4	0.287	0.343	0.398	0.455	0.510
0.45	0.260	0.302	0.345	0.388	0.430
0.5	0.234	0.266	0.298	0.331	0.362
0.55	0.210	0.234	0.257	0.280	0.304
0.6	0.187	0.205	0.222	0.239	0.255

*From interpolated *Y* values

Table 4.2 (*continued*)

Izod energy calibration factor

	ϕ				
a/D	$L/D=4$	$L/D=6$	$L/D=7$	$L/D=9$	$L/D=11$
0.05	1.738	1.990	2.125	2.500	—
0.1	1.060	1.165	1.230	1.360	1.570
0.15	0.748	0.834	0.874	0.953	1.050
0.2	0.600	0.642	0.670	0.730	0.810
0.25	0.511	0.540	0.555	0.600	0.679
0.3	0.452	0.480	0.489	0.519	0.587
0.35	0.410	0.438	0.447	0.473	0.525
0.4	0.387	0.410	0.420	0.441	0.478
0.45	0.370	0.391	0.398	0.419	0.440
0.5	0.360	0.379	0.385	0.399	0.411

SEN tension energy calibration factor

	ϕ			
a/D	$L/D=4$	$L/D=8$	$L/D=16$	$L/D=24$
0.04	12.50	24.99	49.96	74.93
0.08	5.93	11.83	23.62	35.41
0.1	4.57	9.09	18.14	27.18
0.2	1.77	3.46	6.83	10.21
0.3	0.87	1.64	3.18	4.71
0.4	0.48	0.84	1.55	2.27
0.5	0.28	0.44	0.76	1.08
0.6	0.18	0.25	0.38	0.51
0.8	0.12	0.13	0.15	0.17

This result for $n=4$, pure bending, is also shown in Fig. 4.4. The nature of the solution is illustrated in Fig. 4.5(i) where ϕ is plotted as a function of L/D for $x=0.5$. To compensate for the additional rotation at the clamp, it is clear that an addition of 8 must be made to the L/D value for the Izod data. When this is done, they lie on a single line, extrapolating to the pure bending result when $L/D=0$. When the same data are plotted versus D/L (Fig. 4.5(ii)), the line

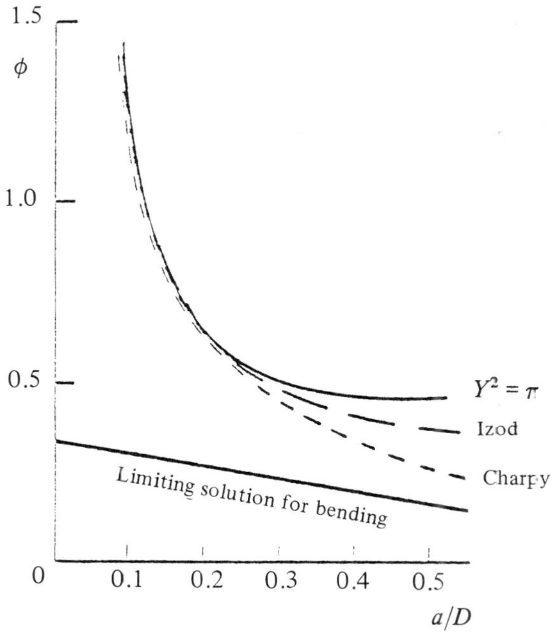

Fig. 4.4 Energy calibration factors at $L/D=6$

extrapolates to the simple tension value for $D/L=0$. An approximation to $\phi=\frac{1}{2}(1-x)$ is obtained in the Charpy geometry when $L/D=5$.

4.5 Loaded crack faces

Some useful results may be obtained for K_I for a crack of length $2a$ in an infinite sheet loaded with symmetrical splitting forces P a distance b from the centre, as shown in Fig. 4.6. The result is [7]:

$$K_I=\frac{2}{\sqrt{\pi}}\frac{P}{B}\frac{\sqrt{a}}{\sqrt{a^2-b^2}} \tag{4.18}$$

and there is a similar result in mode II with shearing forces. For a point load at the centre, we have $2P\equiv P$ and $b=0$, so that:

$$K_I=\left(\frac{P}{B}\right)\frac{1}{\sqrt{\pi a}}$$

(4.19)

and:

$$G=\frac{1}{E}\left(\frac{P}{B}\right)^2\left(\frac{1}{\pi a}\right)$$

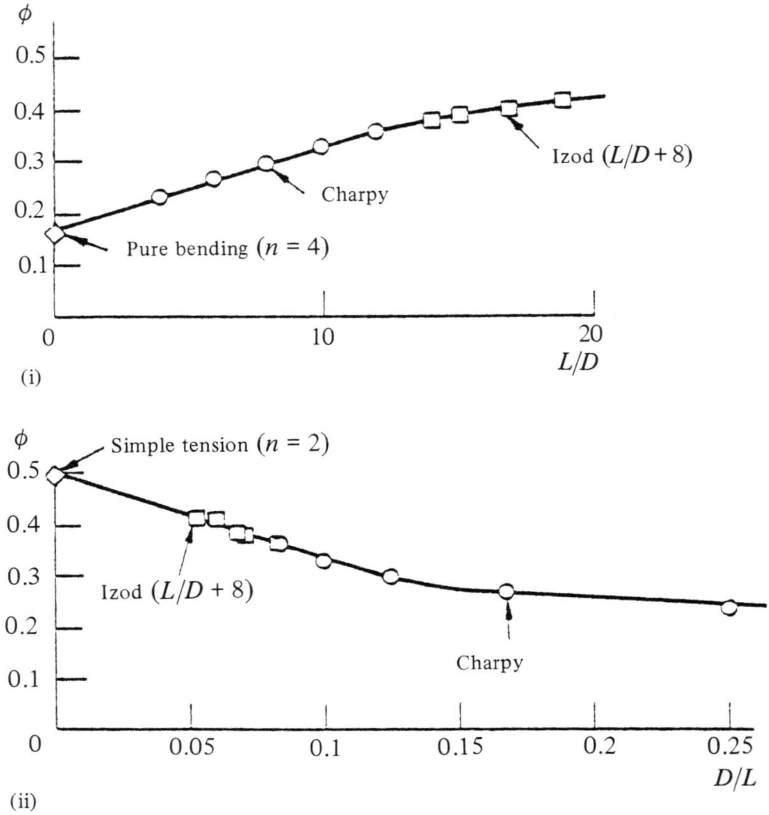

Fig. 4.5 ϕ as a function of L/D for $x = 0.5$

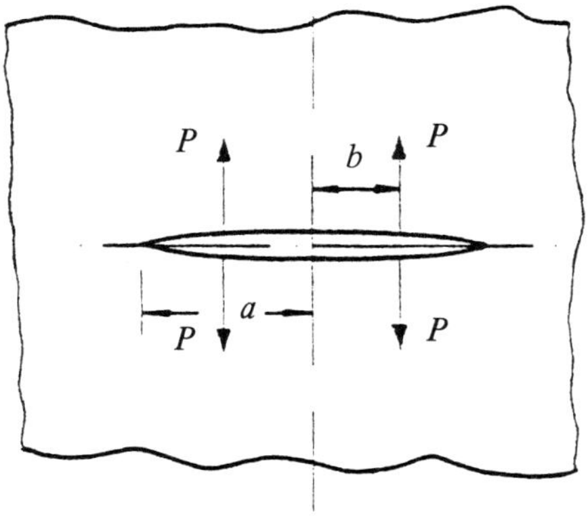

Fig. 4.6 A crack in an infinite sheet with symmetrical splitting forces P

Solutions for distributed symmetrical loads on the crack faces may be obtained by expressing P as due to a local pressure p_ξ acting at a distance ratio $\xi = b/a$ from the centre. Thus, $P \equiv B\, p_\xi\, db$ and the increment of K_I at the crack tip due to this force is:

$$dK_I = \frac{2p_\xi}{\pi} \frac{d\xi}{\sqrt{1-\xi^2}} \sqrt{\pi a}$$

and for a distribution from ξ_1 to ξ_2, we have:

$$K_I = \sqrt{\pi a}\, \frac{2}{\pi} \int_{\xi_1}^{\xi_2} \frac{p_\xi\, d\xi}{\sqrt{1-\xi^2}} \tag{4.20}$$

For $p_\xi = p$ constant, we have $K_I = p\sqrt{\pi a}$ which is an important result since it shows that a uniform pressure over the crack faces is exactly equivalent to the same value of uniform stress remote from the crack. For a body subjected to a uniform pressure which also acts within the crack, the nett K_I value is zero since the external pressure gives a K_I value of $-p\sqrt{\pi a}$ while that within the crack gives $p\sqrt{\pi a}$.

If a uniform stress p_c is applied over a zone of length c at the crack tip, as shown in Fig. 4.7, then:

$$K_I = p_c\sqrt{\pi a}\left[\frac{2}{\pi} \int_{1-c/a}^{1} \frac{d\xi}{\sqrt{1-\xi^2}}\right]$$

and if we let $\xi = \cos\theta$ and $\theta_c = \cos^{-1}(1-c/a)$, then this becomes:

$$K_I = p_c\sqrt{\pi a}\left(\frac{2}{\pi}\theta_c\right)$$

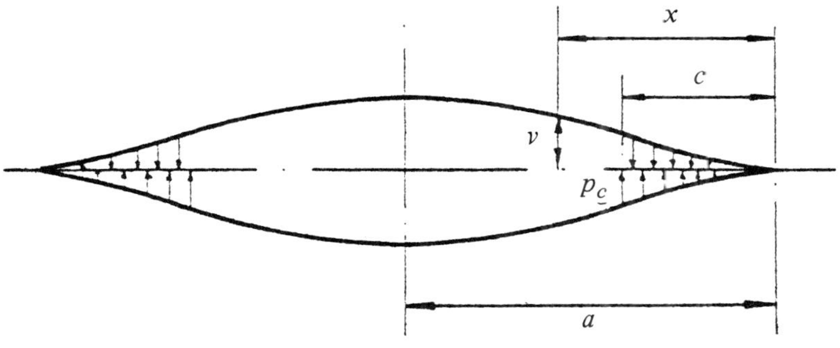

Fig. 4.7 Schematic displacements for crack with a crack tip cohesive stress zone

This result is used in the cohesive zone theory of fracture since the singularity due to a remote stress p may be put to zero by a cohesive stress p_c acting over the zone c such that the K_I values are equal and opposite. If $K_I = p\sqrt{\pi a}$, then we have:

$$\frac{c}{a} = 1 - \cos\frac{\pi}{2}\frac{p}{p_c} \simeq \frac{\pi^2}{8}\left(\frac{p}{p_c}\right)^2 \text{ for } \frac{p}{p_c} \ll 1$$

i.e.

$$c \simeq \frac{\pi}{8}\frac{K_I{}^2}{p_c{}^2} \tag{4.21}$$

Thus, $K_I = p_c\sqrt{2\pi\,(4c/\pi)}$ and on comparing with equation (3.22), we have a further possible interpretation of p_0 and l_0.

The displacement in the crack may be derived by a complex number stress function solution [8], giving the result for plane stress of:

$$\frac{\pi\, E\, v}{p_c\, a} = \cos\theta \ln\left[\frac{\sin(\theta_c - \theta)}{\sin(\theta_c + \theta)}\right]^2 + \cos\theta_c \ln\left[\frac{\sin\theta_c + \sin\theta}{\sin\theta_c - \sin\theta}\right]^2 \tag{4.22}$$

The form of this profile may be expressed more simply for the case of a small zone for which $x = b - a \ll a$. For this case, equation (4.22) may be expressed as:

$$\frac{\pi}{8}\frac{E}{p_c}\frac{(2v)}{c} = \sqrt{x/c} - \frac{1}{2}\left(1 - \frac{x}{c}\right)\ln\left[\frac{1+\sqrt{x/c}}{1-\sqrt{x/c}}\right] \tag{4.23}$$

since $\sin\theta \simeq \sqrt{2x/a}$ and $\sin\theta_c \simeq \sqrt{2c/a}$, and the resulting cusp-shaped profile is shown in Fig. 4.8, together with the original profile for K_I alone. An important feature of this result is that there is a distinct displacement at the end of the zone which can be regarded as the crack tip when a zone of cohesive force is present. This is termed the crack opening displacement (COD) shown as $\delta = (2v)_{x=c}$, and in the general case is given, from equation (4.22), as:

$$\frac{\pi}{8}\frac{E\,\delta}{p_c\, a} = -\cos\theta_c \ln\cos\theta_c,\ \cos\theta_c = 1 - \frac{c}{a} \tag{4.24}$$

and for the special case of $x/c \ll 1$, we have, from equation (4.23) when $x = c$:

$$\delta = \frac{8}{\pi}\frac{p_c}{E}c = \frac{K_I{}^2}{p_c\, E} \tag{4.25}$$

It should be noted that:

$$G = \frac{K_I{}^2}{E} = \delta\, p_c$$

which is consistent with equation (3.26), since the constant p_c value is equivalent to $n = 0$.

4.6 Mixed mode fracture

A practically important problem where a local crack tip stress analysis is essential is that of fracture under combined loading in both mode I and mode II. All the G calculations discussed previously assumed colinear crack propagation,

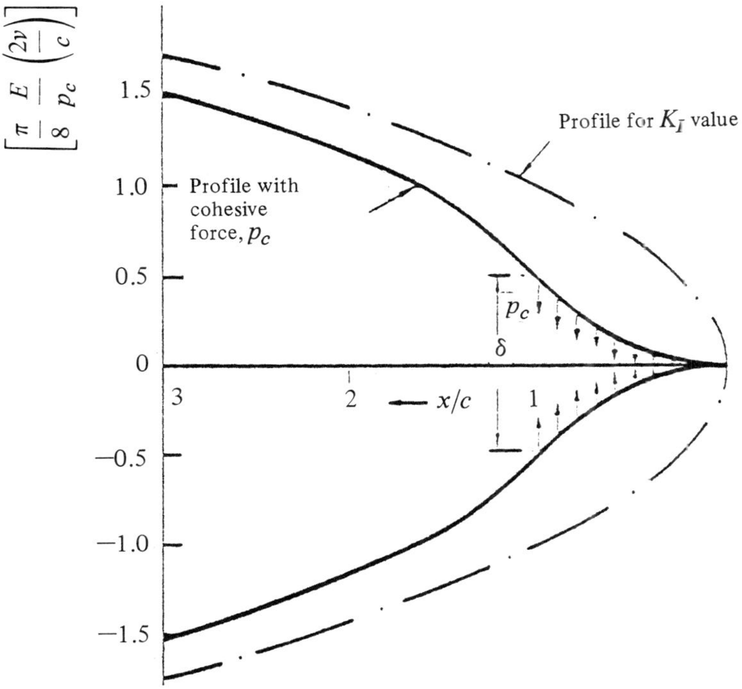

Fig. 4.8 Crack tip displacement profile for a cohesive zone

which does not occur when mode II is present. In pure mode II, one could postulate a K_{IIc} condition of fracture, but experience shows that if such a condition exists, it is a very high value and much greater than K_{Ic}. K_{IIc} values for colinear crack extension have been measured for highly anisotropic materials such as wood [9], where $K_{IIc} \simeq 5K_{Ic}$, but they are extremely rare in practice and it is more useful to assume that all fractures are locally in mode I. If this is so, then the direction of propagation will be when p_θ is a maximum close to the crack tip to give a sensible extension of the crack. The condition $dp_\theta/d\theta = 0$ is given when $p_{r\theta} = 0$ which, from equations (3.11), gives:

$$K_I \sin\frac{\bar{\theta}}{2}(1+\cos\bar{\theta}) = K_{II}\cos\frac{\bar{\theta}}{2}(1-3\cos\bar{\theta})$$

where $\bar{\theta}$ is the angle of crack propagation with respect to the original crack line, as shown in Fig. 4.9. The fracture condition is determined by the expression for p_θ, from which we have:

$$K_{Ic} = K_I\frac{1}{2}\cos\frac{\bar{\theta}}{2}(1+\cos\bar{\theta}) - K_{II}\frac{3}{2}\sin\frac{\bar{\theta}}{2}(1+\cos\bar{\theta})$$

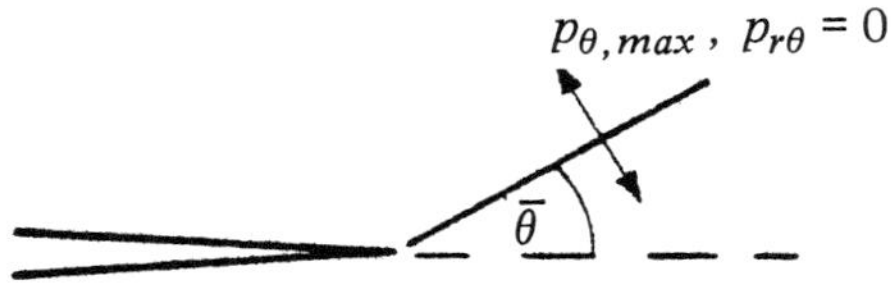

Fig. 4.9 Crack propagation under mixed mode loading

and by suitable rearrangement, we have the fracture condition defined as:

$$\frac{K_I}{K_{Ic}} = \frac{3 \cos \bar{\theta} - 1}{\cos \bar{\theta}/2 \, (1 + \cos \bar{\theta})}$$

$$\frac{K_{II}}{K_{Ic}} = -\frac{2 \sin \bar{\theta}/2}{(1 + \cos \bar{\theta})} \qquad (4.26)$$

For $K_{II} = 0$, we have the usual mode I result of $\bar{\theta} = 0$, as expected, while in pure mode II with $K_I = 0$:

$$\cos \bar{\theta} = \tfrac{1}{3}, \ \bar{\theta} \simeq -70.3°$$

(Note: The negative angle is necessary to give positive K_{II} values).

and $K_{II} = \sqrt{3} \, 2 \, K_{Ic}$, i.e. an apparent K_{IIc} value of $\sqrt{3} \, 2 \, K_{Ic}$. The locus of K_I and K_{II} for fracture, together with appropriate $\bar{\theta}$ values, is shown in Fig. 4.10.

A frequently used test to-study mixed mode loading is a plate in simple tension with a crack inclined at an angle β to the vertical axis, as shown in Fig. 4.11. The stress normal to the crack is now $p \sin^2 \beta$ and the shear stress $p \sin \beta \cos \beta$, giving:

$$K_I = p \sqrt{\pi a} \sin^2 \beta \text{ and } K_{II} = p \sqrt{\pi a} \sin \beta \cos \beta$$

On substituting into equations (4.26), we have:

$$\frac{K_{II}}{K_I} = \cot \beta = -\frac{\sin \bar{\theta}}{3 \cos \bar{\theta} - 1} \qquad (4.27)$$

and the resulting curve for $\bar{\theta}$ as a function of β is shown in Fig. 4.12, together with the line for propagation normal to the applied stress for which $\beta = (\pi/2) - \bar{\theta}$. At $\beta = 0$, we have a vertical crack giving the shear result, whereas, in reality, the propagation would be normal to the crack direction, i.e. at 90°, so that the curve rises vertically at that point as shown. These angles apply only at the very first stage of propagation and as the crack grows, it does tend to the direction normal to the applied load. This may be understood from Figs. 4.11 and 4.12 where $\bar{\theta}$ tends to increase the effective β value so that subsequent $\bar{\theta}$ values tend to zero. An approximate model for the crack path may be obtained by noting that the change in crack length, S, may be expressed in terms of the change in effective β

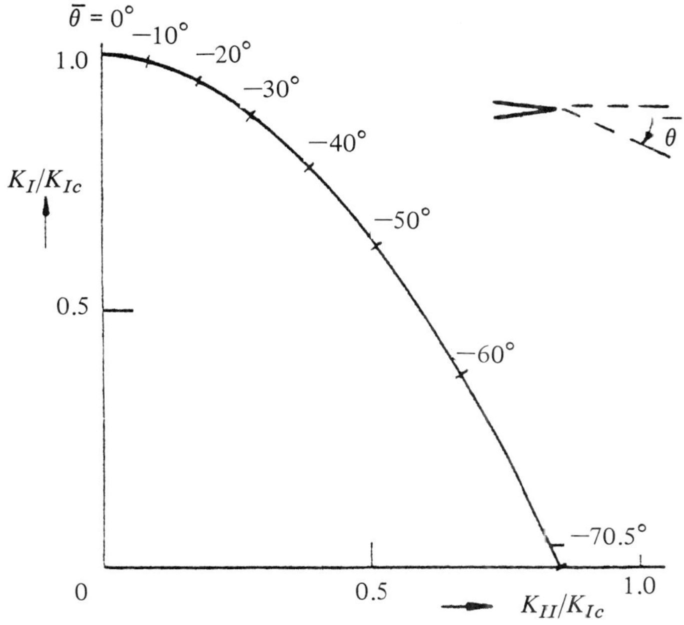

Fig. 4.10 Mixed mode failure criteria – loaded with K_I and K_{II} and failing at K_{Ic} at $\theta = \bar{\theta}$

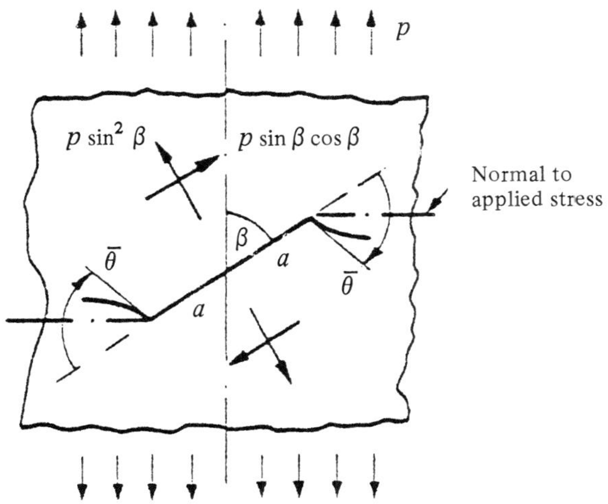

Fig. 4.11 The angled crack

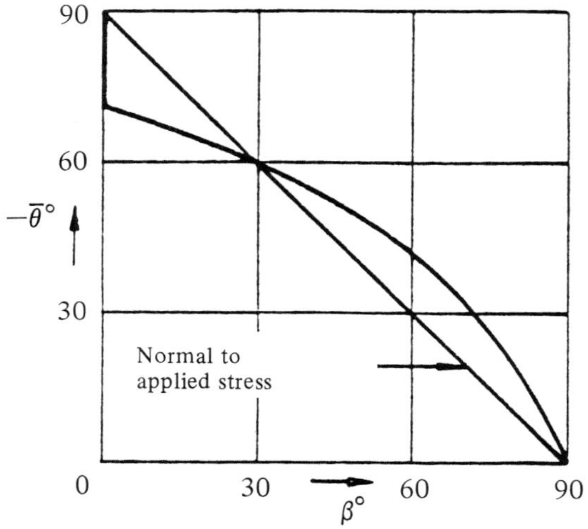

Fig. 4.12 Crack propagation direction for the angled crack

such that:

$$a\,d\beta = -dS \sin\bar{\theta}$$

as shown in Fig. 4.13.

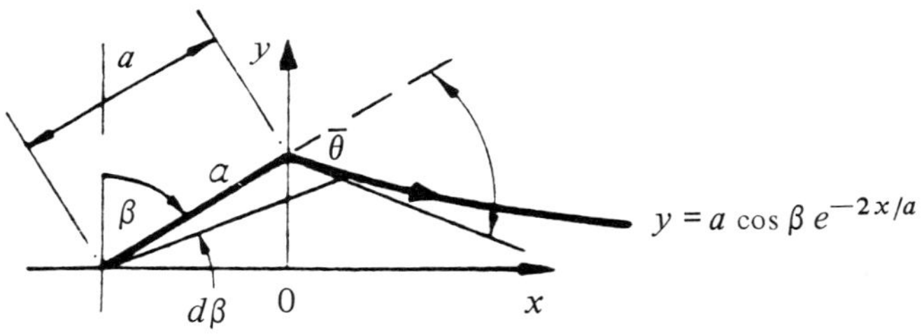

Fig. 4.13 Approximate crack propagation profile

On substituting for $d\beta/d\bar{\theta}$ using equation (4.27), we have:

$$\frac{dS}{d\bar{\theta}} = -\frac{a}{2\sin\bar{\theta}}\left[\frac{3-\cos\bar{\theta}}{(2\cos\bar{\theta}-1)^2+\cos\bar{\theta}}\right] \tag{4.28}$$

which may be integrated to give S as a function of $\bar{\theta}$. For small values of $\bar{\theta}$, this tends to:

$$\frac{dS}{d\bar{\theta}} \to -\frac{a}{2\bar{\theta}}$$

which may be rewritten as: $$\frac{d^2y}{dx^2} = -\frac{2}{a}\frac{dy}{dx}$$

giving a final exponential crack profile of the form:

$$y = a \cos \beta \, e^{-2x/a} \tag{4.29}$$

A further case of interest is that of a crack under tension with a small oscillating shear stress applied. Such systems are used in high speed crack propagation studies where stress waves are applied to a loaded specimen and sometimes occur in practical cases such as pipe rupture. Here, we have:

$$K_{II} = K_{II0} \sin \omega t$$

where ω is the frequency, and if the crack is assumed to be moving with a velocity V, then this may be written in terms of the crack propagation direction x such that:

$$K_{II} = K_{II0} \sin \frac{\omega x}{V}$$

On substituting into equation (4.26), we have:

$$\left(\frac{K_{II0}}{K_{Ic}}\right) sin \frac{\omega x}{V} = -\frac{2 \sin \bar{\theta}/2}{(1+\cos \bar{\theta})}$$

For a small perturbation, $\bar{\theta} \ll \pi/2$ and may be approximated to dy/dx giving:

$$\frac{dy}{dx} = -2\left(\frac{K_{II0}}{K_{Ic}}\right) \sin \frac{\omega x}{V}$$

and a final sinusoidal crack profile of:

$$y = 2\left(\frac{K_{II0}}{K_{Ic}}\right) \frac{V}{\omega} \left(\cos \frac{\omega x}{V} - 1\right) \tag{4.30}$$

4.7 Cracks in two dimensions

In all the foregoing analysis, it has been assumed that a crack will propagate in a manner which can be predetermined. Thus, in the analysis of Chapter 2, the change of area was assumed to occur in a self similar manner from the original shape so that it could be defined in terms of a single parameter. The simplest case, and the most widely used, is of a through-crack in a uniform thickness sheet, for which the crack length alone defines the area change. Implicit in this assumption is that the local fracture condition is reached simultaneously along the whole crack front and this is frequently the case for simple geometries. There are, however, circumstances for which this condition does not hold. These may be grouped under three main headings:

1. Stresses varying along the crack front giving rise to a variation in K_I (e.g. lateral bending of a sheet).
2. Crack front curvature giving rise to a variation in K_I, even for uniform loading (e.g. an elliptical crack).

3. Variations in K_{Ic} along a crack front (e.g. plane stress effects at specimen surfaces giving rise to tough skins (see Section 5.3)).

In any combination of these three cases, the situation occurs in which $K_I \geqslant K_{Ic}$ over some parts of the crack front but not in others, and thus growth occurs in a non-self similar manner. In principle, there is no difficulty in dealing with this situation since an elastic analysis of the original profile AD will determine K_I as shown in Fig. 4.14. In the region BC, $K_I > K_{Ic}$ so that the crack will grow to establish a new profile between B and C as shown, such that $K_I = K_{Ic}$ along this new section. In fact, the growth may lead to some redistribution of K_I so that B and C may move and the final $K_I < K_{Ic}$ distribution may be altered. The determination of the new crack front profile is a free boundary problem and the K_I calculations involve the solution of a three-dimensional elasticity problem. Such a combination is of considerable difficulty and there are few analytical or numerical solutions available. Some useful results for special cases may be derived, however, at least approximately, and these will be considered further here.

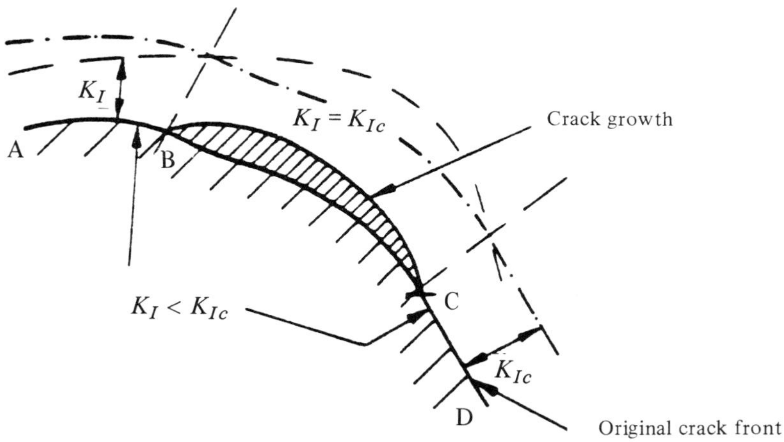

Fig. 4.14 Crack growth under non-uniform conditions

4.7.1 *Circular cracks under symmetrical loading*

Although we are dealing with a curved crack front here, it is, in reality, a self similar case where the crack growth profile is predetermined. It does, however, illustrate the effects of crack front curvature and in Fig. 4.15 we have a penny-shaped, embedded crack subjected to a symmetrical internal pressure $p(r)$. The section defined by the angular increment $d\theta$ may be treated by the point load result of Section 4.5 since, by symmetry, K_I on the crack front $dS = a\,d\theta$ must derive from the forces arising from the pressure over the subtended element.

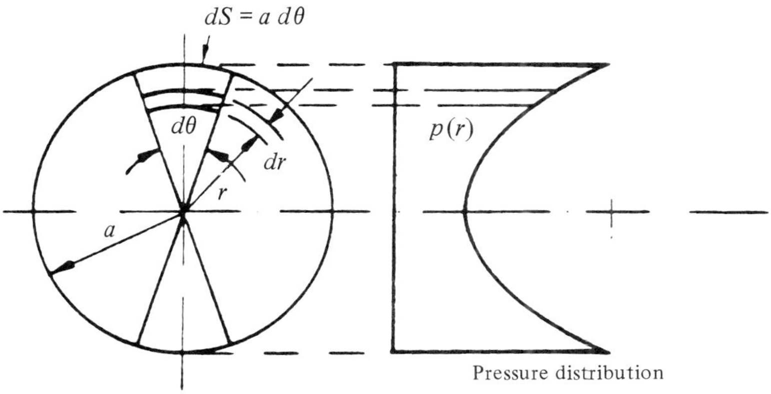

Fig. 4.15 The penny-shaped crack

Thus, we may write:

$$(a\, d\theta)\, dK_I = \frac{2r}{\pi}\, d\theta\, dr\, \frac{p_r \sqrt{\pi a}}{\sqrt{a^2 - r^2}}$$

Rewriting in terms of $\xi = r/a$ and integrating between loading limits, we have:

$$K_I = \sqrt{\pi a}\, \frac{2}{\pi} \int_{\xi_1}^{\xi_2} \frac{p_\xi\, \xi\, d\xi}{\sqrt{1-\xi^2}} \tag{4.31}$$

For constant loading, $p_\xi = p$ and we have:

$$K_I = \left(\frac{2}{\pi}\right) p \sqrt{\pi a} \tag{4.32}$$

so that K_I is smaller than for the equivalent uniform section case since the load taken by a given crack front length is less. For a cohesive zone stress p_c over a length c at the crack front, we have:

$$K_I = p_c \sqrt{\pi a} \left[\frac{2}{\pi} \int_{1-c/a}^{1} \frac{\xi\, d\xi}{\sqrt{1-\xi^2}} \right]$$

Making the substitution $\xi = \cos\theta$ with $\theta_c = \cos^{-1}(1 - c/a)$, we have:

$$K_I = p_c \sqrt{\pi a} \left(\frac{2}{\pi} \sin\theta_c \right) \tag{4.33}$$

For small zones, $c/a \simeq \pi^2/8\, (p/p_c)^2$, as in the parallel section case.

The same method may be applied to the circumferentially notched bar diameter D as shown in Fig. 4.16. If the crack root diameter is d, then the

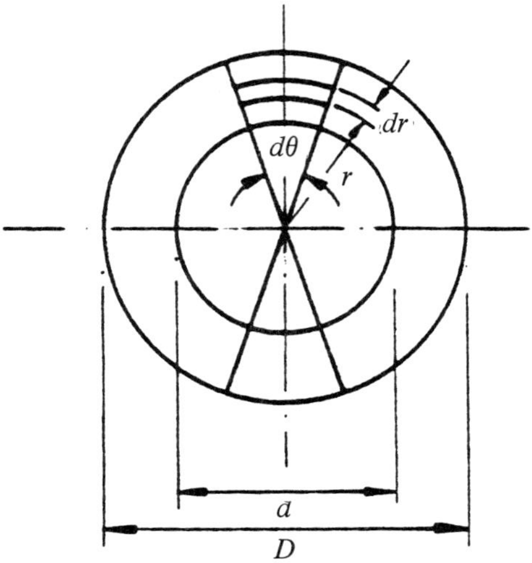

Fig. 4.16 Circumferentially notched bar

equation for K_I becomes:

$$\left(\frac{d}{2}\,d\theta\right)dK_I=\frac{2}{\pi}\frac{p_r\,r\,d\theta}{\sqrt{((D-d)/2)^2-((D/2)-r)^2}}\sqrt{\pi\,(D-d)/2}$$

If we now let $\xi=(1-2r/D)/(1-d/D)$, then this may be written as:

$$K_I=\sqrt{\pi\,((D-d)/2)}\,\frac{2}{\pi}\left[\frac{D}{d}\int_{\xi_1}^{\xi_2}\frac{p_\xi\,d\xi}{\sqrt{1-\xi^2}}-\left(\frac{D}{d}-1\right)\int_{\xi_1}^{\xi_2}\frac{p_\xi\,\xi\,d\xi}{\sqrt{1-\xi^2}}\right] \tag{4.34}$$

For a uniform stress p with $\xi_1=0$ and $\xi_2=1$, this becomes:

$$K_I=p\sqrt{\pi\,((D-d)/2)}\left[\frac{D}{d}-\frac{2}{\pi}\left(\frac{D}{d}-1\right)\right] \tag{4.35}$$

In this case, there is a significant finite width effect because of the large nominal stress. To include this, we may use the approximate method of Section 4.3 using:

$$p_m=p\left[\frac{D}{d}-\frac{2}{\pi}\left(\frac{D}{d}-1\right)\right]\text{ and }p_n=p\left(\frac{D}{d}\right)^2$$

and:

$$Y^2=\frac{\pi}{p^2}\left(\frac{3p_m+p_n}{4}\right)^2$$

This result is shown in Table 4.3 and agrees closely with other published values for this case [10].

This geometry is often used in rotating bending fatigue tests which may be regarded as a symmetrical linear stress distribution of the form:

$$p_{\xi}=p\left[1-\left(\frac{D/d-1}{D/d}\right)\xi\right]$$

where p is the maximum bending stress in the usual way. Substituting in equation (4.34) gives the results:

$$p_m=p\left[\frac{D}{d}-\frac{4}{\pi}\left(\frac{D}{d}-1\right)+\frac{1}{2}\frac{(D/d-1)^2}{(D/d)}\right] \tag{4.36}$$

and:

$$p_n=p\left(\frac{D}{d}\right)^3$$

The computed Y^2 values for this case are also given in Table 4.3.

Table 4.3

Circumferentially notched bar Y^2 values

D/d	Tension	Bending
1.1	3.66	3.57
1.25	4.59	4.57
1.4	5.72	6.16
1.6	7.58	9.51
1.8	9.93	14.93
2.0	12.85	23.58

4.7.2 *Elliptical cracks*

For an elliptical crack, or indeed any non-circular shape, the load from the crack faces is distributed around the crack front depending upon the local curvature. For an ellipse of minor and major axis a and b, respectively, as shown in Fig. 4.17, Irwin [11] deduced the K_I distribution for uniform loading by taking crack face displacement solutions for such a crack and equating it at the crack tip to the displacement-K_I relationship for a planar crack. For an ellipse of the form:

$$\frac{x^2}{a^2}+\frac{y^2}{b^2}=1$$

it is usually most convenient to write expressions in terms of a parametric angle defined as:

$$x=a\cos\phi \text{ and } y=b\sin\phi$$

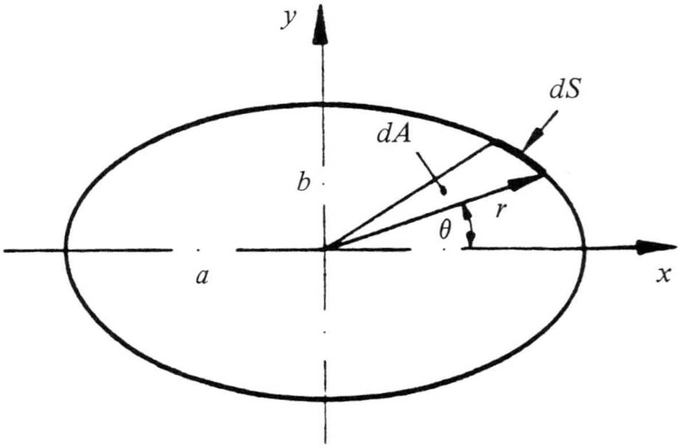

(i) Elliptical crack

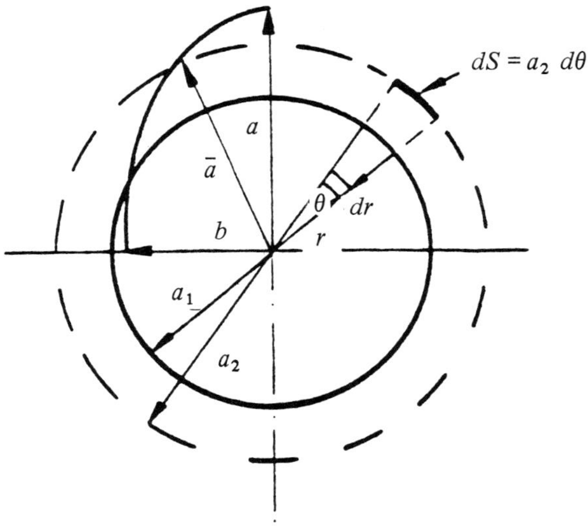

(ii) Equivalent circular crack

Fig. 4.17

and the real angle θ, as shown in Fig. 4.17(i), is related to ϕ via:

$$\tan\theta = \left(\frac{b}{a}\right)\tan\phi$$

Irwin's result is:

$$K_I = \frac{p\sqrt{\pi}}{E_2}(b^2\cos^2\phi + a^2\sin^2\phi)^{1/4}\left(\frac{b}{a}\right)^{1/2} \tag{4.37}$$

where:

$$E_2 = \int_0^{\pi/2}\left[1-\left(1-\frac{b^2}{a^2}\right)\sin^2\phi'\right]^{1/2} d\varphi'$$

the complete elliptic integral of the second kind. This function, together with its complementary form E_1, is tabulated in Table 4.4. For $b/a=0$, $E_2=1$ and $E_1\to\infty$ and at $b/a=1$, $E_2=E_1=\pi/2$.

Table 4.4

Complete elliptic integrals

b/a	0	0.1	0.2	0.3	0.4	0.5	0.6	0.7	0.8	0.9	1.0
E_1	∞	3.702	3.019	2.625	2.358	2.157	1.995	1.863	1.751	1.555	1.571
E_2	1.000	1.016	1.051	1.097	1.151	1.211	1.277	1.346	1.418	1.493	1.571

$$E_1 = \int_0^{\pi/2}\left[1-\left(1-\frac{b^2}{a^2}\right)\sin^2\phi'\right]^{-1/2} d\phi'$$

$$E_2 = \int_0^{\pi/2}\left[1-\left(1-\frac{b^2}{a^2}\right)\sin^2\phi'\right]^{+1/2} d\phi'$$

An alternative approach is to pursue an analogy with the circular crack and define two circles; one with the same area as the ellipse of radius a_1, and one with the same circumference of radius a_2. Thus:

$$\pi\, a_1^{\,2} = \pi\, a\, b,\ a_1 = \sqrt{a\, b} \tag{4.38}$$

and:

$$2\pi\, a_2 = 4E_2\, a,\ a_2 = \frac{2}{\pi}E_2\, a$$

The effective crack length at any point on the ellipse is obtained from the expression for the ratio dS/dA, which is given by:

$$\frac{dS}{dA} = \frac{2}{a\,b}(b^2\cos^2\phi + a^2\sin^2\phi)^{1/2}$$

On the equivalent circles, $dA = \frac{1}{2}a_1^{\,2}\, d\theta$ and we may write:

$$dS = \bar{a}\, d\theta$$

where $\bar{a}$ is now the effective crack length. On substitution, we have:

$$\bar{a}=(b^2 \cos^2 \phi + a^2 \sin^2 \phi)^{1/2} \tag{4.39}$$

If we now integrate the pressure over the circle of equivalent area, then:

$$(a_2\, d\theta)\, dK_I = \frac{2}{\pi} r\, d\theta\, dr \frac{p_r \sqrt{\pi \bar{a}}}{\sqrt{{a_1}^2 - r^2}}$$

and on defining $\xi = r/a_1$ and substituting from equations (4.38) and (4.39), we have:

$$K_I = \sqrt{\pi \bar{a}} \left[\sqrt{b/a} \frac{1}{E_2} \int_{\xi_1}^{\xi_2} \frac{p_\xi\, \xi\, d\xi}{\sqrt{1-\xi^2}} \right] \tag{4.40}$$

This reduces to the circular crack case of equation (4.31) for $b=a$ since $\bar{a}=a$ and $E_2=\pi/2$. For constant pressure, $p_\xi = p$ with $\xi_1=0$ and $\xi_2=1$, so that equation (4.37) is recovered.

It should be noted that any shape may be dealt with in this way. As shown in Fig. 4.18, the centre of the circle a_1, 0 is fixed by the equivalent area condition, a_2 may be found from S and at any point ds/dA may be found so that, in general:

$$a_1 = \sqrt{A/\pi},\ a_2 = \frac{S}{2\pi},\ \bar{a} = \frac{A}{2\pi}\left(\frac{ds}{dA}\right)$$

and:

$$K_I = \sqrt{\pi \bar{a}} \left[\frac{2}{\pi} \frac{a_1}{a_2} \int_{\xi_1}^{\xi_2} \frac{p_\xi\, \xi\, d\xi}{\sqrt{1-\xi^2}} \right] \tag{4.41}$$

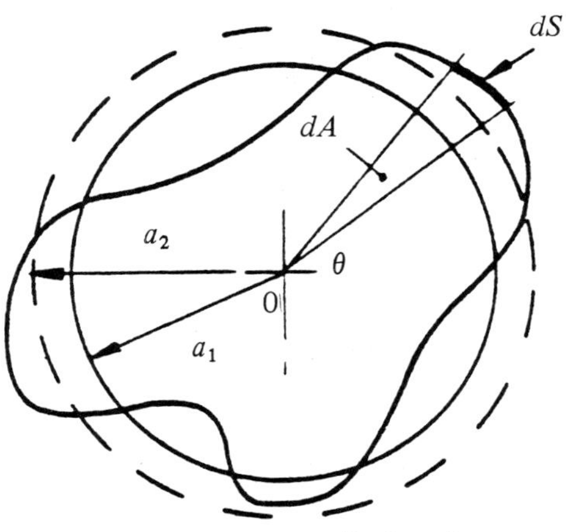

Fig. 4.18 Approximating a general crack shape to an equivalent ellipse

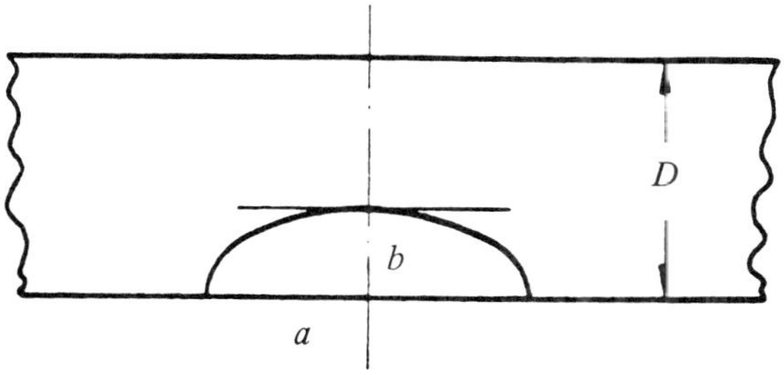

Fig. 4.19 Part-through elliptical surface crack

The most important practical case is the surface flaw, as shown in Fig. 4.19. The maximum K_I value is at the minor axis and is given by:

$$K_I = \frac{1}{E_2} p \sqrt{\pi b}$$

for uniform tension. There is no bending effect induced as in SEN and finite width effects can be approximated by scaling the factor $\pi/E_2{}^2$ by the DEN Y^2 factor. A similar calculation may be performed in bending, since:

$$p_\xi = p\left(1 - \frac{2b}{D} \sin\phi\, \xi\right) \text{ with } \xi_1 = 0 \text{ and } \xi_2 = 1$$

and on integrating equation (4.40), we have:

$$K_I = \frac{p\sqrt{\pi}}{E_2}(b^2\cos^2\phi + a^2\sin^2\phi)^{1/4}\left(\frac{b}{a}\right)^{1/2}\left(1 - \frac{\pi}{2}\frac{b}{D}\sin\phi\right) \quad (4.42)$$

where p is the maximum bending stress. The maximum of this function is always at $\phi = \pi/2$, the minor axis of the ellipse. Thus, at this point, we have:

$$K_I = \frac{1}{E_2} p \sqrt{\pi b}\left(1 - \frac{\pi}{2}\frac{b}{D}\right) \quad (4.43)$$

and finite width effects may be approximated using the pure bending result given in Table 4.1.

A practically useful result may be derived for a crack at the inner wall of a thick-walled cylinder. Here, the stress distribution is given by the Lamé equations and if we assume that the internal pressure acts inside the crack, then:

$$p_\xi = \frac{p_0 K^2}{K^2 - 1}\left[1 + \frac{1}{\left(1 + \frac{b\sin\phi}{a_0}\xi\right)^2}\right]$$

where K is the outside to inside radius ratio of the cylinder, p_0 is the applied pressure, and a_0 is the bore radius. The final result is:

$$K_I = \frac{1}{E_2}\left(\frac{p_0\, 2K^2}{K^2-1}\right)\sqrt{\pi\, \bar{a}}\left(\frac{b}{a}\right)^{1/2} f(\alpha)$$

where:
$$f(\alpha) = \frac{1}{2}\left(\frac{2-\alpha^2}{1-\alpha^2}\right) - \frac{\alpha \tan^{-1}\sqrt{(1-\alpha)/(1+\alpha)}}{(1-\alpha^2)^{3/2}}, \alpha < 1 \tag{4.44}$$

$$f(\alpha) = \frac{1}{2}\left(\frac{2-\alpha^2}{1-\alpha^2}\right) + \frac{\alpha}{2}\ln\left[\frac{1+\sqrt{(\alpha-1)/(\alpha+1)}}{1-\sqrt{(\alpha-1)/(\alpha+1)}}\right], \alpha > 1$$

and $f(1) = 11/16$ with $\alpha = b \sin\phi/a_0$. At the minor axis of the crack, $\phi = \pi/2$ and $\alpha = b/a_0$. For $\alpha \ll 1$, the result may be approximated by:

$$f(\alpha) \simeq 1 - \frac{\pi}{4}\frac{b}{a_0}$$

4.7.3 *G for elliptical cracks*

The computation of G for an elliptical crack illustrates one of the difficulties arising from the varying K_I situations. Since $K_I \neq K_{Ic}$ at every point, there cannot be a true equivalent between an overall G and K_{Ic}. One may, of course, define a local G from K_I and for the elliptical crack this becomes, from equation (4.37):

$$G_{local} = \frac{K_I{}^2}{E} = \frac{p^2}{E\, E_2}\pi\frac{b}{a}(b^2\cos^2\phi + a^2\sin^2\phi)^{1/2}$$

If we now assume that the growth is everywhere the same, then the global G may be computed by simply integrating this expression around the ellipse, i.e.

$$\begin{aligned} G_{global} &= \frac{p^2\,\pi}{E\, E_2{}^2}\frac{b}{a}\left[\frac{2}{\pi}\int_0^{\pi/2}(b^2\cos^2\phi + a^2\sin^2\phi)^{1/2}\, d\phi\right] \\ &= \frac{2p^2\, b}{E\, E_2} \end{aligned} \tag{4.45}$$

However, unlike a straight crack, the area term in the definition of G (dU/dA) can cover a range of possibilities and still remain elliptical, i.e. since $A = \pi\, a\, b$:

$$dA = \pi\,(b\, da + a\, db)$$

From Irwin's displacement expression, we may calculate U as:

$$U = \frac{4\pi}{3}\frac{p^2\, a\, b^2}{E\, E_2}$$

and thus:
$$G = \frac{4\pi}{3}\frac{p^2}{E}\frac{d}{dA}\left(\frac{a\, b^2}{E_2}\right)$$

On performing the necessary differentiation, this becomes:

$$G=\frac{2p^2\,b}{E\,E_2}\left[\frac{2}{3}\;\frac{1+2\dfrac{db}{da}\dfrac{a}{b}}{1+\dfrac{db}{da}\dfrac{a}{b}}+\frac{2}{3}\;\frac{1-\dfrac{db}{da}\dfrac{a}{b}}{1+\dfrac{db}{da}\dfrac{a}{b}}\;\frac{(b/a)^2}{1-(b/a)^2}\left(\frac{E_1}{E_2}-1\right)\right] \tag{4.46}$$

where:

$$E_1=\int_0^{\pi/2}\left[1-\left(1-\frac{b^2}{a^2}\right)\sin^2\phi'\right]^{-1/2}d\phi'$$

the complete elliptic integral of the first kind, and is given in Table 4.4. For $db/da=b/a$, we have self-similar propagation and we recover equation (4.45). Since K_I is greatest at the minor axis, the most realistic estimate of G within the limitations of the elliptical solution would be to assume no growth along the major axis, i.e. $da=0$, and we have:

$$G_{da=0}=\frac{2p^2\,b}{E\,E_2}\left[\frac{4}{3}-\frac{2}{3}\frac{(b/a)^2}{1-(b/a)^2}\left(\frac{E_1}{E_2}-1\right)\right] \tag{4.47}$$

For the penny-shaped crack, $b=a$, but $(E_1/E_2-1)\rightarrow\frac{1}{2}(1-(b/a)^2)$, so that:

$$G\rightarrow\frac{4p^2\,a}{\pi\,E}$$

as expected from equation (4.32).

4.7.4 *Cracks in plates and shells*

A three-dimensional problem for which solutions do exist is that of a thin sheet with a through-crack subjected to a biaxial moment M_0, as shown in Fig. 4.20.

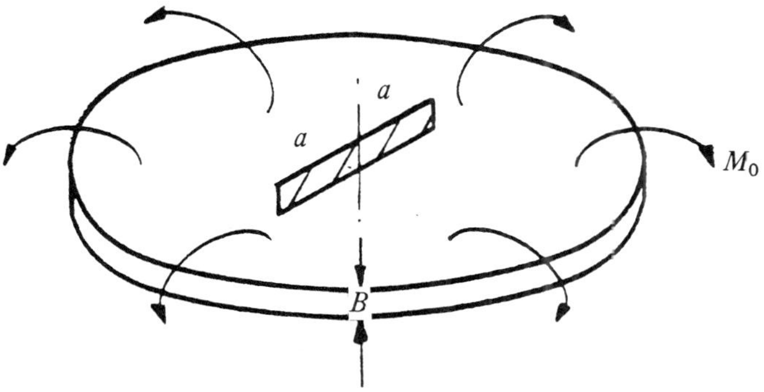

Fig. 4.20 Cracked plate subjected to a uniform bending moment

The solution of the problem is not simple and conventional bending theory gives rise to anomalous results. If higher order terms are included [2], the moments induced at the crack tip have precisely the same angular distribution as the elastic stresses for the general uniformly stressed plate, as given in equation (3.11). For $\theta=0$, we have:

$$M_\theta = M_r = \left(\frac{1+\nu}{3+\nu}\right)\frac{M_0}{\sqrt{2r}}, \quad M_{r\theta}=0 \tag{4.48}$$

If we now define K_I in the usual way and express M_0 in terms of the bending stress $p=6M_0/B^2$, then we have:

$$K_I = \left(\frac{1+\nu}{3+\nu}\right) p\sqrt{\pi a} \tag{4.49}$$

This result is valid only for thin shells in which $a/B \gg 1$ and numerical solutions have shown that the factor $(1+\nu/3+\nu)$ increases for thicker plates [12], tending to unity as $a/b \to 0$, as expected. If a tensile stress is present, then a combined K_I is determined by simple superposition. It should be noted also that the solution assumes stress free crack faces so that narrow cracks in which the compression side faces touch may not be described accurately by this result.

In membrane shells subjected to lateral pressure, the presence of a crack introduces local bending moments and these may be assumed to be dependent on the pressure in the form:

$$M \propto p\,a^2$$

by analogy with uniformly loaded beams. Thus, the bending stress may be written as:

$$p_B \propto \frac{p\,a^2}{B^2}$$

The tensile stress in the shell has the form:

$$p_T \propto \frac{p\,R}{B}$$

where R is the radius of curvature. The total K_I is now:

$$K_I = \left(p_T + \left(\frac{1+\nu}{3+\nu}\right) p_B\right)\sqrt{\pi a}$$

i.e.

$$K_I = p_T\sqrt{\pi a}\left(1 + k\frac{a^2}{R\,B}\right) \tag{4.50}$$

where k is a constant. Some computations have been performed which suggest that for a cylinder with an axial crack $k \simeq 0.83$, for a circumferential crack $k \simeq 0.17$, and for a sphere $k \simeq 0.96$ [13, 14].

References for Chapter 4

[1] Timoshenko, S. P. & Goodier, J. N., *Theory of Elasticity*, 3rd edn, McGraw-Hill (1970).
[2] Koiter, W. T., *Ingenieur Archiv*, **28** (1959) 168–172.
[3] Brown, W. F. & Srawley, J. E., *ASTM STP 410* (1966).
[4] Rooke, D. P. & Cartwright, D. J., *Compendium of Stress Intensity Factors*, HMSO (1976).
[5] Williams, J. G. & Isherwood, D. P., *J. Strain Anal.*, **3** (1968) 17–22.
[6] Plati, E. & Williams, J. G., *Polym. Eng. Sci.*, **15**(6) (1975) 470–477.
[7] Sih, G. C. & Liebowitz, H., Chapter 2 in *Fracture: Volume II* (ed. H. Leibowitz), Academic Press (1968).
[8] Goodier, J. N. & Field, F. A., in *Fracture of Solids* (eds. D. C. Rucher & J. J. Gilman), Wiley–Interscience, New York (1963) 103–118.
[9] Barrett, J. D. & Foschi, R. O., *Eng. Fract. Mech.*, **9** (1977) 371–378.
[10] Bueckner, H. F., in *ASTM STP 381* (1966) 82–83.
[11] Irwin, G. R., *J. Appl. Mech.*, **29** (1962) 651–654.
[12] Knowles, J. K. & Wang, N. M., *J. Math. Phys.*, **39** (1960) 223–236.
[13] Folias, E. S., *Int. J. Fract. Mech.*, **1**(1) (1965) 20–46.
[14] Folias, E. S., *Int. J. Fract. Mech.*, **1**(2) (1965) 104–133.

5

Systems with energy dissipation

5.1 Introduction

In the preceding three chapters, we have developed the analysis of the growth of brittle cracks in elastic bodies in some detail since it is a very useful description of many practical situations. It is, however, based on distinct assumptions and the limitations imposed by them must be recognised when applying the analysis. Firstly, the deformation is assumed to be controlled solely by the elastic response of the material and crack growth results in an elastic energy release. This is balanced (in a static case) by the sole energy dissipation process in the body which is embodied in R, the dissipation rate, which is local to the crack tip and on such a small scale that it does not affect the elastic response. This is the 'crack tip specific' energy dissipation and it is a natural assumption to take this as a constant, independent of crack growth, as used in the definition of G_c. The elastic analysis will, as outlined in Section 2.3, cope with a varying R within this set of assumptions, but such a condition is unlikely for brittle cracks.

In many cases, however, the analysis breaks down because the energy dissipation is not confined to a local zone on a scale which can be deemed crack tip specific. The dissipation then influences the load-deflection behaviour of the system and we cannot assume that all deflections are the result of elastic responses. We can, of course, analyse the elastic response of the system and so it is always possible, at least in principle, to determine G. It is also a reasonable assumption to infer that this energy release is *always* taken up by the crack tip specific part of the energy dissipation. This is because the very nature of G requires a local, non-continuum type of response, which can only be characterised in this way. The additional energy dissipation in R is thus computable, at least in principle, from the usual continuum analyses and is supplied by the external work done on the system. In some cases, it may be appropriate to continue to characterise a fracture using G_c, as for example in stability considerations. For the initiation of fractures which are stable and require the continuous supply of external work, a further definition of R is often more suitable since it is some assessment of total work input which is more sensible. This leads naturally to the adoption of the J integral to be discussed later in this chapter.

Before proceeding with this analysis, it is useful to define some basic parameters used in the description of dissipative processes. All materials, with the possible exception of the most brittle solids, undergo some form of flow or yield process at some stress level. Such deformations are irreversible and lead to energy dissipation and are normally characterised by the stress at which they occur in simple tension or compression, p_Y. Because these are shear determined constant volume flow processes, they are controlled by invariant functions of the deviatoric stresses and are independent of hydrostatic stress. Thus, the stress level at which yield will occur is a function of the state of stress pertaining and is most commonly defined in terms of the Lévy–von Mises yield criterion which may be expressed in terms of the three principal stresses at a point as:

$$2p_Y^2 = (p_1 - p_2)^2 + (p_2 - p_3)^2 + (p_3 - p_1)^2 \tag{5.1}$$

In general, the applied stress, say, p_1, may be elevated to any value by p_2 and p_3, but the condition of most importance here is when one of the stresses is determined by the plane strain condition, i.e.

$$p_3 = \nu(p_1 + p_2)$$

If we now define a constraint factor M to describe the elevation of the applied stress at yield, we have:

$$M = \frac{p_1}{p_Y} = \left[\left(1 + \frac{p_2}{p_1}\right)^2 (1 - \nu + \nu^2) - \frac{3p_2}{p_1}\right]^{-1/2} \tag{5.2}$$

In the crack tip region along the crack line, $p_2 = p_1$ and the stress state is controlled by the elastic field, at least for small-scale yielding, and we have:

$$M = (1 - 2\nu)^{-1/2} \tag{5.3}$$

An approximate idea of the size scale of the yielded zone at the tip of a crack for the linear elastic case may be obtained from equation (3.11), since:

$$p_\theta = p_r = \frac{K_I}{\sqrt{2\pi r}} \text{ at } \theta = 0$$

and if we assume that $p_\theta \equiv p_1$, then the size scale for yielding l will be given by:

$$l = \frac{K_{Ic}^2}{2\pi M^2 p_Y^2} \tag{5.4}$$

If we now use the plane stress case ($M = 1$) as a reference, a material may now be characterised by any two parameters of K_{Ic} (or G_{Ic}), p_Y or $l = K_{Ic}^2/(2\pi p_Y^2)$.

Use may now be made of l to divide fractures into four convenient classes, as shown in Fig. 5.1. Here, we have a plate of crack length a, width D and thickness

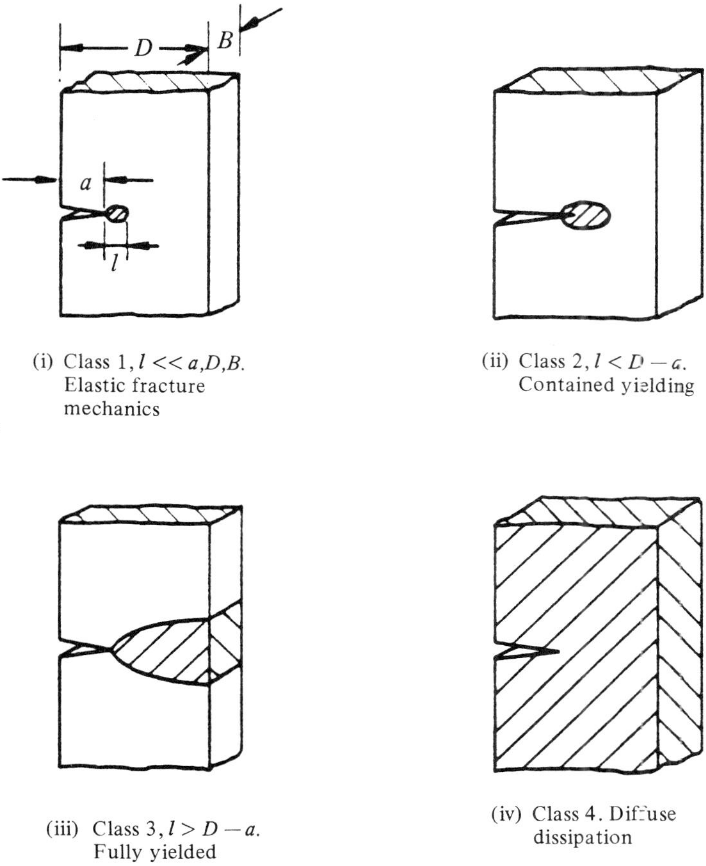

(i) Class 1, $l << a,D,B$. Elastic fracture mechanics

(ii) Class 2, $l < D - a$. Contained yielding

(iii) Class 3, $l > D - a$. Fully yielded

(iv) Class 4. Diffuse dissipation

Fig. 5.1 Types of fracture

B and if $l \ll a,D,B$, then we have a class 1 situation which is termed elastic fracture mechanics and is covered by our previous G_c and K_{Ic} analysis. If $l < D - a$, the ligament length, but can be significant compared with a or B, then we have class 2 which will be termed contained yielding. In class 3, $l > D - a$ and the whole ligament has yielded so that the deformation is controlled by the fully yielded or plastic collapse situation. Class 4 involves diffuse dissipation throughout the body and is usually associated with a general deformation other than elasticity, such as creep or viscoelasticity. It would be possible to have class 4 combined with any of the other three types of behaviour.

This classification will be a useful framework in which to study non-elastic fractures and will form the basis of this chapter. It is first necessary, however, to return to the basic definition of R in order to consider appropriate criteria.

5.2 Fracture criteria for dissipative systems [1]

Let us consider the loading of a cracked body, in which energy dissipation occurs over and above that which may be defined as crack tip specific. The resistance to crack propagation R may now be assumed to contain the local part, characterised by G_c, plus the additional contribution which occurs in the rest of the body. If we assume that we have stable crack propagation, then a possible criterion for initiation is to define a total energy dissipation rate from the load-deflection diagram, as shown in Fig. 5.2. Let OB be the loading curve for a body with a crack area A and assume that fracture occurs at B with A increasing by δA at B'. If we now assume that loading the body with an initial crack of $A+\delta A$ would give the loading line OB' with fracture occurring at B',

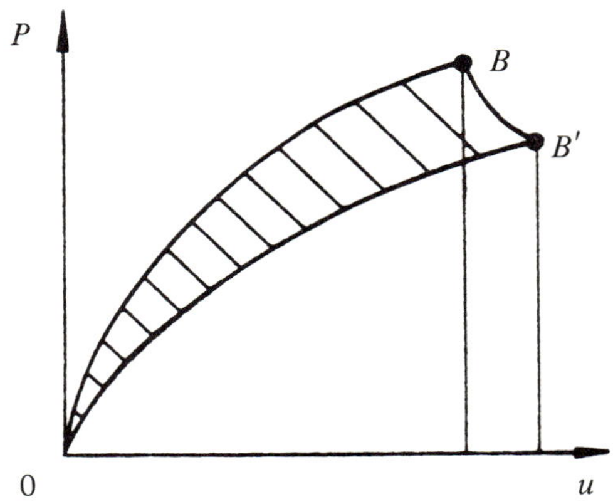

Fig. 5.2 Load-deflection diagram for a dissipative system

then the shaded area between the two curves represents the total energy necessary to propagate the crack δA, and in the limit we may write:

$$R=\frac{dU}{dA}=P\frac{du}{dA}-\frac{d}{dA}\int_0^u P\,du \tag{5.5}$$

This is precisely the definition of G for the elastic case, but here we are not assuming just an elastic response but are including energy dissipation as well. We shall term this parameter J and note that it is entirely dissipative and that for the elastic case:

$$J=G_c \tag{5.6}$$

For elastic systems, the definition is precise since the deformations are history-independent and the assumptions of the loading lines are correct. For dissipative systems, this is not necessarily so since the deformations may be history-dependent but, if the loading is monotonic and increasing, then path dependence will not occur and J will be an accurate representation. Under such circumstances, the use of a critical value of J, J_c, to characterise crack initiation is sensible.

Because J is defined in the same way as G, the various G derivations may be used for J if it is assumed that the monotonic, dissipative, loading curves have the same form as elastic lines. Thus, we would expect path independence of J and we may evaluate it as a global parameter from load deflection diagrams or from local crack tip stress fields using contours. An important result is that for power-law elasticity, as outlined in Section 3.5, for $p \propto e^n$ which is a reasonable representation of plastic stress–strain curves of different degrees of work hardening. Here we have, from equation (3.26):

$$J = 2u_0\, p_0 \left[\frac{((n\,\pi)/1+n)}{\sin\,((n\,\pi)/1+n)} \right]$$

and for non-work hardening, $n=0$, we have:

$$J = 2u_0\, p_0 \tag{5.7}$$

p_0 may be regarded as Mp_Y, the yield stress at the crack tip, and $2u_0$ as the local displacement, usually termed the crack opening displacement (COD) and defined as δ. It is therefore possible to define J_c in terms of a critical COD, δ_c, which, as a local criterion representing crack tip stretching, is, perhaps, a more physically appealing parameter than J_c. The conversion of δ_c to a local strain requires the addition of a further length parameter into the analysis.

The J_c and δ_c criteria frequently lead to very similar results and, indeed, in many cases the same ones, but not in all cases (see Section 5.7). The nature of the problem analysed and the type of dissipation involved will determine how convenient it is to evaluate J and where problems of path dependence and time occur, the use of δ may be more appropriate.

5.3 Small-scale yielding (Class 1)

A first approximation to yielding behaviour around the crack tip may be obtained by determining where the elastic stress field reaches the yield criterion. This will not be an exact solution since the local flow will result in stress redistribution, but some useful results are obtained. From equations (3.11), expressions for the principal stresses near the crack tip may be obtained, since:

$$p_{1,2} = \left(\frac{p_\theta + p_r}{2}\right) \pm \sqrt{\left(\frac{p_\theta - p_r}{2}\right)^2 + p_{r\theta}^{\,2}}$$

and hence:

$$p_{1,2} = \frac{K_I}{\sqrt{2\pi\, r}} \cos\frac{\theta}{2}\left(1 \pm \sin\frac{\theta}{2}\right) \tag{5.8}$$

for mode I loading. If we now assume plane strain at the crack tip, then the transverse strain $e_z = 0$, and we have:

$$p_z = p_3 = \nu\,(p_1 + p_2)$$

Using this in conjunction with equations (5.8) and (5.1), we have an expression for the size and shape of the plastic zone:

$$r_p = \frac{1}{2\pi}\left(\frac{K_I}{p_Y}\right)^2 \cos^2\frac{\theta}{2}\left[4(1-\nu+\nu^2) - 3\cos^2\frac{\theta}{2}\right] \tag{5.9}$$

Fig. 5.3 shows the shapes of the zones for $\nu = 1/2$ and $1/3$ and also for $\nu = 0$, which is the plane stress case. The plane stress zone is clearly much larger than those achieved in plane strain because of the higher constraint, and thus effective yield stress, of plane strain. There is also a distinct change in form from the double lobes of plane strain to the almost circular form for plane stress.

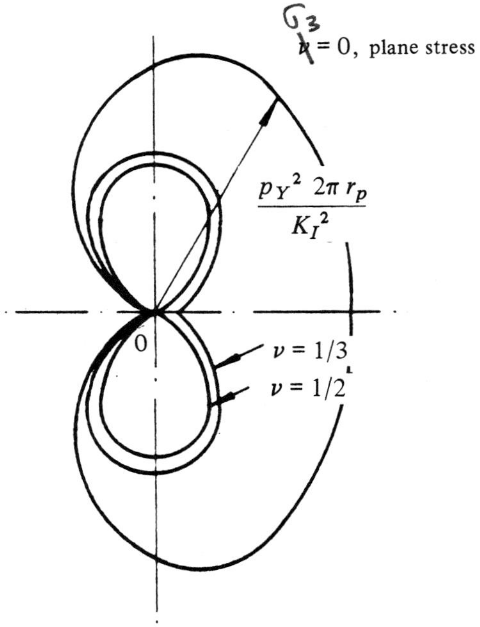

Fig. 5.3 Plastic zone shapes

The fracture toughness, G_c, will be closely related to the zone size and would be expected to be a minimum when the zone is at its smallest. This condition is achieved with maximum constraint under full plane strain conditions and represents a lower bound to the material toughness. This is an important concept since this value, G_{c1} (or K_{c1}), is of particular value in both design and material evaluations. As the constraint decreases, G_c may be expected to increase and an upper limit, for conditions when the plastic deformation is still local, will be the plane stress case, G_{c2} (or K_{c2}). The phenomenon is of considerable importance in edge notched sheets, as illustrated in Fig. 5.4, where the central region gives rise to plane strain conditions while the surfaces are in plane stress since $p_3 = 0$. This will result in a distribution of K_c values across the section, as shown in Fig. 5.5, and any measured value obtained will be an average. In

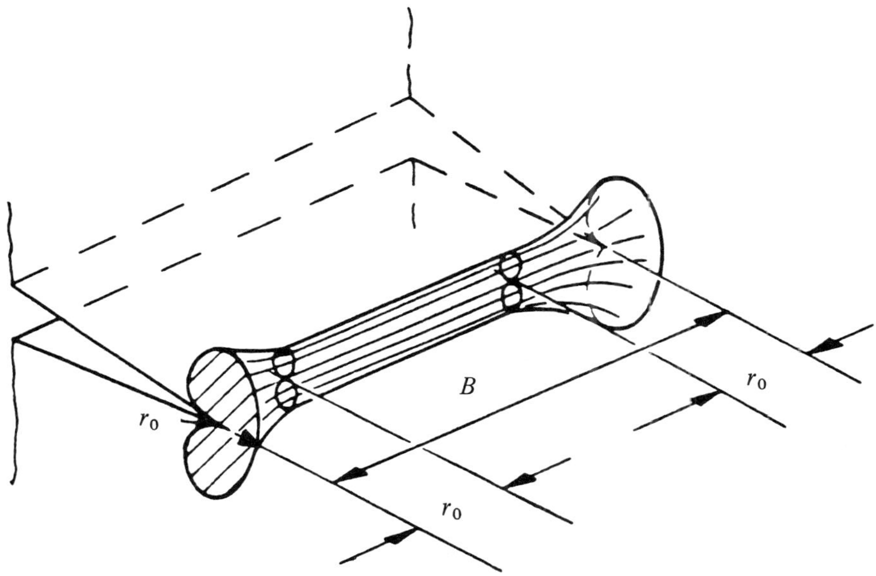

Fig. 5.4 Plastic zone shape across a plate thickness

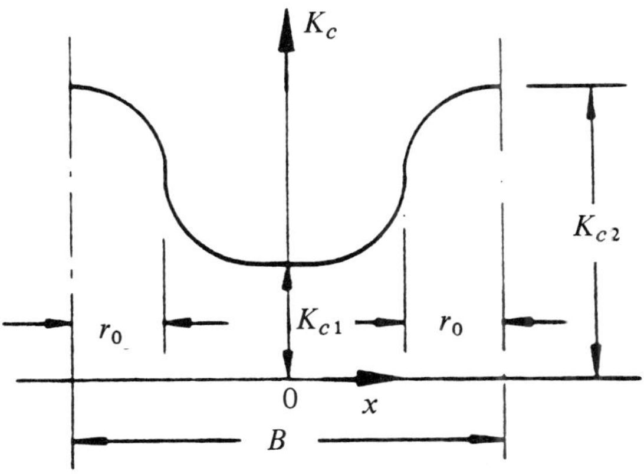

Fig. 5.5 K_c distribution across a crack front

reality, of course, fracture will occur first in the centre, giving crack front bowing which reduces K (see Section 4.6). Finally, complete fracture will occur when the K_{c2} condition is achieved at the outer surface by redistribution of the load and an average will be measured:

$$K_c' = \frac{2}{B} \int_0^{B/2} K_c \, dx$$

A useful approximation to the distribution is obtained by assuming a bimodal distribution with K_{c2} over surface zones over a thickness r_0, where:

$$r_0 = \frac{1}{2\pi} \left(\frac{K_{c2}}{p_Y} \right)^2 \qquad (5.10)$$

Thus:

$$K_c' = \frac{2}{B} \left[K_{c2} \, r_0 + K_{c1} \left(\frac{B}{2} - r_0 \right) \right]$$

i.e.

$$K_c' = K_{c1} + \frac{K_{c2}^2}{\pi \, B \, p_Y^2} (K_{c2} - K_{c1}) \qquad (5.11)$$

Thus, the measured K_c' value is inversely proportional to B and tends to K_{c1} for large thicknesses. When $B = 2r_0$, the fracture will be entirely plane stress and there will be a transition to plane stress, ductile failure at the condition:

$$\hat{B} = \frac{1}{\pi} \frac{K_{c2}^2}{p_Y^2} \qquad (5.12)$$

This is frequently expressed in terms of K_{c1} since this is usually known, and if we assume that $K_c^2 \propto r_p \, (\theta = 0)$, then:

$$\hat{B} = \frac{1}{\pi} \left(\frac{K_{c2}}{K_{c1}} \right)^2 \left(\frac{K_{c1}}{p_Y} \right)^2 = \frac{1}{\pi \, (1 - 2\nu)^2} \left(\frac{K_{c1}}{p_Y} \right)^2$$

For $\nu \simeq 1/3$, we have $\hat{B} \simeq 2.9 \, (K_{c1}/p_Y)^2$, which is somewhat more severe than the ASTM criteria [2], $\hat{B} \simeq 2.5 \, (K_{c1}/p_Y)^2$.

For $B < \hat{B}$, the K_c value decreases since the thickness becomes the controlling parameter and $K_c' \propto B$ and since $K_c' = K_{c2}$ at $B = \hat{B}$, we have:

$$K_c' = p_Y \sqrt{\pi \, B} \qquad (5.13)$$

Fig. 5.6 shows a schematic diagram of K_c' as a function of inverse thickness, indicating the various types of behaviour.

It should be noted that this analysis has averaged the K_c values along the crack front on the basis that the stresses are redistributed. It might be argued that there is an energy redistribution so that p^2 is averaged, giving expressions in K_c^2 (or G_c). In practice, differences in the two forms are difficult to discern and can only be decided on the basis of experimental data.

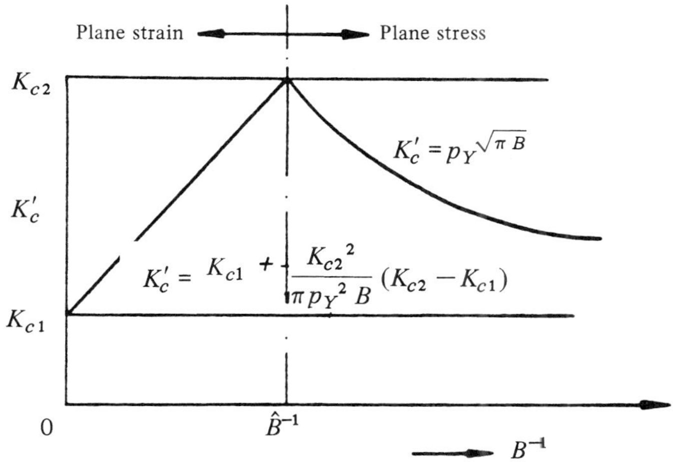

Fig. 5.6 Variation of K_c', the observed K_c value, with plate thickness, plane stress-plane strain effects

5.4 Contained yielding (Class 2)

The assumption that the occurrence of yielding does not affect the crack tip stress distribution is clearly less reasonable as the zone size increases and a correction to the zone size equation is needed. A simple analysis, which is due to Irwin [3], is illustrated in Fig. 5.7, where the original stress distribution for $K_I(a)$ is shown cut off at $p = Mp_Y$. This clearly does not satisfy equilibrium along the crack line and a first approximation correction may be obtained by considering an apparent crack tip a distance Δa from the real one, as shown. There is thus an elastic distribution for $K_I(a + \Delta a)$ now also cut off at $p = Mp_Y$, but the lost load, given by the upper shaded area, is compensated by the lower shaded area.

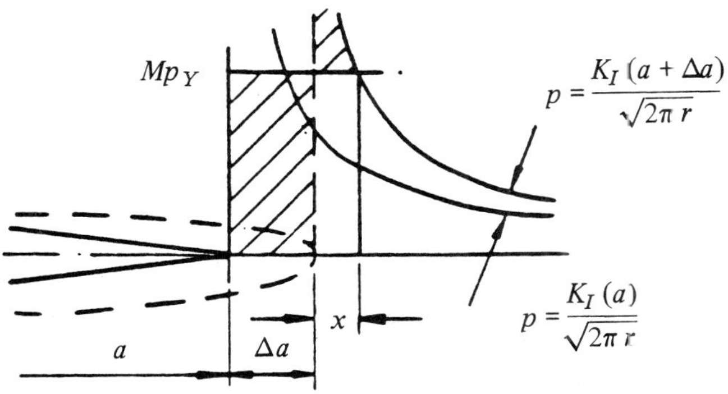

Fig. 5.7 Plastic zone correction

Equating these two areas, we have:

$$M\, p_Y\, \Delta a = \int_0^x \frac{K_I(a+\Delta a)}{\sqrt{2\pi\, r}}\, dr - M\, p_Y\, x$$

Noting that:

$$x = \frac{1}{2\pi}\frac{K_I{}^2(a+\Delta a)}{(M\, p_Y)^2}$$

we have:

$$M\, p_Y\,(\Delta a + x) = \frac{K_I(a+\Delta a)}{\sqrt{2\pi}}\, 2x^{1/2} = M\, p_Y\,(2x)$$

i.e.

$$x = \Delta a$$

Thus, the effective zone size is given by:

$$r_p = x + \Delta a = \frac{1}{\pi}\frac{K_I{}^2(a+\Delta a)}{(M\, p_Y)^2} \tag{5.14}$$

and if $\Delta a \ll a$, this is twice the original value. There is also an expression for the displacement at the real crack tip, the COD, since from equations (3.12):

$$\delta = 2u_{\theta=\pi} = \frac{4}{\pi}\frac{K_I(a+\Delta a)}{E}\sqrt{2\pi\, x}$$

i.e.

$$\delta = \frac{4}{\pi}\frac{K_I{}^2(a+\Delta a)}{E\, M\, p_Y} \tag{5.15}$$

If we now compute $G = K_I{}^2(a+\Delta a)/E$, we do, in fact, have an expression for J, and not G, since the correction constitutes the additional work term inherent in J. For an infinite plate, we have:

$$J = \frac{\pi\, p^2}{E}(a+\Delta a) = \frac{\pi\, p^2}{E}\left(a + \frac{1}{2\pi}\frac{E\, J}{(M\, p_Y)^2}\right)$$

i.e.

$$J = \frac{\pi\, p^2\, a}{E}\left[1 - \frac{1}{2}\left(\frac{p}{M\, p_Y}\right)^2\right]^{-1}$$

G for this system is still given by $(\pi\, p^2\, a)/E$, so we write finally:

$$\frac{J}{G} = \left[1 - \frac{1}{2}\left(\frac{p}{M\, p_Y}\right)^2\right]^{-1} \tag{5.16}$$

This model is an approximation to the more precise line zone analysed in Section 4.4 for cohesive stresses. If we replace p_c by $M\, p_Y$, then the same concept may be used giving:

$$r_p = a\left(\sec\frac{\pi}{2}\frac{p}{M\, p_Y} - 1\right) \tag{5.17}$$

where a is now $(a-c)$ in the previous analysis. For $p/(M\,p_Y) \ll 1$, this reduces to:

$$r_p = \frac{\pi}{8}\frac{K_I^2}{(M\,p_Y)^2}$$

which is a factor of $\pi^2/8 \simeq 1.23$ different from equation (5.14), the approximate result. The COD comes from equation (4.24), from which:

$$\delta = \frac{8}{\pi}\frac{(M\,p_Y)}{E}\,a \ln \sec \frac{\pi}{2}\frac{p}{M\,p_Y} \tag{5.18}$$

As shown in equation (4.21), this reduces to $\delta = K_I^2/(E\,M\,p_Y)$ for $p/(M\,p_Y) \ll 1$, which is a factor of $\pi/4 \simeq 0.78$ different from equation (5.15). The exact expression for J may be deduced from equation (5.18) since $J = \delta\,M\,p_Y$ so that:

$$\frac{J}{G} = \frac{8}{\pi^2}\left(\frac{M\,p_Y}{p}\right)^2 \ln \sec \frac{\pi}{2}\left(\frac{p}{M\,p_Y}\right) \tag{5.19}$$

This is compared with the approximation given by equation (5.16) in Fig. 5.8.

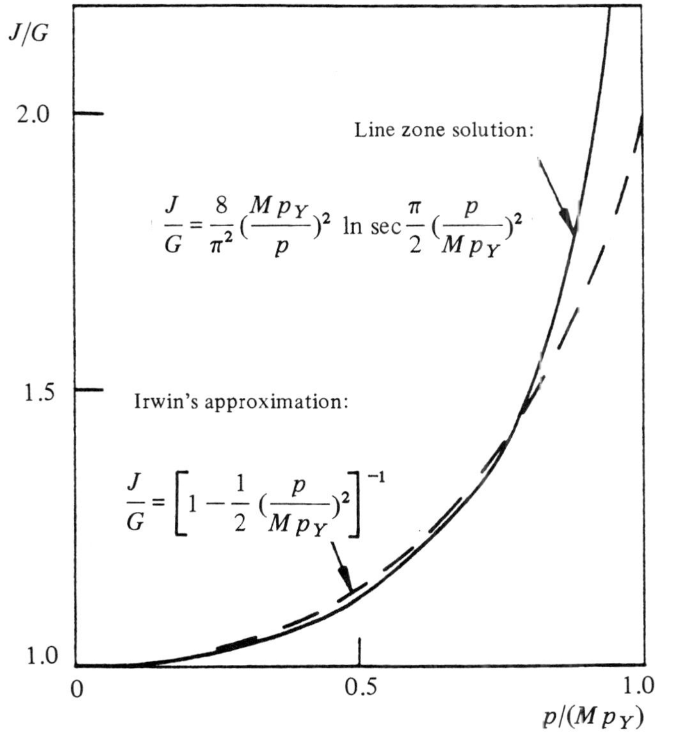

Fig. 5.8 Comparison of the line zone and approximate solutions for J/G

The results agree to within about 10% up to $p/(M\,p_Y)=0.9$, but the line zone solution gives $J/G\rightarrow\infty$ at $p/(M\,p_Y)=1$ which the model requires.

Finite width effects may be corrected for, as in the elastic case, and here it is useful to rewrite equation (5.8) as:

$$J^* = \frac{\pi}{4}\frac{E\,J}{(M\,p_Y)^2\,D} = \frac{2a}{D}\ln\sec\frac{\pi}{2}\left(\frac{p}{M\,p_Y}\right) \tag{5.20}$$

This is a 'pseudo' finite width solution and is shown in Fig. 5.9, where a bound for full yielding in a centre-notched plate in tension (CT) is drawn since, from symmetry, the limiting case is when:

$$\frac{\bar{p}}{M\,p_Y} = 1 - \frac{a}{D}$$

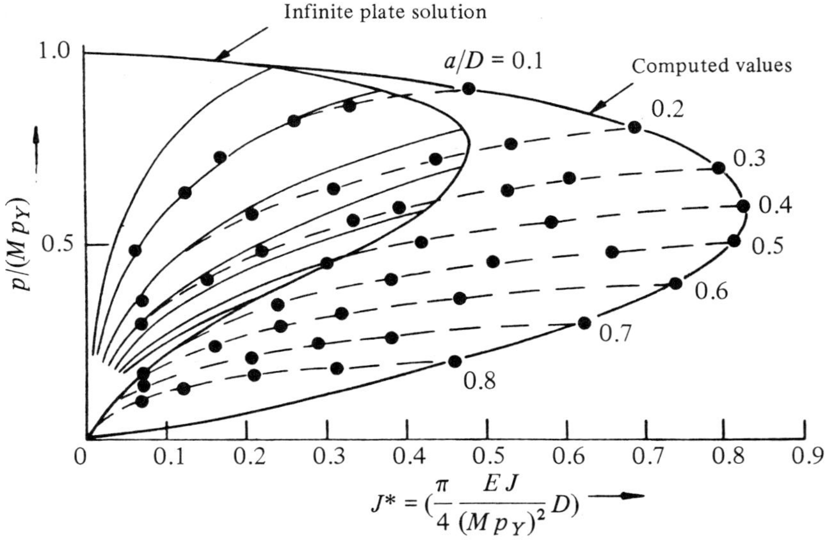

Fig. 5.9 J^* as a function of p/Mp_Y—CN

Numerical results are available for some cases [4] and those for CN are also shown in Fig. 5.9. At low a/D and $p/(M\,p_Y)$, the results converge on the infinite plate case, but the differences are large for $a/D>0.1$. It is interesting to note that there is a limiting value J^* from which a minimum D, $\bar{D}$, may be defined, below which full plasticity will always precede fracture. In this case:

$$\frac{\pi}{4}\frac{E\,J}{(M\,p_Y)^2\,\bar{D}} = 0.825$$

If fracture occurs at $J=J_c$, then:

$$\bar{D} = 1.05\frac{E\,J_c}{(M\,p_Y)^2} \equiv 2.7\,r_p$$

where $r_p = \pi/8\ (E\ J_c)/(M\ p_Y)^2$, the small-scale yielding zone size.

The line zone model of plasticity provides a way of approximating plasticity effects in determining J in practical cases. Of course, the plastic zones may not be true line zones, but it is a useful first approximation. A convenient representation is to define a plastic version of Y^2 as used to represent finite width effects in elasticity and to do this we may write:

$$Y_p^{\,2} = \frac{E\ J}{p^2\ a} = \frac{4}{\pi}\left(\frac{M\ p_Y}{p}\right)^2\left(\frac{D}{a}\right) J^*$$

from equation (5.20). For low stresses and short cracks, $Y_p^{\,2} = Y^2$, but in general $Y_p^{\,2} > Y^2$ since $J > G$. Figs. 5.10, 5.11, 5.12 and 5.13 show such curves deduced

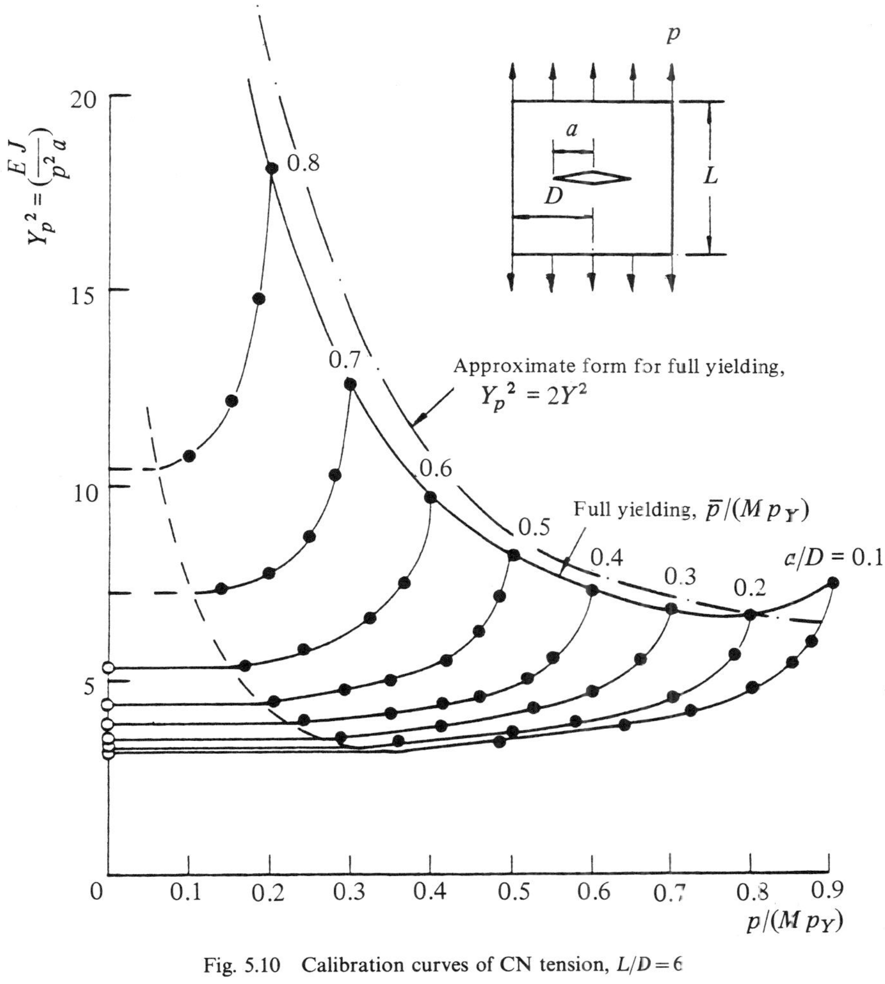

Fig. 5.10 Calibration curves of CN tension, $L/D = 6$

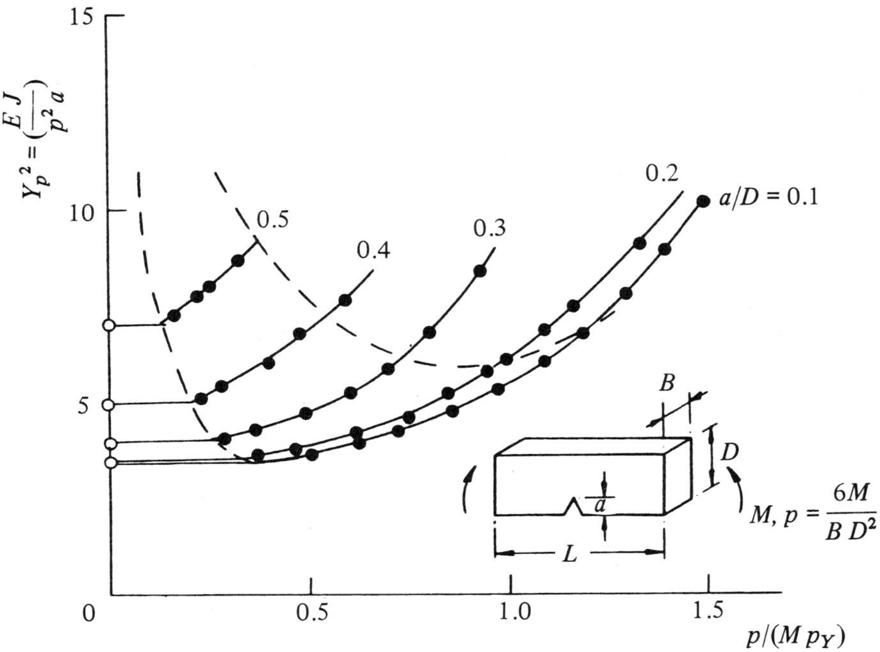

Fig. 5.11 Calibration curve for SEN pure bending, $L/D=12$

from computed results [4] for four practically important geometries, i.e. CN, SEN pure bending, SEN tension and a circumferentially notched round bar. The Y^2 values taken from Chapter 4 are also shown and the lower broken line is an approximate limit for $p/M\,p_Y$, for which $Y_p{}^2=Y^2$. For most cases of a/D, $p/M\,p_Y<0.3$, the elastic solution is adequate and $J \simeq G$, but outside this range there are appreciable differences. It should be noted that SEN tension shows larger plasticity effects compared with other loadings because of induced bending and that Y^2 for deep notches in the round bar case shows some discrepancies. It should also be noted that for $a/D<0.18$ in SEN bending, $p/M\,p_Y>1$.

The upper broken lines are estimates of the onset of full plastic deformation across the ligament. For the tensile cases, this condition is easily defined, since:

$$\frac{\bar{p}}{M\,p_Y}=1-\frac{a}{D}\ \text{for CN, tension} \tag{5.21}$$

and:

$$\frac{\bar{p}}{M\,p_Y}=\left(\frac{d}{D}\right)^2\ \text{for the circumferentially notched bar} \tag{5.22}$$

For the bending case and SEN tension, in which there is induced bending, the situation is more difficult to define since compressive yielding can be induced on the back face. However, as a first approximation, the fully plastic limit loads

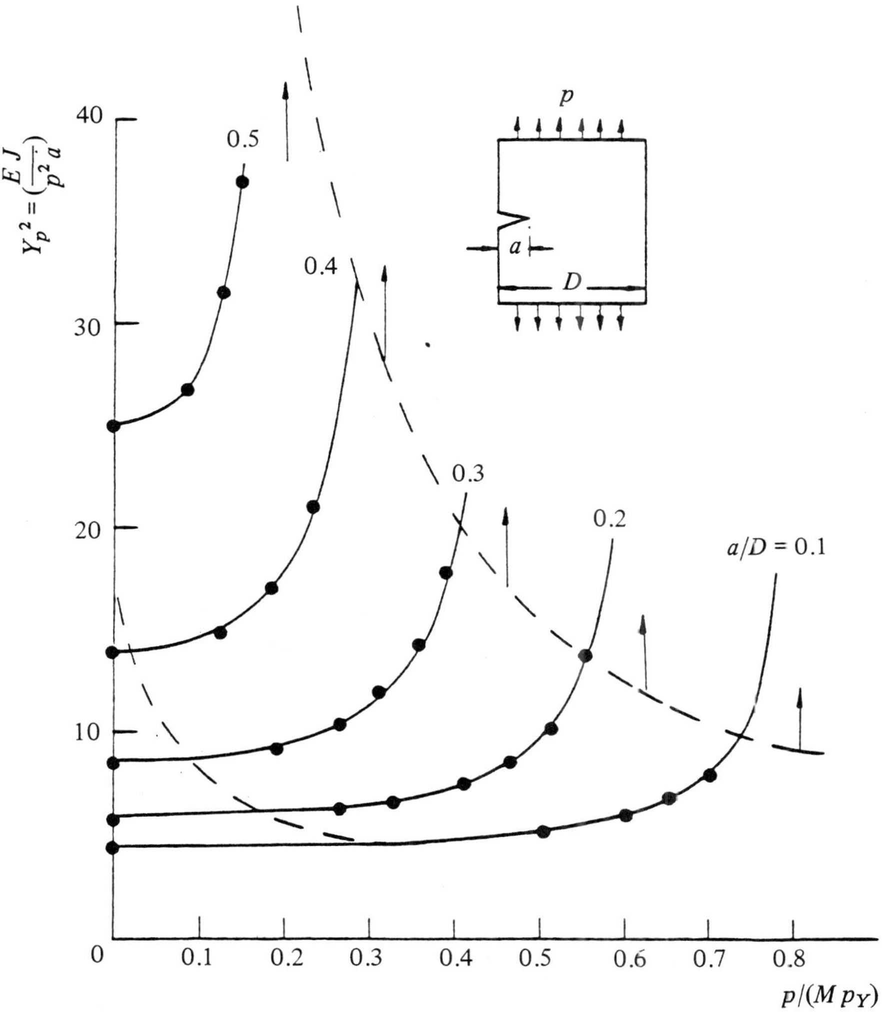

Fig. 5.12 Calibration curves for SEN tension, $L/D = 8$

have been used, i.e.

$$\frac{\bar{p}}{M\,p_Y} = \frac{3}{2}\left(1 - \frac{a}{D}\right)^2 \text{ for SEN, bending} \tag{5.23}$$

and:

$$\frac{\bar{p}}{M\,p_Y} = \sqrt{1 - 2\frac{a}{D} + 2\left(\frac{a}{D}\right)^2} - \frac{a}{D} \text{ for SEN, tension} \tag{5.24}$$

The lines in Figs. 5.10 to 5.13 may be used directly in estimating J, but it is frequently sufficient if an approximation can be made. This can be done using

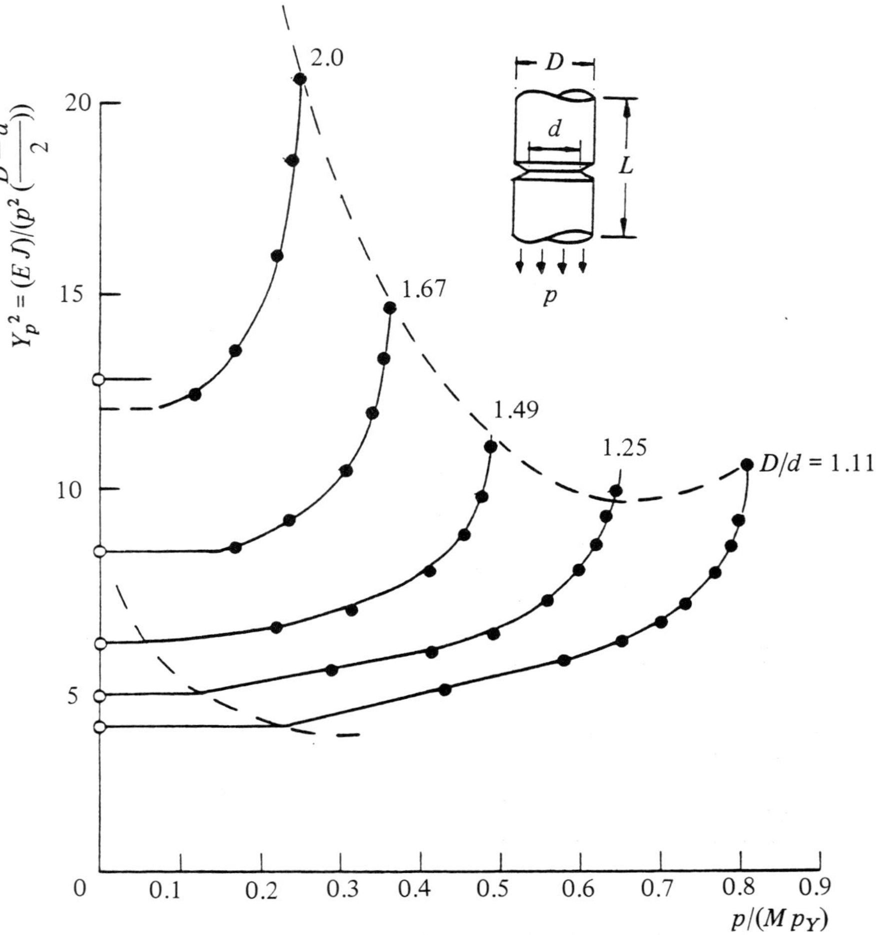

Fig. 5.13 Calibration curves for circumferentially notched round bars, $L/D = 3$

the Irwin approximation in equation (5.16), in which the collapse condition is scaled by replacing $M\,p_Y$ by $\bar{p}$. Thus:

$$Y_p^2 \simeq Y^2\left[1 - \frac{1}{2}\left(\frac{p}{\bar{p}}\right)^2\right]^{-1} \tag{5.25}$$

This is preferable to using the full line zone form, since the limiting case is always finite in the computation as it is determined by the elastic field. The upper limit for this approximation is always:

$$Y_p^2 = 2Y^2 \text{ (i.e. } J = 2G) \tag{5.26}$$

for $p = \bar{p}$, the fully plastic collapse condition. The chain dotted line in Fig. 5.10 shows this approximation for the CN case.

5.5 Fully yielded failure (Class 3)

In this situation, $l > D - a$ and the controlling length parameter is the ligament in the specimen. The loads on the body are determined by the fully yielded condition and, as outlined in the previous section, may be determined from a limit analysis. For non-work hardening materials, some additional information may be gleaned from slip line field solutions since kinematically admissible displacement fields may be worked out which give upper bounds, i.e. loads greater than the exact solution. Solutions satisfying equilibrium alone give lower bounds, i.e. loads less than the true solution. Centre notched plates, as shown in Fig. 5.14(a), have a simple slip system since slip lines as shown give acceptable

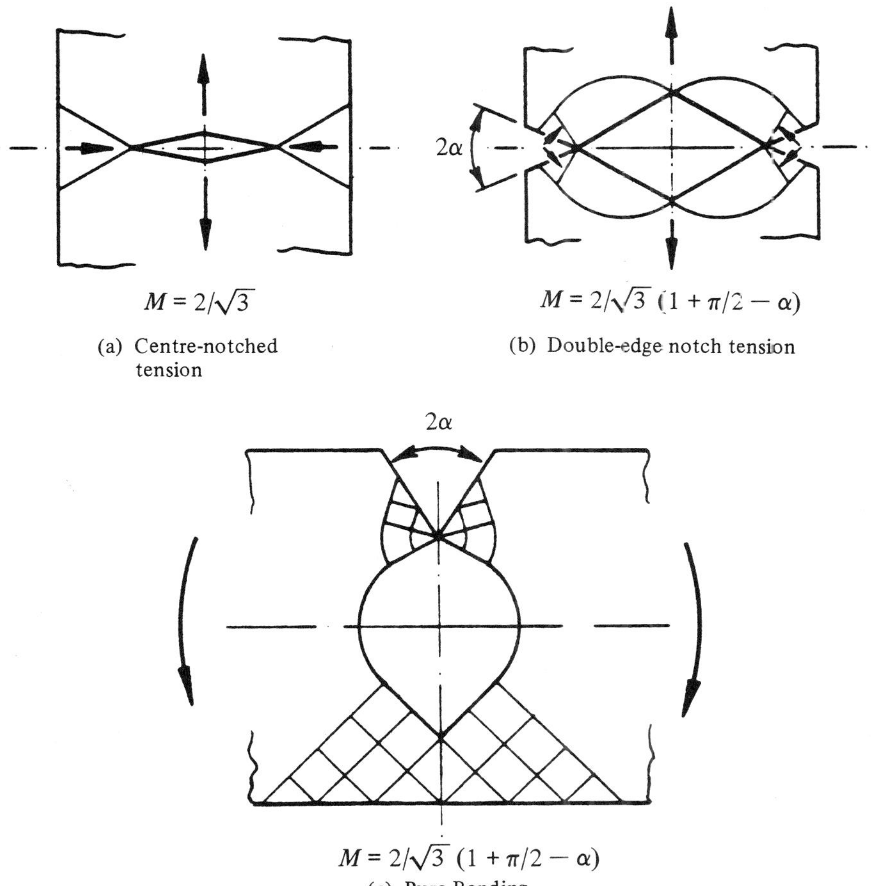

$M = 2/\sqrt{3}$

(a) Centre-notched tension

$M = 2/\sqrt{3}\,(1 + \pi/2 - \alpha)$

(b) Double-edge notch tension

$M = 2/\sqrt{3}\,(1 + \pi/2 - \alpha)$

(c) Pure Bending

Fig. 5.14 Slip line fields for fully plastic ligaments, M values for plane strain, $M = 1$ for plane stress kinematics. These result in the plane strain yield stress ($\nu=\frac{1}{2}$) being achieved on the nett section so that $M=2/\sqrt{3}$ for slip lines at $\pm 45°$. For double edge notch tension, as shown in Fig. 5.14(b), a more complex slip system is necessary for plane strain and this leads to a nett section stress in which $M=(1+(\pi/2)-\alpha)\,2/\sqrt{3}$, where α is the included angle of the notch [1]. Thus, for $\alpha=0$, we have $M=2.97$. For pure bending, shown in Fig. 5.14(c), a similar fan pattern exists near the crack tip and gives the same constraint factor [5]. For all cases where plane stress pertains, local necking occurs and $M=1$.

For failures which occur beyond full yielding, J cannot be defined in terms of load since this is constant and thus J must be expressed in terms of the displacement in the fully plastic state, u_p. A useful general formulation is to express J in terms of a power-law behaviour for which a plastic compliance may be defined as:

$$C_{p,n}=\frac{u_p{}^n}{P}$$

and because of the analogy with G, we may write, from equations (2.21):

$$J=\frac{1}{1+n}P^{(1+n)/n}\,C_{p,n}{}^{(1/n)-1}\frac{dC_{p,n}}{dA}=\frac{1}{1+n}\frac{u_p{}^{1+n}}{C_{p,n}{}^2}\frac{dC_{p,n}}{dA}$$

$$=U_3\frac{1}{C_{p,n}}\frac{dC_{p,n}}{dA} \qquad (5.27)$$

If we now confine our attention to the non-work hardening case, $n=0$, we may analyse the centre-notched plate, for which:

$$P=\frac{2}{\sqrt{3}}p_Y\,B(D-a)\,2 \text{ (D is the half width)}$$

so that: $$C_{p,0}=\sqrt{3}\frac{1}{4\,p_Y\,B(D-a)} \text{ and } A=2B\,a$$

Using these results in equation (5.27) gives:

$$J=u_p\left(\frac{2}{\sqrt{3}}p_Y\right)=\frac{U_3}{2B(D-a)} \qquad (5.28)$$

Thus, J is a linear function of u_p and is given, in this case, by the simple total energy, U_3, divided by the ligament area.

A more general point arises if it is not assumed that M remains constant. In this case:

$$C_{p,0}=\frac{1}{2\,M\,p_Y\,B(D-a)}$$

and: $$J=u_p\left[M-\frac{1}{M}\frac{dM}{da}(D-a)\right]p_Y=U_3\left[\frac{1}{2B(D-a)}-\frac{dM/da}{2\,M\,B}\right] \qquad (5.29)$$

so that J depends on the variation of the constraint factor as the crack grows.

The total energy form gives rise to a convenient factor in terms of which J may be expressed [6]. For the general case, we may introduce the ligament area, A_l, and obtain:

$$J=\left[\frac{U_3}{A_l}\right]\eta_p$$

where:

$$\eta_p=\frac{A_l}{C_{p,n}}\frac{dC_{p,n}}{dA} \tag{5.30}$$

For the centre-notched plate just discussed, $\eta_p=1$ so that J comes directly from the energy per unit ligament area when M is constant.

Bending is of considerable practical importance and for SEN bending, we have:

$$P=M\,p_Y\frac{B}{L}(D-a)^2$$

giving:

$$J=u_p\,M\,p_Y\left[\frac{2}{L}\right]\text{ and }\eta_p=2 \tag{5.31}$$

For SEN tension, we have a combination of tension and bending with the results:

$$J=u_p\,M\,p_Y\left[\frac{1-2a/D}{\sqrt{1-2a/D+2(a/D)^2}}+1\right]$$

and:

$$\eta_p=\frac{1-a/D}{\sqrt{1-2a/D+2(a/D)^2}}\left[\frac{1-a/D}{\sqrt{1-2a/D+2(a/D)^2}-a/D}+1\right] \tag{5.32}$$

The η_p values for the three loading systems are shown in Fig. 5.15, where it can be seen that the SEN tension values are somewhat greater than two with a maximum at $a/D=0.5$ of 2.415.

For a linear elastic–perfectly plastic material, the load-deflection curve would be as shown in Fig. 5.16(a), where the broken line is a possible elastic–plastic region as the zone grows. The resulting J versus u_p curve shown in Fig. 5.16(b) has a parabolic form for the elastic region since $n=1$ and then becomes linear for full plasticity.

5.6 J and δ as fracture criteria

As discussed in Section 5.2, J and δ may be postulated as fracture criteria for any system, but one set of conditions is of particular importance. The toughness of a material is determined by M at the crack tip and that for plane strain which defines the minimum is of special interest. It may be postulated that for any degree of plasticity, if M is unchanged then J (or δ if a line model is used) remains constant. Thus, if the failure occurs in classes 1, 2 or 3, the fracture may be

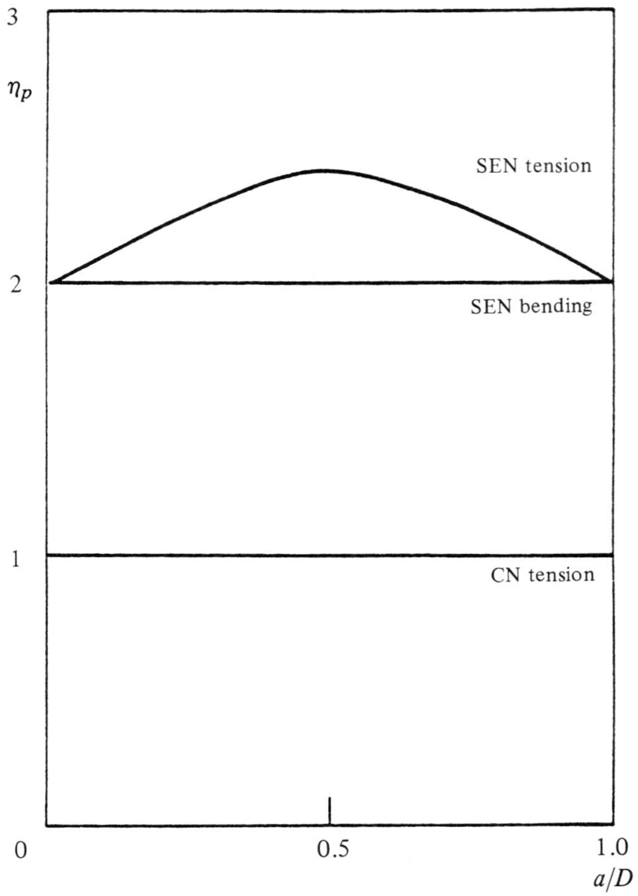

Fig. 5.15 Calibration factor η_p to determine J from the total energy

characterised by a constant J, *providing* that M remains the same. It will be noted that for class 1 the elasticity controlled $M \simeq 3$ and that for full plasticity in edge-notched tension and SEN bending again $M \simeq 3$. It is thus reasonable to assume that $J = J_c$, a constant at crack initiation, for any degree of plasticity. For a class 2 failure, J may be computed from the stress and Y_p^2 as given in Section 5.4, while for class 3 J may be found either from a set of P, u diagrams of various a values (the Begley & Landes method [7]) or from η_p and the total energy. This latter is particularly useful since it requires only one specimen, but with $a/D \simeq 0.5$ to ensure full plasticity.

It should be noted that if M changes, as for example in CN tension going from class 1 ($M \simeq 3$) to class 3 ($M \simeq 1.15$), then J_c would also be expected to change.

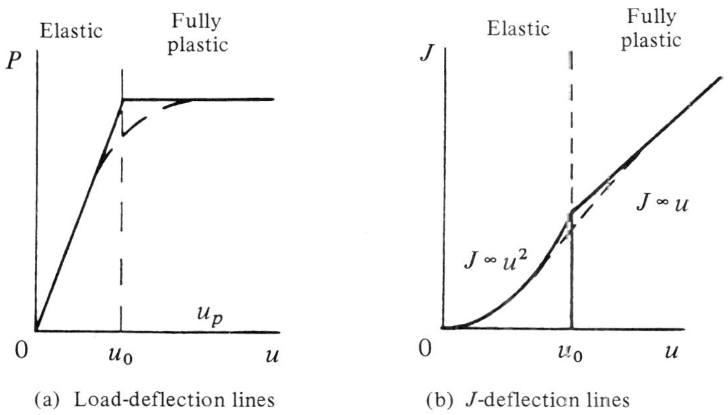

Fig. 5.16 J-deflection diagrams for an elastic-perfectly plastic body

K_{c2} and G_{c2} defined in Section 5.3 are J_c values for $M=1$, such that:

$$J_{c2}=G_{c2}=\frac{K_{c2}^{2}}{E} \tag{5.33}$$

5.7 Visco-elastic fracture (Class 4)

In this section, we will deal with fracture in bodies which exhibit time-dependence so that any loading cycle results in some dissipation of energy. It is possible to compute stresses and strains or loads and displacements so that time-dependent forms of the various fracture criteria may be determined. As in the previous sections, it is convenient to consider both global and local energy analyses.

5.7.1 *Overall energy analysis*

Most of the relevant factors may be explored using the assumption that the material is linearly visco-elastic (non-linear effects will be considered in Section 5.8). This means that the stress–strain relations, for example in simple tension, may be expressed as convolution integrals of the form:

$$e(t)=\int_0^t C(t-\tau)\frac{dp(\tau)}{d\tau}\,d\tau$$

(5.34)

and:

$$p(t)=\int_0^t M(t-\tau)\frac{de(\tau)}{d\tau}\,d\tau$$

where $C(t)$ is the creep compliance function, and $M(t)$ is the relaxation modulus with $1/C=M=E$ for the elastic case. It is usually more convenient to deal with such relationships in terms of their Laplace transforms defined as:

$$\bar{f}(s) = \int_0^\infty f(t)\, e^{-st}\, dt \tag{5.35}$$

where s is the transformed variable and the bar denotes the transformed function. Equations (5.34) now become:

$$\left.\begin{aligned} \bar{e}(s) &= \bar{C}(s)\, \bar{p}(s)\, s \\ \bar{p}(s) &= \bar{M}(s)\, \bar{e}(s)\, s \\ \text{and:} \quad s\, \bar{M}(s) &= \frac{1}{s\, \bar{C}(s)} \end{aligned}\right\} \tag{5.36}$$

For any loading history, energy dissipation occurs throughout the body and, although the stress–strain relationships, and thus load-deflection diagrams, are linear for fixed times, they are non-linear for a given history of loading. An important general point is that for a stationary crack at initiation, the problem is termed variables separable [8] and results in a solution of the same form as the elastic one with the appropriate time-dependent modulus included. In terms of loads and deflections, we have:

$$J = -\frac{1}{B}\frac{d}{da}\left|\int_0^u P\, du\right|_{u\ const}$$

from equation (2.14) where, in general: (5.37)

$$u(t),\ P(t) \text{ and } u = P\, C(t)$$

with either u or P being the input function. Thus, $J(t)$ is derived in the usual way from the load-deflection diagrams for a and $a + da$ for a constant u.

As an example of the method, let us consider a DCB specimen made from a standard linear solid, consisting of a modulus E_1 in a series with a modulus E_2 in parallel with a viscosity η. For this material [8]:

$$\left.\begin{aligned} C(t) &= \frac{E_1 + E_2}{E_1\, E_2}\left(1 - \frac{E_1}{E_1 + E_2}\, e^{-t/\tau_2}\right),\ \tau_2 = \frac{\eta}{E_2} \\ \text{and:} \quad M(t) &= \frac{E_1\, E_2}{E_1 + E_2}\left(1 + \frac{E_1}{E_2}\, e^{-t/\tau_1}\right),\ \tau_1 = \frac{\eta}{E_1 + E_2} \end{aligned}\right\} \tag{5.38}$$

For a constant displacement rate, $\dot{u}$, from the second of these equations, we have:

$$P(t) = \frac{B\, D^3}{8a^3}\, u\, \frac{E_1\, E_2}{E_1 + E_2}\left[1 + \frac{E_1}{E_2}\frac{\tau_1}{t}\left(1 - e^{-t/\tau_1}\right)\right] \text{ with } u = \dot{u}\, t$$

by using equations (5.34) and variables separable conditions. We may now write:

$$P(u) = \frac{B\,D^3}{8a^3}\frac{E_1\,E_2}{E_1+E_2}\,u\left[1+\frac{E_1}{E_2}\frac{\dot{u}\,\tau_1}{u}(1-e^{-u/\dot{u}\tau_1})\right]$$

giving the energy expressions:

$$U_3 = \int_0^u P\,du = \frac{B\,D^3}{8a^3}\frac{E_1\,E_2}{E_1+E_2}\left[\frac{u^2}{2}+\frac{E_1}{E_2}\dot{u}\,\tau_1\,(u+\dot{u}\,\tau_1\,e^{-u/\dot{u}\tau_1}-\dot{u}\,\tau_1)\right]$$

and:
$$J = \frac{3D^3}{16a^4}u^2\frac{E_1\,E_2}{E_1+E_2}\left[1+\frac{2E_1}{E_2}\left(\frac{\dot{u}\,\tau_1}{u}\right)^2\left(\frac{u}{\dot{u}\,\tau_1}+e^{-u/\dot{u}\tau_1}-1\right)\right]$$

from equation (5.37). As $\dot{u}\to\infty$ $(t\to 0)$, we may determine G for the current P with the short-term modulus E_1 so that we may write finally:

$$\left(\frac{J}{G}\right)_{\dot{u}} = (1+R)\frac{[1+2R\,(\tau_1/t)^2((t/\tau_1)+e^{-t/\tau_1}-1)]}{[1+R\,(\tau_1/t)(1-e^{-t/\tau_1})]^2} \tag{5.39}$$

where $R = E_1/E_2$. Similarly, for constant loading rate:

$$\left(\frac{J}{G}\right)_{\dot{P}} = (1+R)\left\{1-\frac{2R}{1+R}\left(\frac{\tau_2}{t}\right)^2\left[\frac{t}{\tau_2}-(1-e^{-t/\tau_2})\right]\right\} \tag{5.40}$$

For creep loading, P is constant and:

$$u(t) = \frac{8a^3\,P}{B\,D^3}\frac{E_1+E_2}{E_1\,E_2}\left(1-\frac{E_1}{E_1+E_2}e^{-t/\tau_2}\right)$$

The energy is given by $P\,u$, but it should be noted that G is twice the usual result here, so that:

$$U_3 = P^2\frac{8a^3}{B\,D^3}\left[\frac{E_1+E_2}{E_1\,E_2}\left(1-\frac{E_1}{E_1+E_2}e^{-t/\tau_2}\right)\right]$$

and finally:
$$\left(\frac{J}{G}\right)_P = 1+R\,(1-e^{-t/\tau_2}) \tag{5.41}$$

In relaxation, the displacement is constant and so the energy input is a constant at the initial elastic value and J is constant, i.e.

$$J = \frac{3}{16}\frac{D^3}{a^4}u^2\,E_1$$

However, G is calculated on the current load:

$$P(t) = \frac{u\,B\,D^3}{8a^3}\frac{E_1\,E_2}{E_1+E_2}\left(1+\frac{E_1}{E_2}e^{-t/\tau_1}\right)$$

so that:
$$\left(\frac{J}{G}\right)_u = \left[\frac{1+R}{1+R\,e^{-t/\tau_1}}\right]^2 \tag{5.42}$$

A final example is for a moving crack of constant velocity $\dot{a}$ such that $a=\dot{a}\,t$ and constant load P. Here, it is possible to solve the DCB problem using the curvature relationship [8]:

$$\frac{d^2v}{dx^2}=\frac{12\,P\,x}{B\,D^3}\frac{E_1+E_2}{E_1\,E_2}\left(1-\frac{E_1}{E_1+E_2}\,e^{-(t-x/\dot{a})/\tau_2}\right)$$

and the boundary conditions $v=dv/dx=0$ at $x=a$ and $v=u/2$ at $x=0$. The displacement expression becomes:

$$u=\frac{8\,P\,a^3}{B\,D^3}\frac{E_1+E_2}{E_1\,E_2}\left\{1+\frac{6E_1}{E_1+E_2}\left[\left(\frac{\tau_2}{t}\right)^2-\frac{1}{2}\left(\frac{\tau_2}{t}\right)-\left(\frac{\tau_2}{t}\right)^3(1-e^{-t/\tau_2})\right]\right\}$$

and the final result is:

$$\left(\frac{J}{G}\right)_{\dot{a},p}=1+R\left\{1+6\left[\left(\frac{\tau_2}{t}\right)^2-\frac{1}{2}\left(\frac{\tau_2}{t}\right)-\left(\frac{\tau_2}{t}\right)^3(1-e^{-t/\tau_2})\right]\right\} \tag{5.43}$$

The major point here is that an expression for J may be computed for a visco-elastic material and, in general, it will be a function of time. The time-dependence is different for each history and any criterion based on, say, a fixed J would result in the prediction of times to failure.

The standard linear solid is not particularly realistic and more useful results come from a simple power-law dependence of the form:

$$C(t)=\frac{(t/\tau_0)^n}{E_0} \tag{5.44}$$

although the G solution for $t=0$ is not retrievable. Expressions for J may, however, be found and for the DCB we have:

$$\left.\begin{aligned}
\left(\frac{J}{J_0}\right)_p &= \left(\frac{t}{\tau_0}\right)^n \\
\left(\frac{J}{J_0}\right)_{\dot{p}} &= \left(\frac{2!}{2+n!}\right)\left(\frac{t}{\tau_0}\right)^n \\
\left(\frac{J}{J_0}\right)_{\dot{a},p} &= \left(\frac{2!}{2+n!}\right)\left(\frac{t}{\tau_0}\right)^n,\ t=\frac{a}{\dot{a}} \\
\left(\frac{J}{J_0}\right)_{\dot{u}} &= \left(\frac{n\,\pi}{\sin n\,\pi}\right)\frac{2!}{(2-n)!}\left(\frac{t}{\tau_0}\right)^n
\end{aligned}\right\} \tag{5.45}$$

where $J_0=(12\,P^2\,a^2)/(B^2\,D^3\,E_0$. (Note: J_u has no meaning for this model).

5.7.2 *Local crack tip solutions*

For a stationary crack, the eigen solution for the local stress field as given in Section 3.2 may be repeated in terms of transformed variables. Since neither r

nor θ change with time, they are not changed and we have an identical solution with $m=\frac{1}{2}$ giving the dominant term near the crack tip with stresses of the form:

$$p(t)=\frac{K(t)}{\sqrt{2\pi\, r}}f(\theta) \tag{5.46}$$

while the displacements are given by:

$$u(t)=\frac{\sqrt{2\pi\, r}}{2\pi}\,g(\theta)\int_0^t C(t-\tau)\frac{dK(\tau)}{d\tau}\,d\tau \tag{5.47}$$

where $f(\theta)$ and $g(\theta)$ are as in equations (3.11) and (3.12). An expression for $J(t)$ may now be deduced as for G in equation (2.15), but here we must use:

$$J(t)=\frac{1}{\Delta a}\int_0^{\Delta a}\int_0^t [-2u_\theta(\pi,s,t)]\,\frac{d}{dt}\,[p_\theta(0,\,\Delta a-s,\,t)]\;dt\;ds$$

On rearranging and using the appropriate forms, we have:

$$J(t)=2\int_0^t\left[\int_0^t C(t-\tau)\frac{dK(\tau)}{d\tau}\,d\tau\right]\frac{dK(t)}{dt}\,dt \tag{5.48}$$

For elasticity, $C(t)=1/E$, giving $J(t)=G=K^2/E$ as before. For creep, $K(t)=K_0$, a constant, and:

$$J(t)=2C(t)\,\mathrm{K}_0{}^2 \tag{5.49}$$

which confirms the results from the global solution given in equations (5.41) and (5.45). For a linear ramp loading, $K=\dot{K}\,t$, $\dot{K}$ a constant, and:

$$J(t)=2\dot{K}^2\int_0^t\int_0^t C(t-\tau)\,d\tau\,dt \tag{5.50}$$

from which equation (5.40) and the equivalent power-law result may be derived.

Moving cracks present particular problems in the analysis but the local field would be expected to be of the same form as the static case. The appropriate time-scale cannot be defined, however, since it is governed by the crack speed which may only be converted to a time using some length factor. The solution for the DCB given in equations (5.43) and (5.45) has no general significance since it overcomes this problem by the use of a, which defines t for the whole beam. On the local level considered here, we must use a *local* length scale to give an appropriate time, and line zones provide a useful vehicle for this purpose.

5.7.3 *Line zone models* [9]

Let us now consider a small-scale line zone at the crack tip with a constant zone stress, p_c. The zone length arises from the superposition of linear solutions and may be extended to the visco-elastic case. The elastic solution given in equation (4.21) now becomes:

$$c(t)=\frac{\pi}{8}\frac{K^2(t)}{p_c^{\,2}} \tag{5.51}$$

If we now take a zone tip displacement profile of $f(x/c)$, then for the elastic case we have:

$$(2v)=\frac{K^2}{p_c E}f\left(\frac{x}{c}\right), f(1)=1$$

and we may write the analogous visco-elastic definition of J:

$$J(t)=p_c\,(2v)=\int_0^t C(t-\tau)\frac{d}{d\tau}\left[K^2(\tau)f\left(\frac{x}{c}\right)\right]d\tau \tag{5.52}$$

For a constant zone stress we have:

$$J=p_c\,(2v)_{x/c=1}$$

and for a constant load and a stationary crack, $K^2(t)=K_0^{\,2}$, and we have:

$$J=p_c\,(2v)_1=C(t)\,K_0^{\,2} \tag{5.53}$$

since x is not a function of time. For the moving crack with constant load, $\dot{x}=\dot{a}$ and putting $t=0$ at $x=0$ so that the whole time history is in the range 0 to $t_1=c/\dot{a}$, we have:

$$J=p_c\,(2v)=K_0^{\,2}\int_0^{t_1} C(t-\tau)\frac{df}{d\tau}\left(\frac{\tau}{t_1}\right)d\tau$$

The expression for f given in equation (4.23) may now be used, i.e.

$$f\left(\frac{x}{c}\right)=\sqrt{\frac{x}{c}}-\frac{1}{2}\left(1-\frac{x}{c}\right)\ln\left(\frac{1+\sqrt{x/c}}{1-\sqrt{x/c}}\right)$$

and the expression for J becomes:

$$J=p_c\,(2v)_1=\frac{K_0^{\,2}}{E_0}\left(\frac{c}{\dot{a}\,\tau_0}\right)^n\int_0^1(1-\xi)^n\,\frac{1}{2}\ln\left(\frac{1+\xi^{1/2}}{1-\xi^{1/2}}\right)d\xi$$

which may be integrated to give:

$$J=\frac{K_0^{\,2}}{E_0}\left(\frac{c}{\dot{a}\,\tau_0}\right)^n\lambda_n \tag{5.54}$$

where:

$$\lambda_n=\frac{\Gamma(1+n)}{(1+n)}\frac{\sqrt{\pi}/2}{\Gamma(\frac{3}{2}+n)}$$

and Γ is the gamma function. If a constant J ($(2v)_1$ in this case) criterion, J_c, is used and since $J_0=K_0^{\,2}/E_0$, then we may write an expression for $\dot{a}$:

$$\dot{a}=\lambda_n^{\,1/n}\left(\frac{J_0}{J_c}\right)^{1/n}\frac{c}{\tau_0} \tag{5.55}$$

for $n \ll 1$, $\lambda_n{}^{1/n} \simeq 0.20 = 1/5$, and $\lambda_n{}^{1/n} = 1/3$ for $n = 1$. In some cases, the expression for f is difficult to integrate and may be approximated by $(x/c)^2$. When this is used here:

$$\lambda_n = \frac{2}{(2+n)!} \text{ and } \lambda^{1/n}_{n_{n\to 0}} \simeq 0.22$$

It should be noted that much of this analysis is concerned with defining λ_n since the form of the result may be found simply from $t = \lambda_n{}^{1/n}$ $(c/\dot{a})$ and the pseudo elastic $J = K_0{}^2/E_0\,(t/\tau_0)^n$.

A crack initiation time, t_i, may be found from:

$$J_c = J_0 \left(\frac{t_i}{\tau_0}\right)^n \tag{5.56}$$

so that an initiation crack speed, $\dot{a}_i$, comes from combining equations (5.54) and (5.55), i.e.

$$\dot{a}_i \simeq \frac{1}{5}\frac{c}{t_i} \tag{5.57}$$

If the crack speed is assumed to be slowly varying, then equation (5.54) may be used to predict slow crack growth in, say, a sheet with a centre crack of length $2a_0$ and a constant stress, p. Thus:

$$J_0 = \frac{p^2\,\pi\,a}{E_0}$$

and equation (5.55) becomes:

$$\dot{a} \simeq \frac{1}{5}\frac{c}{\tau_0}\left(\frac{\pi\,a\,p^2}{J_c\,E_0}\right)^{1/n}$$

i.e.

$$\int_{a_0}^{a_0+\Delta a} a^{-1/n}\,da = \left(\frac{\pi\,p^2}{J_c\,E_0}\right)^{1/n} \frac{c}{5\tau_0}\int_{t_i}^{t} dt$$

and:

$$\frac{n\,a_0}{1-n}\left[1 - \left(1 + \frac{\Delta a}{a}\right)^{1-(1/n)}\right] = \left(\frac{J_0}{J_c}\right)^{1/n} \frac{c}{5\tau_0}(t - t_i) \tag{5.58}$$

For $\Delta a \ll a_0$, we have an expression for the growth:

$$\Delta a \simeq \left(\frac{J_0}{J_c}\right)^{1/n} \frac{c}{5\tau_0}(t - t_i) \tag{5.59}$$

and a time for which $\Delta a \gg a_0$:

$$\hat{t} \simeq \frac{5n}{1-n}\left(\frac{\tau_0\,a_0}{c}\right)\left(\frac{J_c}{J_0}\right)^{1/n} + t_i,\ n \ll 1$$

i.e.

$$\frac{\hat{t}}{t_i} = 1 + \frac{5n}{1-n}\left(\frac{a_0}{c}\right) \tag{5.60}$$

The analysis may be extended to include a time-dependent zone stress. For crack initiation, we may write:

$$C(t)=\frac{\pi}{8}\frac{K^2(t)}{p_c{}^2(t)} \tag{5.61}$$

which gives a zone growth law prior to failure. Equation (5.52) must now be rewritten as:

$$(2v)=\int_0^t C(t-\tau)\frac{d}{d\tau}\left[\frac{K^2(\tau)}{p_c(\tau)}f\left(\frac{x}{c}\right)\right]d\tau \tag{5.62}$$

If we now introduce a time-dependent stress:

$$p_c=p_0\left(\frac{t}{\tau_1}\right)^{-m}$$

then for a constant crack length and fixed K_0, we have:

$$(2v)_{x/c=1}=\frac{K_0{}^2}{p_0\,E_0}\int_0^t\left(\frac{t-\tau}{\tau_0}\right)^n\frac{d}{d\tau}\left(\frac{\tau}{\tau_1}\right)^m d\tau,\ \text{with}\ C(t)=\frac{1}{E_0}\left(\frac{t}{\tau_0}\right)^n$$

i.e.

$$(2v)_1=\frac{K_0{}^2}{p_0\,E_0}m\left(\frac{t}{\tau_1}\right)^m\left(\frac{t}{\tau_0}\right)^n\int_0^1(1-\xi)^n\,\xi^{m-1}\,d\xi$$

giving:

$$(2v)=\frac{K_0{}^2}{p_0\,E_0}\left(\frac{t}{\tau_0}\right)^n\left(\frac{t}{\tau_1}\right)^m\lambda_{nm}$$

where:

$$\lambda_{nm}=\frac{\Gamma(1+n)\,\Gamma(1+m)}{\Gamma(1+n+m)}$$

i.e.

$$J=p_0\left(\frac{t}{\tau_1}\right)^{-m}(2v)_1=J_0\left(\frac{t}{\tau_0}\right)^n\lambda_{nm} \tag{5.63}$$

For the moving crack with $f=(x/c)^2$, we have:

$$(2v)=\frac{K_0{}^2}{p_0\,E_0}\int_0^t\left(\frac{t-\tau}{\tau_0}\right)^n\frac{d}{d\tau}\left[\left(\frac{\tau}{\tau_1}\right)^m\left(\frac{\tau\,\dot{a}}{c}\right)^2\right]d\tau$$

i.e.

$$(2v)=\frac{K_0{}^2}{p_0\,E_0}\left(\frac{t}{\tau_0}\right)^n\left(\frac{t}{\tau_1}\right)^m\left(\frac{t\,\dot{a}}{c}\right)^2\int_0^1(1-\xi)^n\frac{d}{d\xi}\,\xi^{2+m}\,d\xi$$

and:

$$J=p_0\left(\frac{c}{\dot{a}\,\tau_1}\right)^{-m}(2v)_1=\left[\frac{(2+m)!\,\lambda_{nm}}{(2+m+n)!}\right]J_0\left(\frac{c}{\dot{a}\,\tau_0}\right)^n \tag{5.64}$$

c is crack speed dependent here and from equation (5.61), we have:

$$\left(\frac{c}{\dot{a}\,\tau_1}\right)=\left(\frac{c_0}{\dot{a}\,\tau_1}\right)^{1/(1-2m)}\ \text{with}\ c_0=\frac{\pi}{8}\frac{K_0{}^2}{p_c{}^2}$$

The use of J or COD criteria lead to different expressions for $\dot{a}$, and we obtain:

$$\left.\begin{aligned} \dot{a} &= \left[\frac{(2+m)!\ \lambda_{nm}}{(2+m+n)!}\right]^{(1-2m)/n} \left(\frac{J_0}{J_c}\right)^{(1-2m)/n} \left(\frac{\tau_0}{\tau_1}\right)^{2m} \frac{c_0}{\tau_0}, \text{ for } J=J_c \\ \text{and:} & \\ \dot{a} &= \left[\frac{(2+m)!\ \lambda_{nm}}{(2+m+n)!}\right]^{(1-2m)/(n+m)} \left(\frac{\delta_0}{\delta_c}\right)^{(1-2m)/(n+m)} \left(\frac{\tau_0}{\tau_1}\right)^{m(1+2n)/(n+m)} \frac{c_0}{\tau_0} \end{aligned}\right\} \quad (5.65)$$

for $(2v)_1=\delta_c$ with $\delta_0=J_0/p_0$. Similarly, for initiation times from equation (5.63):

$$\left.\begin{aligned} \lambda_{nm}{}^{1/n}\, t_i &= \left(\frac{J_c}{J_0}\right)^{1/n} \tau_0, \text{ for } J=J_c \\ \text{and:} & \\ \left(\frac{\tau_0}{\tau_1}\right)^m \lambda_{nm}{}^{1/(n+m)}\, t_i &= \left(\frac{\delta_c}{\delta_0}\right)^{1/(n+m)} \tau_0, \qquad \text{for } (2v)_1=\delta_c \end{aligned}\right\} \quad (5.66)$$

As $m=n\rightarrow 0$, $\lambda_{nm}{}^{1/n}=\lambda_{nm}{}^{1/(n+m)}\rightarrow 1$, and $[((2+m)!\ \lambda_{nm})/(2+m+n)!]^{(1-2m)/n}\rightarrow 0.22$ as for λ_n and when the same function is raised to the power $(1-2m)/(n+m)$, as in the second of equations (5.65), we have $0.477=\sqrt{0.22}$ as expected. Thus, for small values of m and n, we may use the more precise form of λ_n to give:

$$\left.\begin{aligned} &\dot{a}\simeq\frac{1}{5}\left(\frac{J_0}{J_c}\right)^{1/n}\frac{c_0}{\tau_0} \text{ and } \dot{a}\simeq\frac{1}{\sqrt{5}}\left(\frac{\delta_0}{\delta_c}\right)^{1/(n-m)}\frac{c_0}{\tau_0} \\ \text{with:}\quad &t_i\simeq\left(\frac{J_c}{J_0}\right)^{1/n}\tau_0 \text{ and } t_i\simeq\left(\frac{\delta_c}{\delta_0}\right)^{1/(n+m)}\tau_0 \end{aligned}\right\} \quad (5.67)$$

5.7.4 *Non-linear models*

The definition of J as given in equation (5.37) can, of course, be used on any time-dependent system, providing that a load-deflection (or stress–strain) function may be defined for a given loading history. We may, for example, write a load-deflection relationship in the form:

$$u(t)=C\,P\left(\frac{P}{P_0}\right)^m\left(\frac{t}{\tau_0}\right)^n \quad (5.68)$$

where $m=0$ gives the linear case, and C is a compliance. For creep loading:

$$U=P\,u=\frac{C\,P^{2+m}}{P_0{}^m}\left(\frac{t}{\tau_0}\right)^n$$

i.e.

$$J(t)=\frac{1}{1+m}\,\frac{P^2}{B}\,\frac{dC}{da}\left(\frac{t}{\tau_0}\right)^n\left(\frac{P}{P_0}\right)^m \quad (5.69)$$

For some creep problems in metals, $n=1$, and a rate of energy dissipation may be defined:

$$\dot{J}_{n=0}=\frac{1}{1+m}\frac{P^2}{B}\frac{dC}{da}\left(\frac{P}{P_0}\right)^m\frac{1}{\tau_0} \tag{5.70}$$

An alternative is to define a work rate via $\dot{u}$ in equation (5.68):

$$\dot{U}=\int_0^{\dot{u}} P\,d\dot{u}=\left(\frac{1+m}{2+m}\right)\frac{P^{2+m}C}{\tau_0}$$

and:

$$J^*=-\frac{1}{B}\left(\frac{d\dot{U}}{da}\right)_{u\ const}=\frac{1}{2+m}\frac{P^2}{B}\frac{dC}{da}\left(\frac{P}{P_0}\right)^m\frac{1}{\tau_0} \tag{5.71}$$

This form avoids the factor of two which arises in $\dot{J}$ for creep and is directly analogous with J but involves a rate. It should be noted that:

$$\frac{\dot{J}}{J^*}=\frac{2+m}{1+m}$$

and $\dot{J}\rightarrow J^*$ for large m which is frequently the case in metals.

The concept of J^* is equivalent to using an isochronous definition of J which, for the linear case, is:

$$J=\frac{K^2}{E(t)}$$

where $E(t)$ is simply that for the particular time history. When $n\neq 1$, there is no simple physical interpretation possible in terms of rates.

Since the linear visco-elastic problem resulted in non-linear stress–strain diagrams, there is no special problem in extending the line zone results to non-linear cases, providing they can be formulated. It seems likely that crack growth laws formulated in terms of J or δ as given in equation (5.67) will suffice.

References for Chapter 5

[1] Rice, J. R., Chapter 3 in *Fracture: Volume II* (ed. H. Liebowitz), Academic Press (1968).

[2] ASTM Standards, **31** (1969) 1099–1114.
ASTM STP 463 (1970) 249–269.

[3] Irwin, G. R., Proceedings of the 7th Sagamore Conference (1960) IV-36.

[4] Hayes, D. J. & Williams, J. G., *Int. J. Fract. Mech.*, **8**(3) (1972) 239–256.

[5] Green, A. P. & Hundy, B. B., *J. Mech. Phys. Solids*, **4** (1956) 128–144.

[6] Turner, C. E., Chapter 2 in *Post-yield Fracture Mechanics* (ed. D. G. H. Latzko), Applied Science Publishers (1979).

[7] Begley, J. A. & Landes, J. D., ASTM STP 514 (1972) 1–20.

[8] Williams, J. G., *Stress Analyses of Polymers*, 2nd edn, Ellis Horwood, Chichester (1980).

[9] Schapery, R. A., *Int. J. Fract. Mech.*, **11** (1975) 141–155, 369–388, 549–562.

6

Fracture initiation

6.1 Introduction

In this chapter, we will utilise the various concepts developed previously to evaluate the fracture toughness of polymers at crack initiation. Fracture toughness as used in fracture mechanics is a material property in that it is independent of test method and geometry, and the demonstration of this independence will form a major part of the information presented. However, in common with other properties, such as modulus and yield stress, the toughness can be temperature-dependent and, because all polymers exhibit some degree of visco-elasticity, time-dependent. In energy terms, therefore, the characterising parameter should be $J_c(t)$ and not G_c, since the latter assumes elasticity. For these generally linearly visco-elastic materials, it is convenient to express this as $G_c(t)$ and the equivalent $K_c(t)$. All fractures discussed are in mode I and thus the subscript I is omitted in what follows. These parameters refer to both stable and unstable crack initiation, and stability will be considered under a separate heading where appropriate. Although mentioned here, the description of crack propagation will be deferred to Chapter 7.

It should also be noted that toughness values are, in common with all other properties, controlled by the molecular structure of the material and are thus sensitive to both molecular variations and micro-structure. For any given polymer, it is possible to see substantial differences in toughness depending on chemical changes, as in different grades of the same material, and due to micro-structural changes arising from processing differences. This having been said, however, it is still convenient to present the data for polymers grouped under their main chemical designation. The various phenomena which occur and which the analysis has to describe will then be dealt with under the headings of materials where they frequently occur. There are, however, some common features of mechanical behaviour which have a strong influence on fracture and it is convenient to deal with these separately.

6.2 Yielding, microvoiding and crazing

In the discussion of energy dissipation in Chapter 5, attention in yielding phenomena was concentrated on shear yielding, familiar from metals behaviour and described in terms of the von Mises criterion. This was convenient since most of the theory is couched in these terms and the results are useful for polymers. However, it is important to recognise that while shear yielding does occur in polymers, it is not the only energy absorbing process or indeed the major one in many cases. Shear yielding is certainly the dominant process in compressive stress fields but when hydrostatic tension is present, as around loaded crack tips, then microvoids frequently occur. These can initiate from inhomogeneities in the micro-structure such as random density fluctuations in amorphous materials or from ordered regions in semi-crystalline materials. In addition, many polymers contain high proportions of particulate matter, either as deliberate fillers, contamination or as material added for other purposes such as lubricants, flame retardants or stabilisers. Thus, many polymers will whiten at some stress level in tension (sometimes the shear yield stress), indicating the formation of diffuse voiding which absorbs considerable energy. Frequently, this whitening will disappear if the material is subsequently compressed as the voids close. Care should be taken in equating voiding too closely with hydrostatic stresses since shear alone in an inhomogeneous material can induce local tensions and thus voiding, as when voiding and the shear yield condition occur together.

An important phenomenon which arises from microvoiding is that of crazing, in which concentrated bands of these voids develop. The origins of this process lie in the ability of polymers to work harden very rapidly at high strains because of molecular orientation. Thus, a ligament between two adjacent voids will become stiff as the voids develop, thus tending to give preference to ligament deformation around adjacent, less well developed, voids. This process leads to the development of planar groups of voids which act effectively as cracks with ties formed by the ligaments across the surfaces. These are the crazes which form in the bulk of many polymers on loading and if rupture of the ligaments occurs, they are often the precursors of fracture.

When cracks are being formed or loaded, all of these processes can occur and their occurrence can greatly influence the initiation toughness values. The process by which the crack is formed and the crack tip sharpness often strongly influence how the deformation proceeds and thus such details are of major importance in testing procedures. In general terms, the minimum toughness is obtained with plane strain conditions which will require a sharp crack. In polymers which do not craze, such as thermoset resins, this leads to plane strain shear yielded zones, as in metals, but for most thermoplastics crazes occur. The minimum energy condition is when a single craze is formed which subsequently fractures, but in some materials craze bunches or a zone of microvoids are formed.

6.3 Polymethylmethacrylate (PMMA)

PMMA is the most intensively studied polymer as far as fracture is concerned (reference [1] lists a series of studies). This is because it has proved to be particularly convenient, being an amorphous glass below 110°C exhibiting brittle fracture under normal testing conditions below about 85°C. It is also remarkably consistent in properties, even between different suppliers, and shows, for example, almost no molecular weight dependence of toughness above a value of 55,000. A further important characteristic is that it is particularly 'good natured' with regard to crack formation. At 20°C, the craze stress is slightly lower than the shear yield stress at around 100 MPa. If fine saw cuts are extended by carefully pressing new razor blades into the end of the slot, then clear crack tips form as the crack runs slowly ahead of the blade. Even here, care must be exercised and any blunting of the blade will lead to excessive loads and a blunt or multiple crazed crack tip. For edge notches, it is also important to form the crack as near to normal to the sheet faces as possible and, although some curvature is inevitable, a very stiff loading system for the blade is desirable. A side view of a well-formed crack tip is shown in Fig. 6.1 (taken from reference [2]) and the cusp-shaped single craze is clearly in evidence. It will be noted that the zone length, $r_p \simeq 10$ μm (this crack is at subcritical conditions) so that most practical specimens with dimensions of the order of several millimetres will certainly be in plane strain at fracture when zone sizes are $\simeq 30$ μm.

A typical example of fracture data for PMMA is shown in Fig. 6.2 (taken from reference [3]) which is plotted as $p^2 Y^2$ versus a^{-1} for a single-edge notch test with Y^2 taken from Table 4.1. The degree of scatter here is typical of well-notched specimens, being about $\pm 12\%$. The load-deflection diagram is linear up to fracture for the crack lengths used here (5 mm $< a <$ 15 mm in 50 mm wide specimens). Close observation of the specimen at slow loading rates shows that the crack initiates before unstable failure and this is confirmed on the fracture surface where the initial razor notch is followed by a ridged, slow growth region

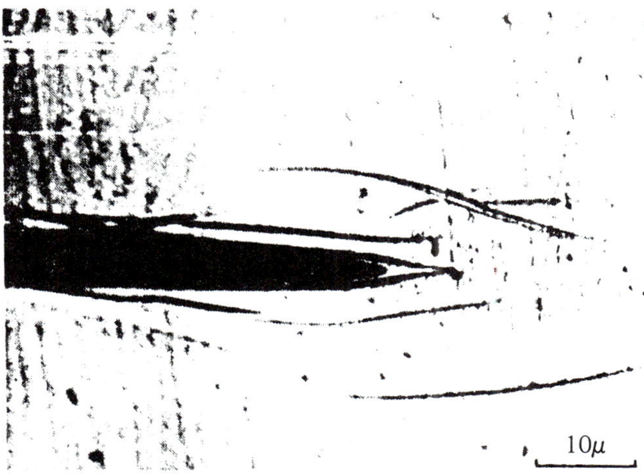

Fig. 6.1 Crack tip in PMMA (from reference [2])

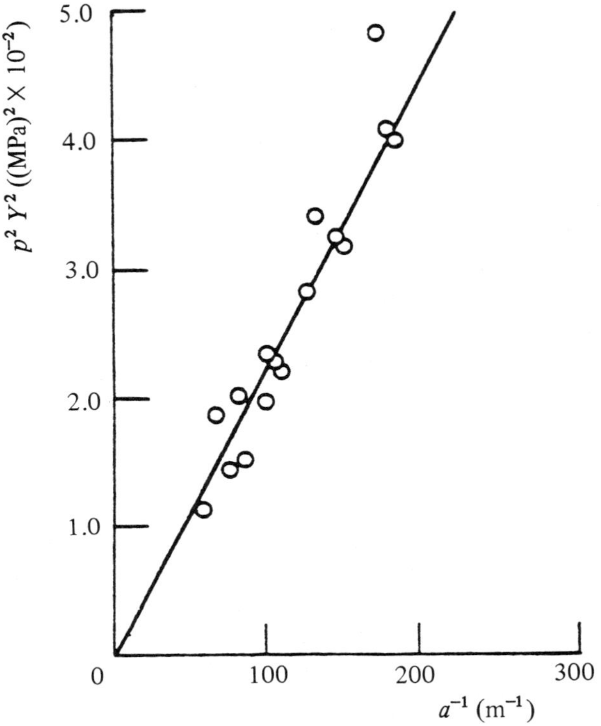

Fig. 6.2 SEN test data for PMMA at 20°C

of crazed polymer terminated by a sudden transition to a mirror surface. The load differences are small but the increased crack length makes a considerable difference to the K_c value. Fig. 6.3 shows the initiation K_c, computed from the initial crack length and the instability value derived from the final value plotted as a function of crosshead rate in 150 mm long specimens (at $\dot{x}=1$ cm/min, $\dot{e}\simeq 10^{-3}$ s^{-1}). The instability value is essentially rate-independent, while that at initiation shows a marked increase with rate. This is a consequence of the strongly visco-elastic β transition in PMMA in the temperature range -60°C to $+40$°C which influences both p_c and E and will be discussed in detail in Chapter 7. The mechanism implied by these data is that the visco-elastically controlled slow crack growth with its attendant crack tip crazing gives rise to an increasing initiation toughness-rate curve which terminates at a particular rate in an instability. Thereafter, there is an apparent decrease in toughness with rate since growth is unstable (see Section 7.2.3) and crazing does not occur with the fast running cracks.

Fig. 6.4 shows the effect of test temperature on the K_c values. The points shown are for instability at a rate of 0.5 cm/min and above about -70°C they

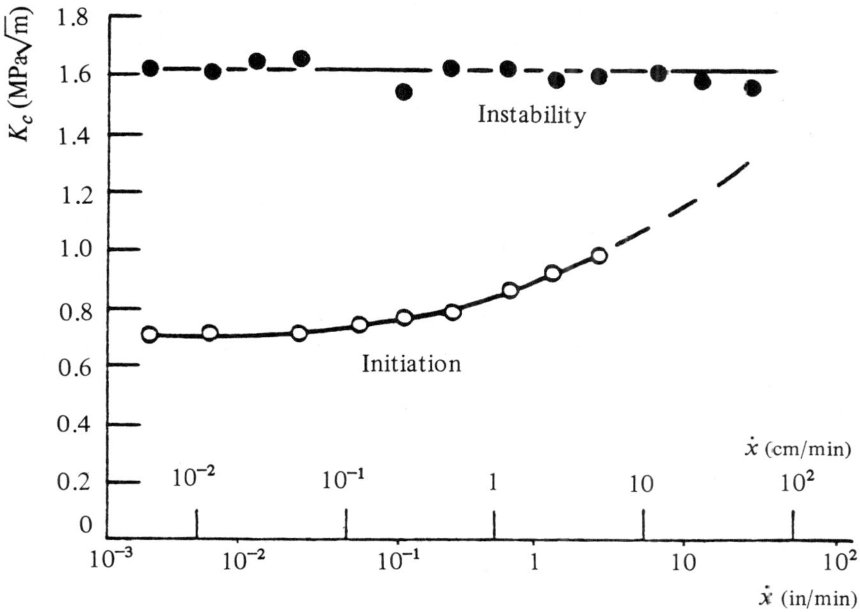

Fig. 6.3 K_c at initiation and instability as a function of crosshead rate at 20°C from SEN tests (from reference [1])

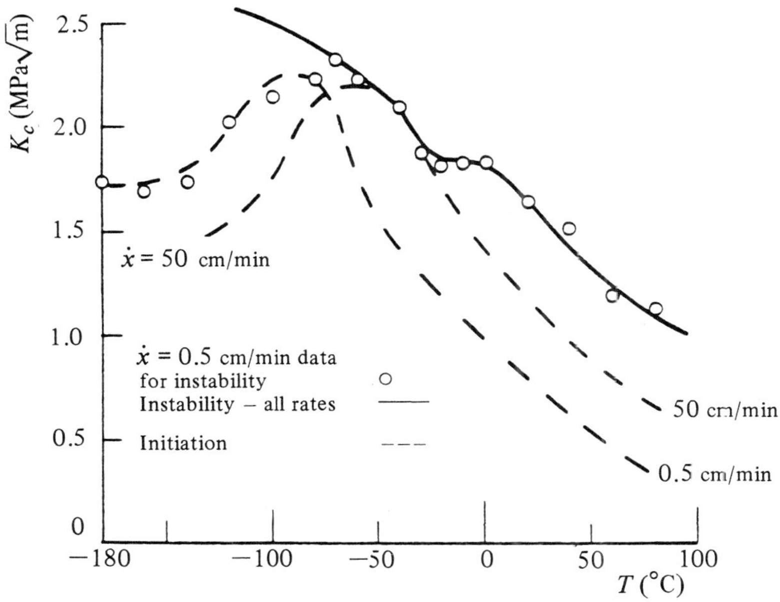

Fig. 6.4 K_c as a function of temperature, SEN tests – PMMA (data replotted from reference [1])

fall on a line common to all rates. The initiation condition is also shown as the lower broken line and demonstrates the pronounced rate effect in this region. Below $-70°C$, the initiation and instability conditions are coincident since there is no stable growth, although substantial craze growth occurs prior to initiation. The initiation line for 50 cm/min ($\dot{e} \simeq 0.05\ s^{-1}$) demonstrates the expected visco-elastic shift which also occurs below $-70°C$, resulting in an inversion of the usual rate effects. This is consistent with a peak in the K_c-rate curve at instability with K_c decreasing for rates above the critical value and that those data shown at low temperatures are all post-instability, while those at higher temperatures are pre-instability. These phenomena will be explored in more detail in Chapter 7 and it is sufficient to note here that PMMA demonstrates clearly the importance of rate in polymers, although it is, in fact, somewhat exceptional in the degree of dependence shown.

The utility of the line zone model for crazes has been demonstrated for PMMA in experiments in which the craze is viewed from the top so that its wedge shaped form produces fringes (see reference [4]). From these, it is possible to compute the craze profile and Fig. 6.5 shows an example and its comparison with equation (4.23). The closeness of agreement gives considerable support for the assumption of a constant craze stress. From a knowledge of the applied K value and the craze length, it is possible to determine the craze stress directly from this method and the data over the range $-30°C$ to $+45°C$ are shown in Fig. 6.6. Also shown are compressive yield stress results obtained from simple compression tests. Above about $-70°C$, upper and lower yield stress values can

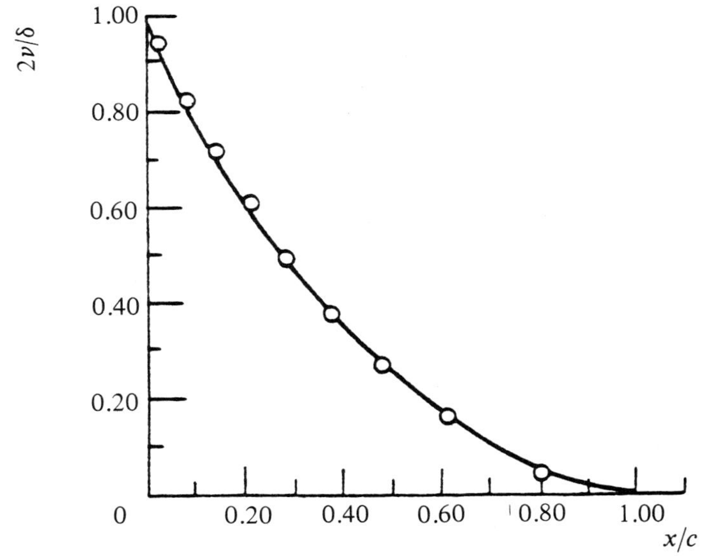

Fig. 6.5 Craze displacement profile; ○experiment, — equation (4.23) (reproduced from reference [4])

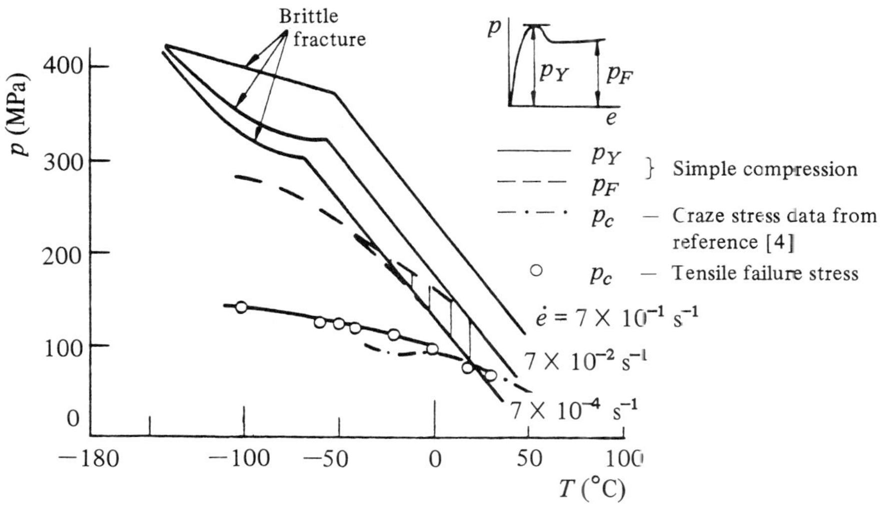

Fig. 6.6 Yield and craze stress data for PMMA

be found and the specimens show substantial deformation. The upper value shows the expected visco-elastic effects since it is approximately proportional to modulus, but the lower value, p_F, shows much less variation. Over the temperature range -30°C to $+20$°C, the yield values at low rates are higher than the craze stresses with a cross-over at about $+20$°C. At temperatures below -60°C, no yielding could be obtained in compression and the specimen shattered. Also shown in Fig. 6.6 are data for unnotched samples tested in tension and there is clearly close agreement with the craze stress values.

A reasonable explanation of these data can be derived in terms of the cohesive stress model demonstrated in Fig. 6.7, where the craze stress is taken as being equivalent to a cohesive stress in equation (5.19), i.e.

$$\frac{r_c}{a} = \ln \sec \frac{\pi}{2} \frac{p}{p_c}$$

where:

$$r_c = \frac{\pi}{8} \frac{E J}{p_c^2}$$

For a constant J (or δ) criterion of fracture, r_c is a constant and we obtain the line shown for the failure stress as a function of crack length. For $a \gg r_c$, the dependence is as for a fixed K criterion, as shown in the broken line. For $a/r_c < 0.2$, the failure stress varies very little and is close to the craze stress value. Thus, for unnotched specimens, the failure stress would be expected to approximate to the craze stress. At low temperatures, the yield stress becomes very high, even when some correction is made for possible hydrostatic stress sensitivity, and the brittle shattering occurs when the stress is approximately $3p_c$.

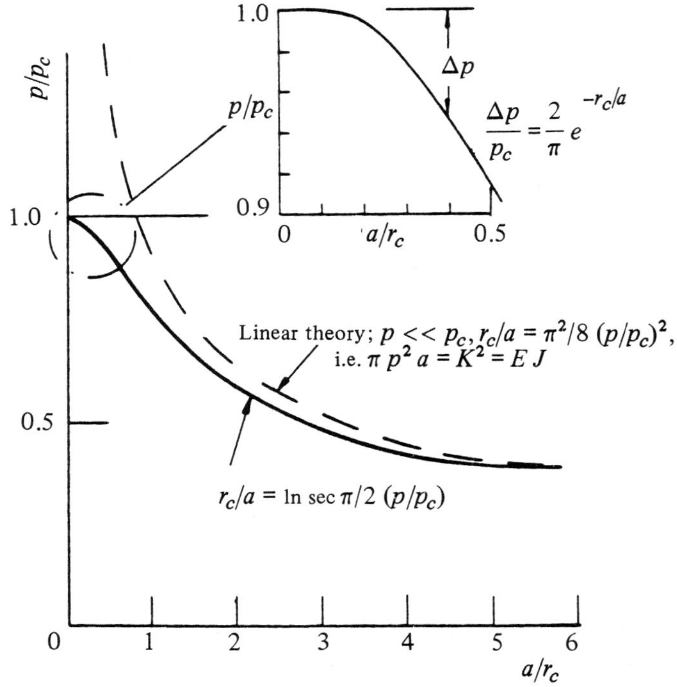

Fig. 6.7 Variation of failure stress with crack length; $r_c = \pi/8\ (E\ J)/p_c^2$, p_c is craze stress

This may represent the achievement of the craze stress at inherent flaws under compressive stresses.

6.4 Polystyrene (PS), unmodified

Polystyrene is used mainly because it has good moulding properties and gives parts with good optical performance. As with PMMA, it is a transparent glass at room temperature but is notorious for being extremely prone to brittle fracture in practice. Early fracture studies were puzzling (see reference [2]) since specimens notched using the same methods that had been successful for PMMA gave K_c values of up to 4 MPa$\sqrt{\text{m}}$, i.e. substantially greater than PMMA, to which it was known to be inferior. The solution lay in the rather low craze stress of PS which resulted in it being very easy to form crazes, so that any attempt at notch sharpening using a wedging action gave craze bundles. This blunted the crack tip and increased the initiation value and, indeed, cracks with craze bundles can be propagated across the whole specimen. Cracks with a single craze can be formed by fatiguing the specimen, by environmental crack growth or by machining in a sharp notch with a very sharp fly cutter. In the last case, tip radii of about 15 μm appear necessary to give satisfactory results.

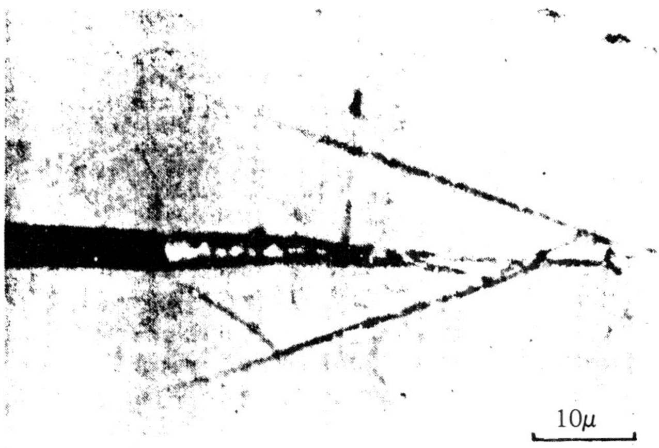

Fig. 6.8 Craze at crack tip in polystyrene. Fatigue notched specimen (from reference [2])

Fig. 6.8 shows a side view of a single craze in PS and a length about 50 μm is apparent, indicative of a lower craze stress than in PMMA since the loading conditions are much the same.

Fig. 6.9 shows $p^2 Y^2$ versus inverse crack depth data for surface-notched specimens and, again, very good linearity is apparent. The cracks were formed with a sharp fly cutter impinging on the specimen surface so that the crack section was a circular arc, as shown in the inset. The Y^2 used was for a semi-elliptical surface crack as discussed in Section 4.7.2 and, although only an approximation, seems quite satisfactory. (D is used for thickness here instead of B since it is the effective width dimension for Y^2). Fig. 6.10 shows some single-edge notch data as a function of crosshead rate at 20°C for fatigue notched specimens where the agreement with the surface notch data should be noted. The rate dependence of K_c is very slight, except above about 10 cm/min ($\dot{e} \simeq 0.01\ s^{-1}$) where there is a sharp rise which came from some craze bunches forming during the shock loading at these high rates. The initial craze could be seen on the fracture surface but there was no evidence of significant slow crack growth. This is consistent with the much less pronounced β transition in PS as compared to PMMA. Also shown in Fig. 6.10 is a craze stress computed from the surface craze length and the K_c value. This value, $\simeq 35$ MPa, is significantly lower than for PMMA and shows some rate dependence and a decrease at higher rates. It is possible that this is an artefact of not being able to distinguish small amounts of slow crack growth and the craze in lengths of the order of 50 μm.

Fig. 6.11 shows the temperature dependence of K_c for this material (designated material 1) and the small change with temperature is again consistent with a smaller β transition effect. Data for another commercially available material is also shown and while it is lower at 20°C, there is a much larger increase with

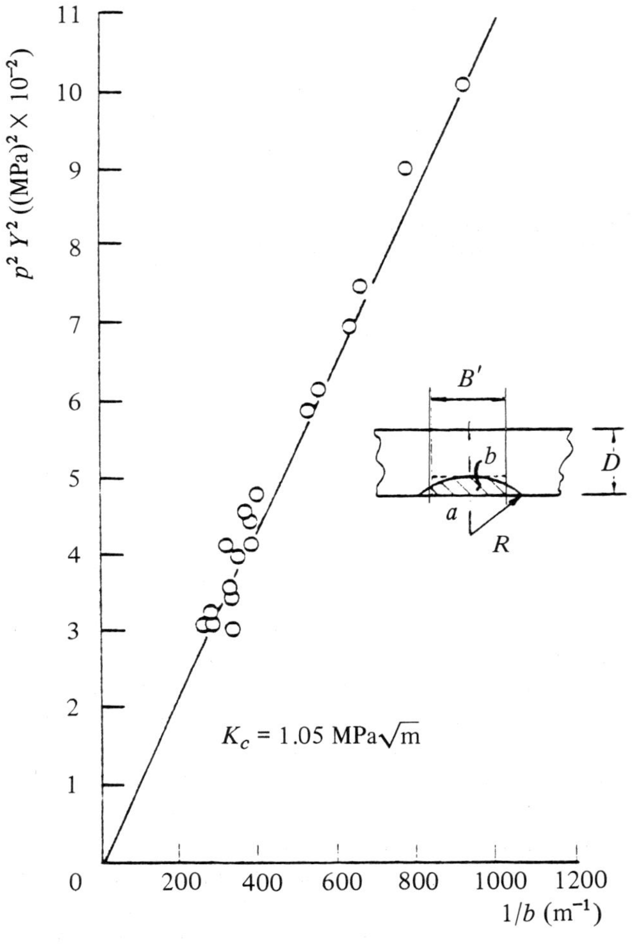

Fig. 6.9 Results of surface notched polystyrene at $\dot{x} = 0.05$ cm/min and 20°C, material 1 (from reference [5])

decreasing temperature. It is believed that this effect arises from mineral oil present in the material and is of relevance here to demonstrate the large differences in different grades of the same basic polymer.

6.5 Polycarbonate (PC)

Polycarbonate is manufactured and sold on the basis of being a high-toughness transparent material suitable, for example, for demanding glazing applications. Like PMMA and PS, it is an amorphous glass at room temperature but is different from them in having a considerably lower shear yield stress. At 20°C, it is about 60 MPa, while for PMMA it is 100 MPa. The result is a material with

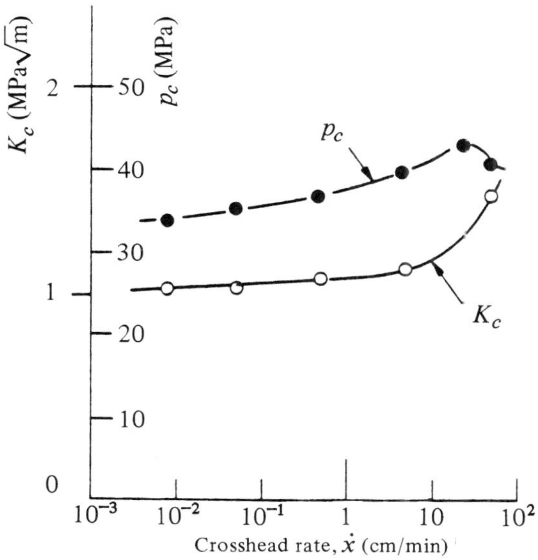

Fig. 6.10 Craze stress and K_c as a function of rate – PS at 20°C (material 1) (from reference [2])

remarkable toughness and ductile behaviour under some circumstances, but a tendency to brittle cracking in others.

This occurs because the toughness stems from the generation of plane stress zones in the surface of the material which are sensitive to the yield stress and produce a marked thickness dependence. The effects are exemplified (see reference [7]) in single-edge notch tests at 20°C at a range of rates on 5 mm thick sheet. At 50 cm/min, almost all the specimens give a brittle fracture but as the rate decreases so does the proportion of brittle failures. Observation of the crack tip shows that an initial crack tip bowing can be observed at loads significantly below those for final fracture. At final failure, some specimens give a complete catastrophic fracture (the brittle ones), some give a 'pop-in' effect where the crack starts but arrests and is followed by ductile tearing, and others go directly to a ductile tearing condition. In all cases, the initial crack bowing gives flat fractures, while subsequent crack growth gives rise to shear lips at each face. These reduce considerably on the fast fracture surface.

Fig. 6.12 shows $p^2 Y^2$ versus a^{-1} data for the 5 mm brittle failures at final fracture for two notch preparations and a range of rates. The crack lengths are corrected for the initial bowing and clearly there is little evidence of any variation with either rate or notching method and a K_c value of 4.0 MPa$\sqrt{m}$ is given. Those specimens showing 'pop-in' give the same value of K_c for this event. Fig. 6.13 shows the same plot for the initial crack bowing or initiation condition and, again, a rate and crack independent value is obtained, this time of

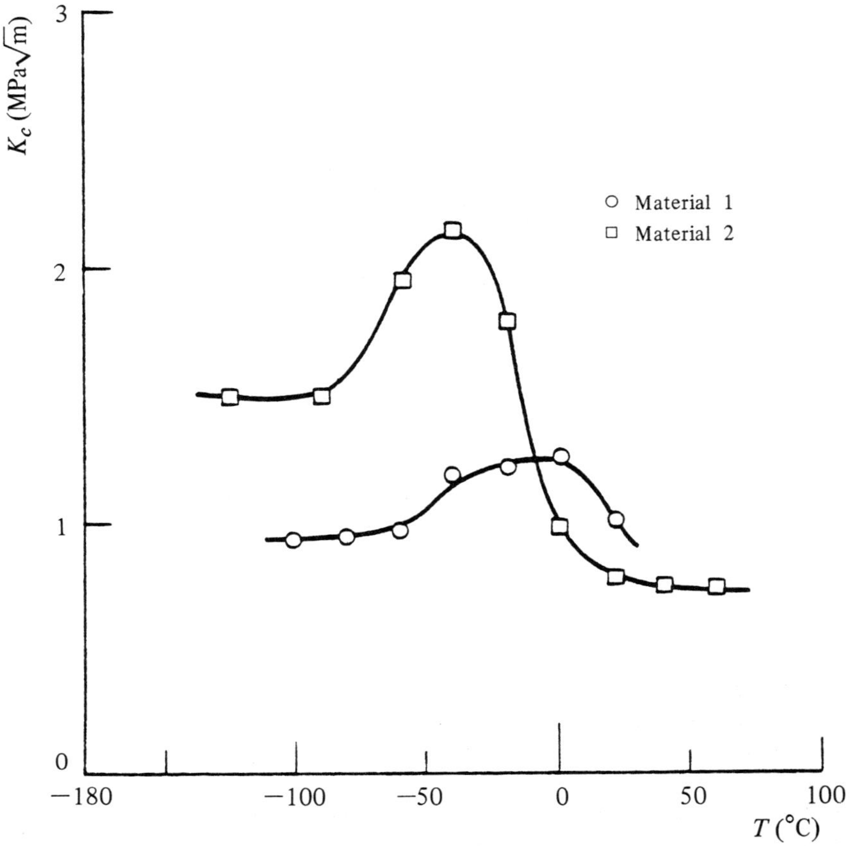

Fig. 6.11 K_c as a function of temperature – PS (○ from reference [5], □ from reference [6])

2.24 MPa$\sqrt{m}$. Also shown on this graph are some points of 3 mm thick specimens which showed the same initiation phenomena, but all of which gave ductile tearing even at high rates.

This type of behaviour may be modelled using the three layer concept given in Section 5.3, and from equation (5.11) we have:

$$K_c' = K_{c1} + \frac{K_{c2}^2}{\pi B p_Y^2}(K_{c2} - K_{c1})$$

where K_{c1} is the plane strain toughness, and K_{c2} is that for plane stress. If we assume that the initial crack bowing corresponds to the plane strain initiation at K_{c1}, then K_{c2} may be deduced using $p_Y \simeq 64$ MPa, $K_{c1} = 2.24$ MPa$\sqrt{m}$ and $B = 5$ mm, giving $K_{c2} = 5.7$ MPa$\sqrt{m}$. The surface zone thickness or shear lip

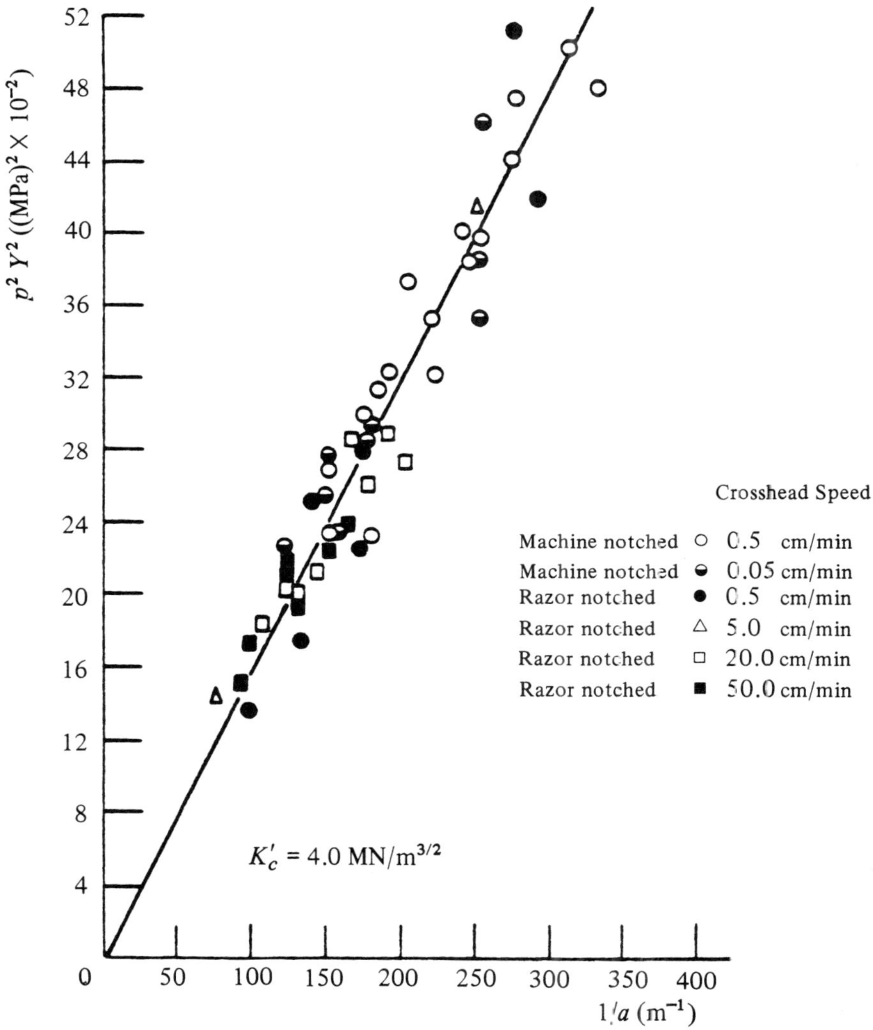

Fig. 6.12 5 mm, single-edge notched specimens at final fracture – PC at 20°C (from reference [7])

width is then given by:

$$r_0 = \frac{1}{2\pi}\left(\frac{K_{c2}}{p_Y}\right)^2 = 1.3 \text{ mm}$$

and ductile failure would be expected for thicknesses $< 2r_0 = 2.6$ mm; a reasonable approximation to the fact that all 3 mm specimens were ductile. The ASTM

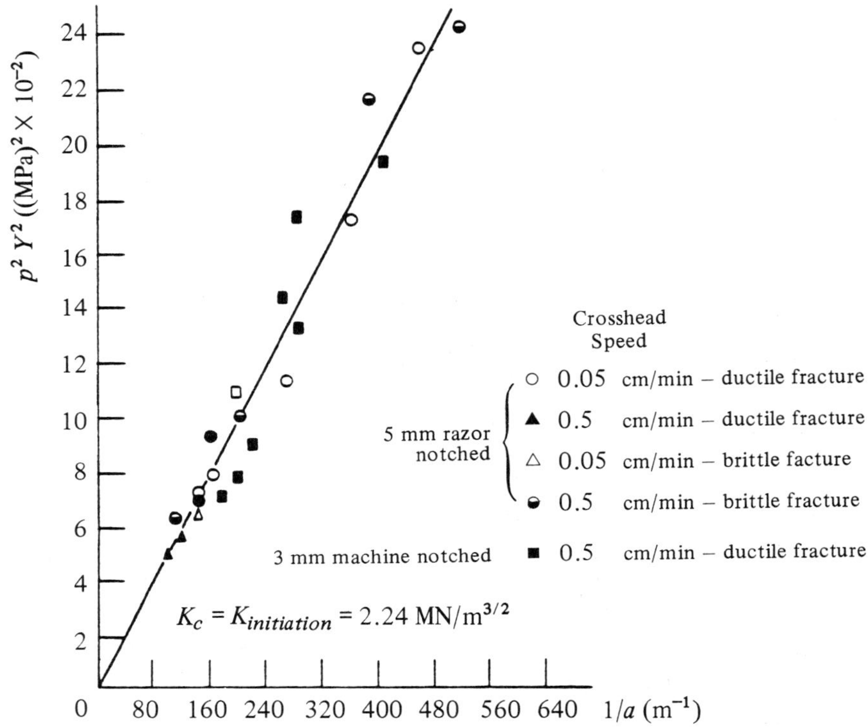

Fig. 6.13 3 mm and 5 mm, single-edge notched specimens at initiation – PC at 20°C (from reference [7])

plane strain limiting thickness criterion is:

$$B > 2.5\left(\frac{K_{c1}}{p_Y}\right)^2 = 3.1 \text{ mm}$$

which appears to be satisfactory for initiation in the tension tests used here.

Some further confirmation of these ideas is given in Fig. 6.14, where 5 mm thick surface notched specimen data are shown as stress versus crack depth for three rates. For a crack depth of greater than 1 mm, the failures are brittle and give $K_c = 2.95$ MPa$\sqrt{\text{m}}$, while for lesser depths the failures become ductile and show the rate dependence of p_Y. The K_c value given is equivalent to a thickness of 12.3 mm and illustrates how surface notches can give more severe conditions than edge notches. A useful model of this effect is to compute an equivalent thickness, B', of a crack of the same area and depth, which, for a cutter of radius R, is:

$$B' \simeq \tfrac{4}{3}\sqrt{a(2R-a)} \text{ (see diagram in Fig. 6.9)} \tag{6.1}$$

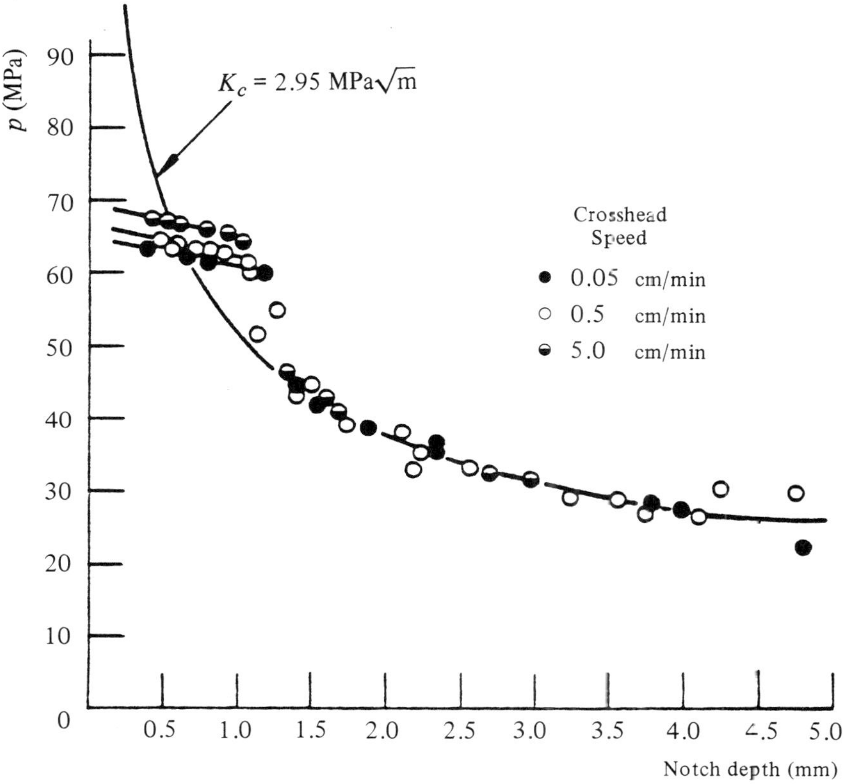

Fig. 6.14 5 mm surface notched results – PC at 20°C (from reference [7])

In this case, $R = 15$ mm and the average $a \simeq 3$ mm, giving $B' = 12$ mm.

It should be emphasised here that the initiation value K_{c1} is a proper toughness characterising parameter, being independent of crack length, specimen size and mode of loading. The higher values at instability are not, even though they are crack length independent, since they depend on specimen thickness. The utilisation of the K_{c2} concept enables the analysis to be extended to embrace this sort of behaviour which may be regarded as an extension of the basic analysis, incorporating the effects of loss of constraint in the specimen surface. The onset of thickness and other size effects would be expected to be controlled by the parameter $(K_{c1}/p_Y)^2$, as used in the ASTM criterion, while the breakdown of the brittle bimodal behaviour to give ductile tearing will be a function of $(K_{c2}/p_Y)^2$.

The criterion for the onset of ductile tearing can be explored using the fact that the fraction of brittle specimens increases as rate is increased, and Table 6.1 shows the relevant data.

Table 6.1

(Data taken from reference [7])

Crosshead Rate (cm/min)	0.05	0.5	5	10	20	50
Yield Stress (MPa)	62.3	64.3	67.5	68.3	69.1	70.1
Number of Specimens	20	20	12	11	11	10
Number of Ductile Fractures	19	11	4	3	2	1

If we assume that K_{c2} varies within a given batch such that:

$$K_{c2}=K_0+\Delta K\left(\ln\frac{1}{x}\right) \tag{6.2}$$

where x is the fraction of specimens which have a value of K_{c2} or greater. K_0 is the minimum value of the cumulative distribution, and ΔK is a measure of its spread. If we now assume a transition to ductile tearing when:

$$\hat{B}\leqslant\alpha\left(\frac{K_{c2}}{p_Y}\right)^2$$

then we may write an expression for the fraction of specimens which exceed this condition for a given thickness and yield stress as:

$$p_Y=\left(\frac{\alpha}{B}\right)^{1/2}K_0+\left(\frac{\alpha}{B}\right)^{1/2}\Delta K\left(\ln\frac{1}{x}\right) \tag{6.3}$$

Fig. 6.15 shows the data in Table 6.1 plotted on this basis and a reasonable line results, except for the single value result at the highest p_Y value. Noting that $B=5$ mm in this case, we have:

$$\alpha^{1/2}\,K_0=4.35\ \text{MPa}\sqrt{\text{m}}\ \text{and}\ \alpha^{1/2}\,\Delta K=0.38\ \text{MPa}\sqrt{\text{m}}$$

so that $\Delta K/K_0=0.09$, i.e. a 10% variation in K_{c2}. If we take the average value as $K_{c2}=5.7$ MPa$\sqrt{\text{m}}$ for $x=0.5$, then $K_0=5.4$ and $\alpha=0.65$. In terms of r_0, this becomes $\hat{B}/r_0=4.1$, which agrees closely with the suggestion by Irwin *et al.* [8] that the flat fracture condition can only be maintained with at least as much material in the centre as in the skins, i.e. $\hat{B}/r_0=4$.

The effect of temperature variations is also governed by changes in the yield stress. Fig. 6.16 shows the variation of the yield stress and directly measured craze stresses (from references [9] and [10]) with temperature, and it should be noted that $p_c>p_Y$ as expected in this case in which shear yielding effects predominate. Figs. 6.17 and 6.18 show K_c values at a rate of 0.5 cm/min for both

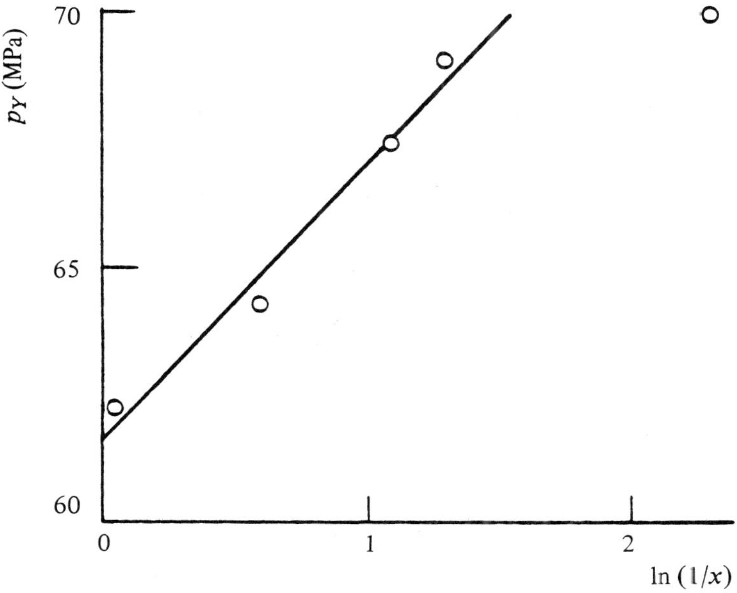

Fig. 6.15 Variation of fraction of ductile failures with yield stress – PC at 20°C (from reference [7])

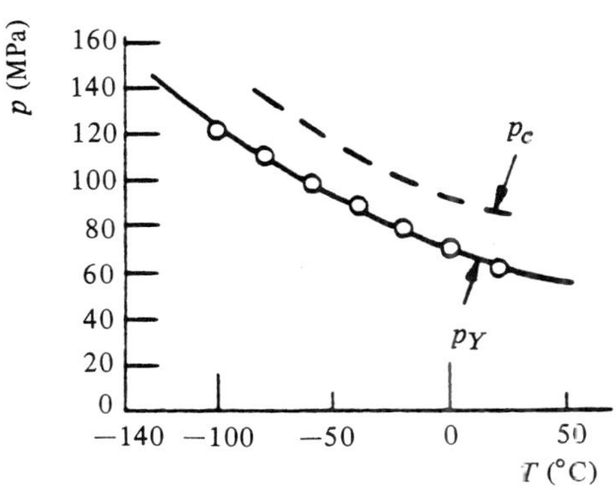

Fig. 6.16 The variation of yield and craze stress with temperature at 0.026 s^{-1} – PC (○ from reference [9], --- from reference [10])

SEN and SN tests in 5 mm and 3 mm sheet, respectively. For 5 mm material, the fractures were mostly brittle in both modes and, as expected, the SEN data are higher than those for SN. For 3 mm sheet, the SN specimens became ductile at 0°C, while for the SEN specimens the transition was at −80°C, although some 'pop-in' values could be detected at −40°C and 0°C. The four curves may be combined using equation (5.11) as shown in Fig. 6.19, where it can be seen that

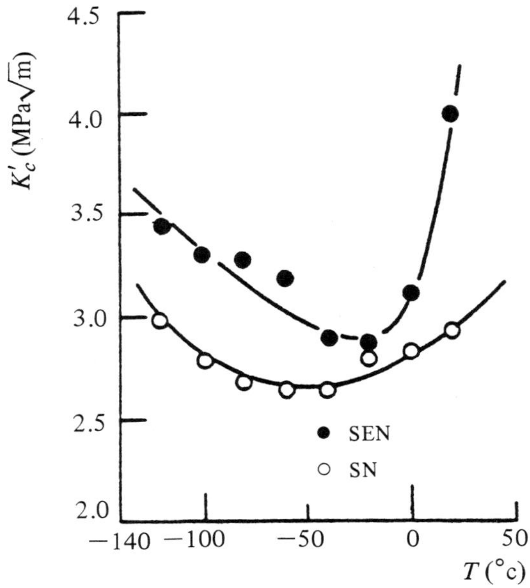

Fig. 6.17 Fracture toughness for 5 mm specimens – PC (from reference [9])

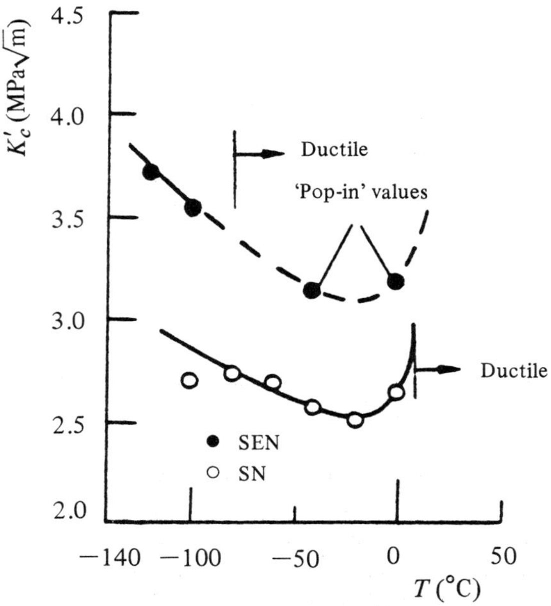

Fig. 6.18 Fracture toughness for 3 mm specimens – PC (from reference [9])

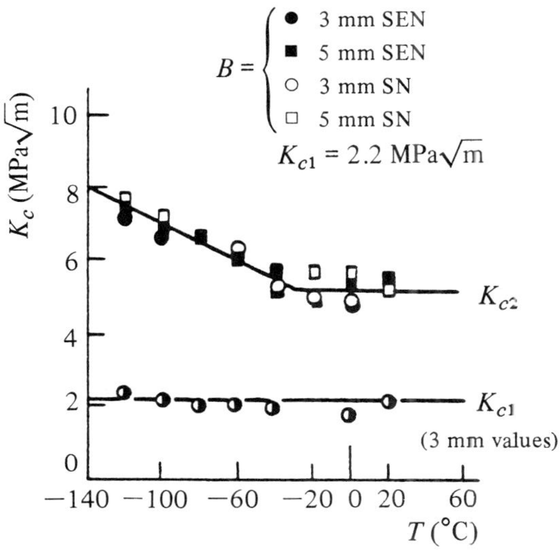

Fig. 6.19 K_{c1} and K_{c2} as functions of temperature – PC (from reference [9])

K_{c1} remains fairly constant while K_{c2} increases below about -40°C, probably reflecting the β transition which occurs in this range. All four tests gave similar values of K_{c2} for the same K_{c1} value as shown. K_{c1} could not be measured directly at low temperatures because of difficulty in seeing the crack tip and it is not possible to define a value precisely from measurements at two thicknesses. There is a slight tendency for K_{c1} to rise at lower temperatures.

Fig. 6.20 shows the thickness for each test plotted versus the transition temperature for ductile tearing. The SEN data are at the appropriate values but for SN the equivalent values are shown, and for 5 mm sheet was 12 mm and was brittle up to 20°C. Also shown is $(K_{c2}/p_Y)^2$ and the transition factor derived from the 5 mm data. The data are very sparse but the general form is sensible.

It is of interest at this stage to compare the toughness of the materials so far considered in terms of G_c or J_c. Taking the data at 20°C, we find that PMMA varies from about 0.22 kJ/m^2 at low rate initiation to 0.51 kJ/m^2 at instability. For PS, the initiation value is about the same but rises to only 0.33 kJ/m^2. It should be noted that the true surface energy of these materials is in the range 2–5 J/m^2 on the assumption of simple bond scission. In normal testing, the molecules are uncoiled and stretched in the craze prior to failure which results in the additional energy being dissipated. For direct loading, it is more reasonable to regard these plane strain values as effective minima since the bonds cannot be loaded to the failure condition without this prior work being done. (Much lower values can be obtained in fatigue (Chapter 7), in which a separate mechanism is involved).

For PC, the plane strain minimum value is 1.51 kJ/m^2 which rises to 10.1 kJ/m^2 under plane stress conditions. These values pertain at initiation for

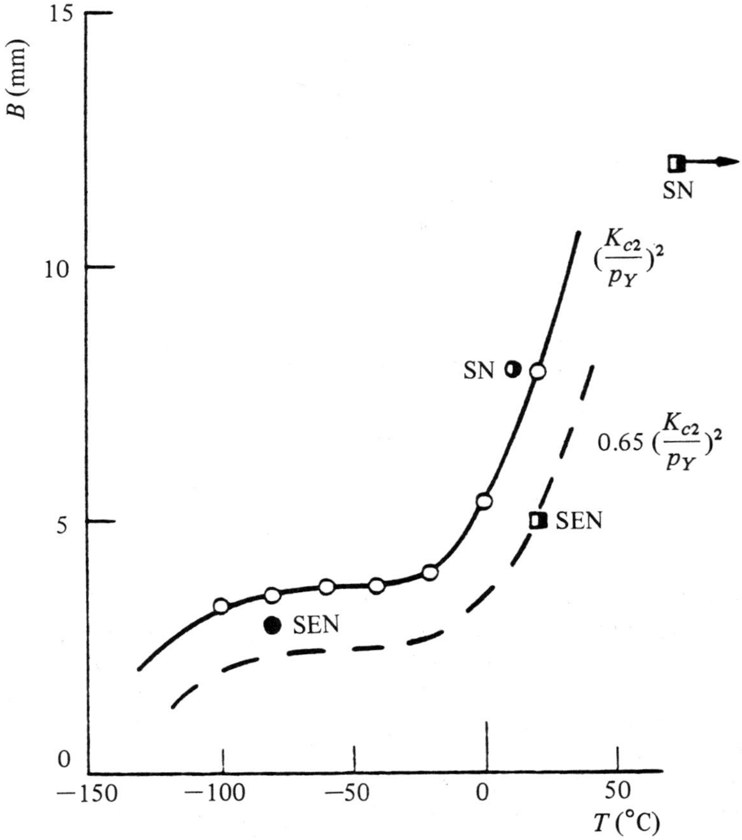

Fig. 6.20 Ductile tearing transitions in PC at $\dot{x} = 0.5$ cm/min and $D = 50$ mm

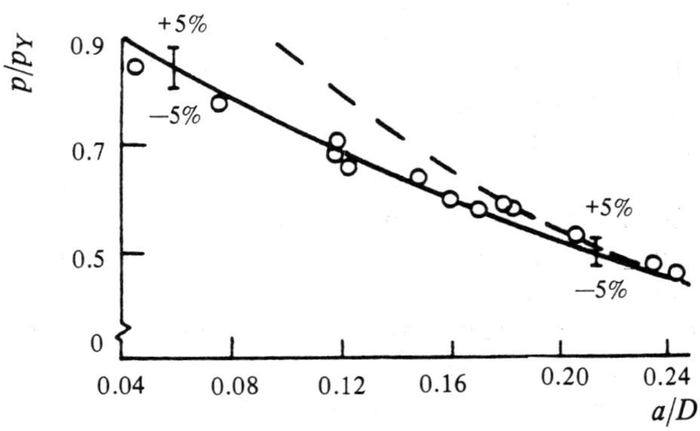

Fig. 6.21 Typical maximum load data for PC in air showing fit achieved with constant COD criterion. Crosshead speed, 0.5 cm/min, $p_Y = 56$ MPa, —— $\delta_c = 0.75$ mm, — — K_c = constant (from reference [11])

brittle fractures and in some geometries go immediately to instability, but in others ductile plane stress tearing can become established which gives stable crack growth with the load maximising and subsequently falling. The increased loads involved lead to extensive plasticity and substantial non-linearity in the load-deflection diagram so that a K_c representation can no longer be used. However, some useful results may be obtained by means of a J_c (or δ) analysis based on maximum load, and Fig. 6.21 shows SEN data in which a fit is given for both a K_c analysis and a constant J_c (or δ) using the SEN solution given in Fig. 5.12. The material was somewhat different here, having $p_Y = 56$ MPa, and a good fit is obtained with $\delta_c = 0.75$ mm so that $J_c = 42$ kJ/m. This increase above the plane stress initiation value represents the rise in J as ductile crack growth occurs (a rising R curve) and corresponds to a crack growth of about 5 mm.

6.6 Rubber modified polystyrene (HIPS)

Rubber modification of polymers is carried out to improve their toughness. Particular emphasis is placed on impact behaviour, giving rise to the general term high impact polystyrene (HIPS) used here. The basic method employed is to produce a material with a dispersed phase of rubber particles of the order of 1 μm in diameter. Originally, this was done by simple mechanical blending when rather small amounts of rubber could be incorporated, say up to 10%, but later the addition was made during polymerisation so that much larger particle contents were achieved. This often involves phase inversion in which PS particles become incorporated in the rubber to give much larger volume fractions of particles than the rubber content would suggest.

The mechanism which was originally imagined was one in which the resilient particles acted as crack stoppers. It is apparent, however, that high stresses give rise to a whitening of the material and this arises from the development of crazes or microvoids around the particles. These act as energy sinks and hence inhibit crack growth. In order to achieve these effects, it is necessary to have good bonding between the matrix and the particles so that a chemically compatible rubber must be used. This is apparent from the fact that holes do not give any enhancement of toughness. It has also been demonstrated that hard particles do not give the required energy absorption. This suggests that the microvoiding or crazing is a consequence of the local stress state around the particle which is in turn strongly dependent on both adhesion and the elastic properties of both phases.

The consequences of the rubber phase are illustrated in Fig. 6.22, in which the yield stress and modulus are shown as a function of rubber content. Both decrease in the same manner, resulting in a constant yield strain, and 6% rubber results in a 37% drop, indicating a large PS included phase. The yield stress is that condition when unnotched tensile bars whiten and the load maximises. This is a consequence of the formation of many crazes and there is little lateral contraction, indicating a volume increase as expected from a voiding process.

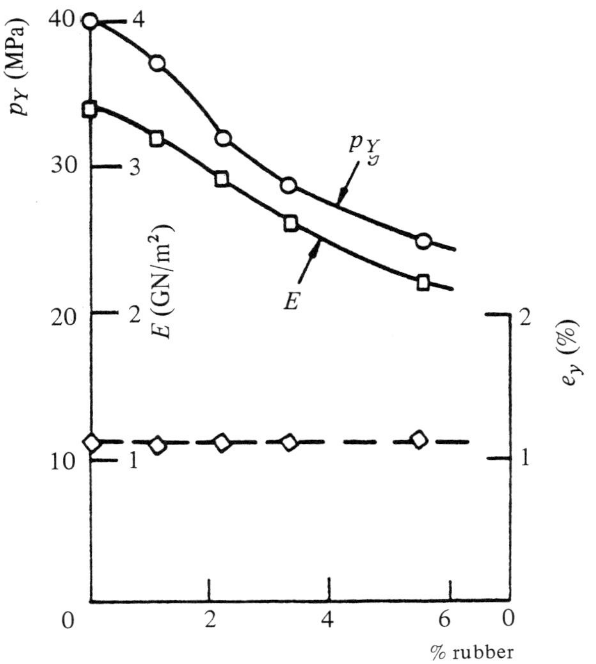

Fig. 6.22 Modulus, yield stress and yield strain as a function of rubber content at 20°C – HIPS (from reference [6])

This does not imply, however, that the process is controlled solely by the hydrostatic stress since considerable local shear may be involved.

In notched samples, the rubber phase induces a large whitened zone around the crack tip which is somewhat similar in form to a plane stress yield zone as seen in PC but without a thickness reduction. Observation of the original crack shows that initiation will occur at values close to or slightly higher than that for the matrix PS in its unmodified form. In SEN, the craze zone development is such that ductile fracture results, but for SN tests a brittle failure occurs at some higher load. Fig. 6.23 shows some SN data at 20°C and demonstrates the enhanced rate dependence of the modified material. K_c values obtained on 6 mm thick SN specimens for which $B' \simeq 12$ mm are shown for 0.5 cm/min in Fig. 6.24 over a temperature range, together with some SEN data. 50 mm wide SEN specimens were only brittle below -60°C but 150 mm wide specimens gave brittle fractures at 20°C as shown. These data may be converted to K_{c2} using the matrix K_c value for K_{c1} and the result is shown in Fig. 6.25.

A pattern of behaviour closely analogous with PC is apparent, in which the rubber modification simply decreases the effective yield stress. At 20°C, $K_{c2} \simeq 2.6$ MPa$\sqrt{\text{m}}$ and $p_Y \simeq 20$ MPa, so that $r_0 = 2.7$ mm and, indeed, surface notches of less than this depth give ductile failure. The SEN specimens used

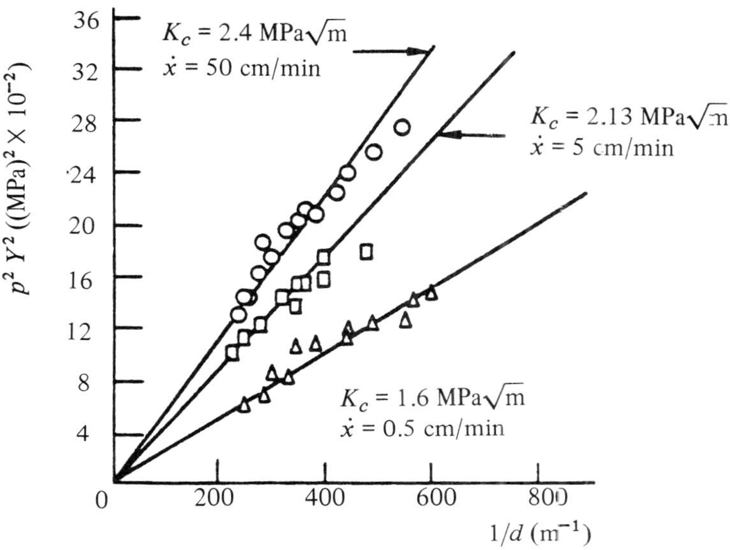

Fig. 6.23 Results for SN rubber modified polystyrene at 20°C (from reference [6])

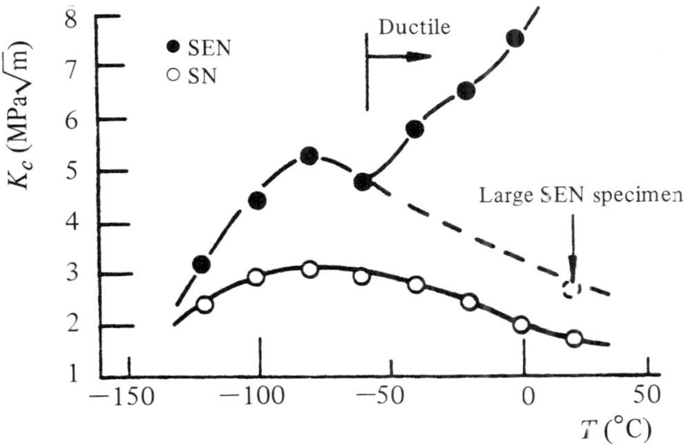

Fig. 6.24 Fracture toughness for rubber modified PS at $\dot{x}$ = 0.5 cm/min (from reference [6])

violate the thickness transition criterion but the absence of lateral contraction appears to prevent any such transition occurring. Those transitions which do occur are apparently controlled by the overall width. That in the SEN data on 50 mm wide specimens occurs when $(K_{c2}/p_Y)^2 = 14.3$ mm (at −60°C) so that:

$$\frac{1}{D}\left(\frac{K_{c2}}{p_Y}\right)^2 \simeq 0.29$$

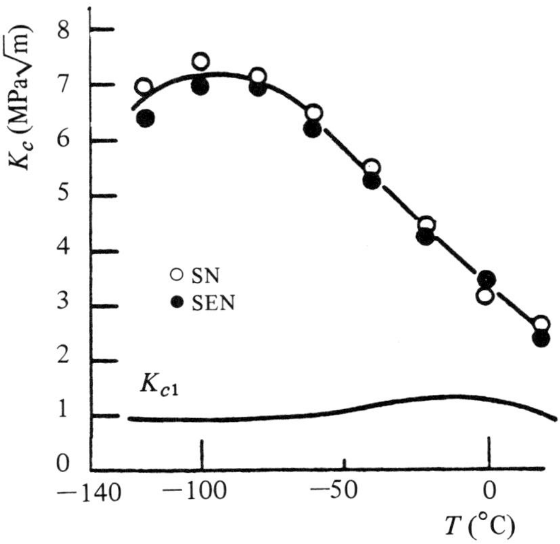

Fig. 6.25 Effect of temperature on K_{c1} and K_{c2} of rubber modified PS at $\dot{x}=0.5$ cm/min (from reference [6])

This is considerably different from the width condition predicted from plastic collapse in Section 5.4, in which the factor is approximately unity. This type of transition will be discussed further in Section 6.14.

Initiation J_{c1} values are very similar to unmodified PS at around 0.3 kJ/m^2 while in plane stress we have $J_{c2}=1.7$ kJ/m^2 at 20°C. Ductile fracture leads to stable tearing and the maximum load data can be modelled quite well using a J_c analysis as for PC, and Fig. 6.26 shows such values as a function of rate. Again, they are considerably greater than J_{c2} and here show a maximum which

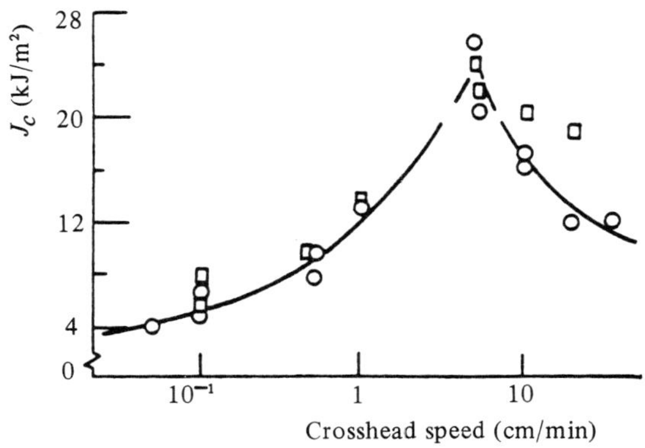

Fig. 6.26 J_c at maximum load, ○ from computed COD, □ from measured COD – HIPS at 20°C (from reference [11])

probably represents a limiting rate for the microvoiding–crazing process. The two types of points shown here are for load-based computations and direct observations of δ (see reference [11]).

The utility of the J_c analysis for this material is demonstrated in Fig. 6.27, which shows some results of four-point bend tests in which both initiation and maximum loads were measured. The computed lines used the SEN tension data which, for initiation, is a K_c value and, for maximum load at the same crack growth, is J_c. The rate is chosen to give comparable strain rates and the agreement is good.

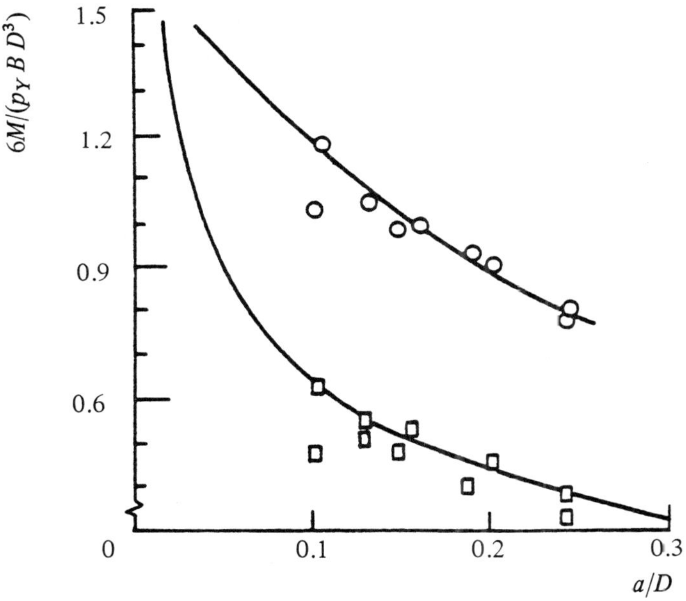

Fig. 6.27 Four-point bend tests results, HIPS tested at 0.5 cm/min and 20°C compared with SEN predictions. ○ SEN data, constant COD = 0.43 mm criterion maximum load; □ SEN data, constant K_c criterion initiation (from reference [11])

6.7 Acrylonitrile-butadiene-styrene (ABS)

ABS is a particularly successful form of rubber modified polystyrene in which the rubber phase is well adhered to the matrix. The material warrants a separate section because it shows remarkable toughness when the interface adhesion is in full operation. A major difference with HIPS is that considerable lateral contraction is observed, together with whitening, indicating substantial shear yielding.

Conventional notched tests do not give brittle failures above −80°C, and crack initiation is followed by arrest and subsequent ductile tearing as reported for PC and HIPS. Below $-80°C$, there is no whitening and the rubber phase is inoperative, giving completely brittle behaviour similar to PS. Fig. 6.28 shows some initiation data at 20°C for bending and tension for two sheet thicknesses and plate widths. The results are obtained from direct observation of the crack tip are difficult in this opaque material so considerable scatter is evident. However, there is a tendency for the thinner sections to give higher K_c values and an analysis on this basis gives $K_{c1} \simeq 1.8$ MPa$\sqrt{m}$ and $K_{c2} \simeq 4.2$ MPa$\sqrt{m}$ for $p_Y = 41$ MPa so that the plane strain value is somewhat higher than that for PS.

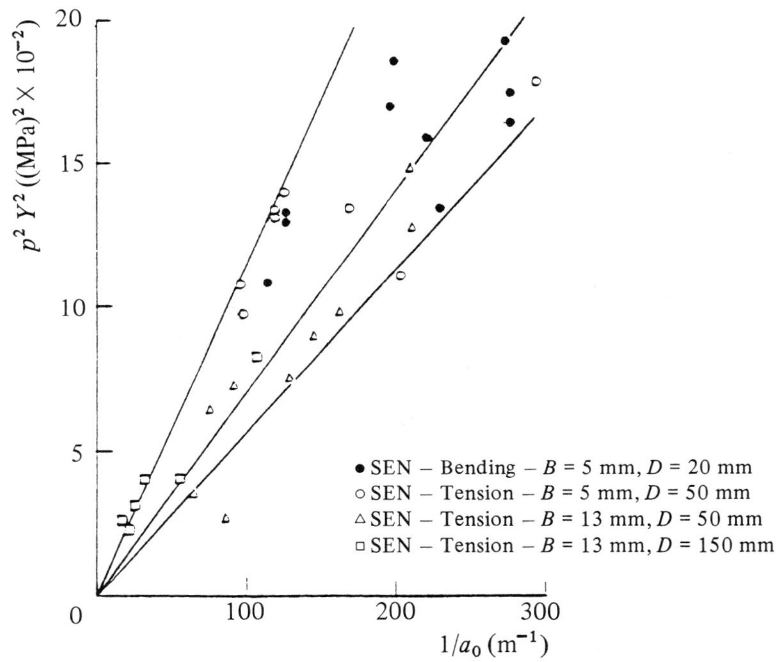

Fig. 6.28 Crack initiation data – ABS at 20°C (from reference [12])

From equation (6.4), the transition thickness for ductile tearing would be:

$$\hat{B} = 0.65 \left(\frac{K_{c2}}{p_Y} \right)^2 = 6.8 \text{ mm}$$

but no transition was observed in the 13 mm sample, suggesting similar behaviour to HIPS in this regard.

The plane strain J values at 20°C are $J_{c1} = 1.5$ kJ/m^2 and $J_{c2} = 8.2$ kJ/m^2. In SEN tension, there is 4.8 mm of slow ductile tearing up to maximum load for $D = 50$ mm. The usual plot of $p^2 Y^2$ versus $a_0{}^{-1}$ is shown in Fig. 6.29 and there is pronounced curvature which is corrected when the final crack length is used, as

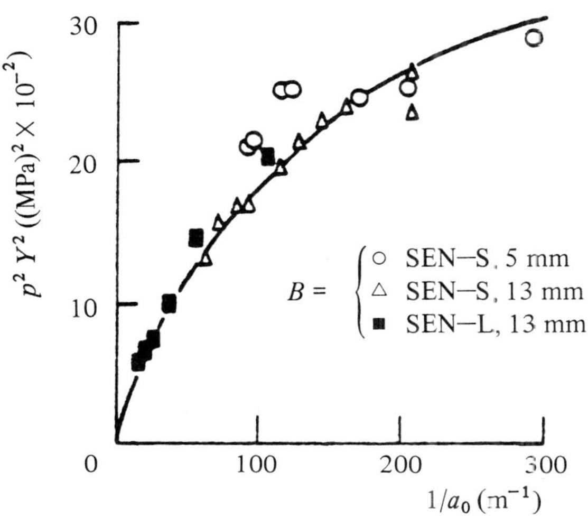

Fig. 6.29 Maximum load data, uncorrected – ABS (from reference [12]) (S, 150 mm × 50 mm; L, 300 mm × 150 mm)

shown in Fig. 6.30. This value of K_c (6.23 MPa$\sqrt{m}$) gives $J_c = 18.0$ kJ/m^2, while in bending the slow growth is 0.8 mm and $J_c = 8.7$ kJ/m^2. This latter figure is very close to J_{c2} which is consistent with the small amount of growth. It also illustrates that, while the R curve may be the same for different geometries, the value of J which is achieved at maximum load can be strongly dependent on the loading system (see Section 7.6).

Fig. 6.31 gives a summary of ABS data as a function of temperature and shows that the discrepancy between tension and bending remains down to −80°C where the microvoiding process ceases and initiation gives immediate failure. The initiation values are for 5 mm sheet and probably somewhat higher than a true K_{c1} value.

6.8 Polyvinyl chloride (PVC)

This is an important commercial material since it can be produced in a form with a good overall performance at a low cost compared to other polymers. It is amorphous but tends to retain a discrete particle structure which forms during manufacture so that great care is needed in processing. Material in which inadequately adhered particles occur is said to have poor gelation and leads to low toughness. An additional complication is that PVC is rarely used in as pure a form as other polymers and is frequently compounded with fillers and toughening agents. It is therefore difficult to give typical data but some insight is provided by that shown in Fig. 6.32.

The results were obtained in SEN tension on four sheet thicknesses for $D = 50$ mm and showed brittle behaviour below −20°C with no evidence of a

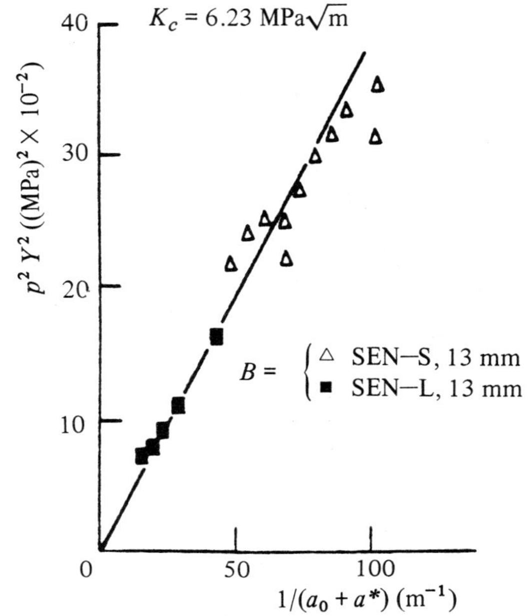

Fig. 6.30 Maximum load data incorporating crack length correction – ABS (from reference [12])

systematic thickness effect. The minimum K_c value is about 4.2 MPa$\sqrt{m}$ down to about -120°C, below which there is a pronounced increase. A rise in this region is observed in other polymers (see Section 6.10) and may represent an effect of the liquid nitrogen used in cooling which condenses in this range. Above -20°C, there is a thickness effect with brittle unstable failures up to 0°C but at 20°C and above ductile tearing occurs. From the thickness data, it is possible to estimate both K_{c1} ($\simeq$4.0 MPa$\sqrt{m}$) and K_{c2} using the yield stress data shown in Table 6.2, and the lines are shown in Fig. 6.32. Also given are the values of 2.5 $(K_{c1}/p_Y)^2$ and it is clear that this provides a reasonable criterion for the onset of the plane stress effects for each thickness. Also shown in Table 6.2 is the length parameter $(K_{c2}/p_Y)^2$ and $4r_0$ which controlled the transition to ductile tearing in PC. The values cover the range used here with no transitions observed and, as observed in HIPS and ABS, D appears to be the controlling variable.

The transition to ductile tearing at 20°C occurs when:

$$\frac{1}{D}\left(\frac{K_{c2}}{p_Y}\right)^2 \simeq 0.25$$

which is a similar value to that observed for HIPS. This is confirmed in tests on 127 mm wide, 3 mm thick sheet which gave brittle fractures up to 0°C.

K_c values for compounded grades of PVC are generally lower than those

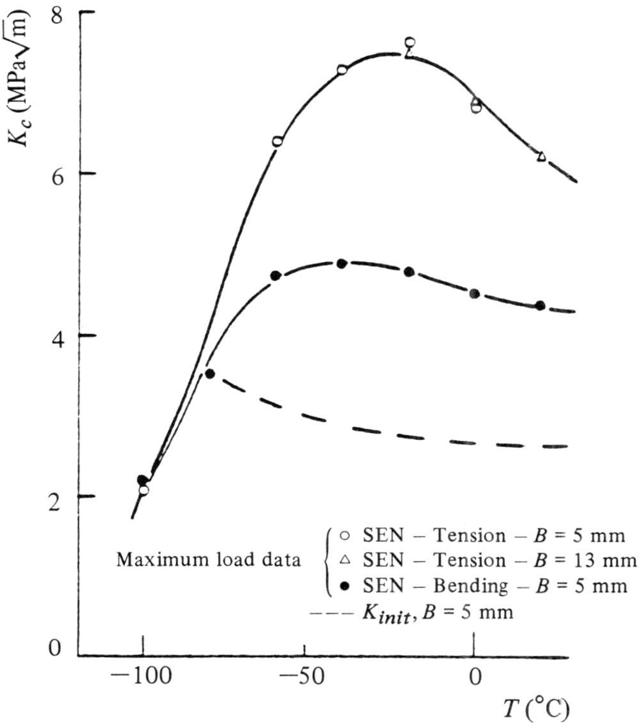

Fig. 6.31 Toughness at maximum load as a function of temperature – ABS (from reference [12])

given here. A water pipe grade gave a consistent brittle failure up to 20°C at 3.2 MPa$\sqrt{m}$, while a rubber modified material shows much tougher behaviour down to −50°C but eventually gives values down to 2 MPa$\sqrt{m}$. This latter figure is also typical of badly gelled material. It seems likely that the higher K_c values are a consequence of microvoiding around the crack tip so that any failure in which this is absent will be at the basic polymer value of the lower figure. Cracks may be seen to initiate at $K_c \simeq 2$ MPa$\sqrt{m}$ but subsequently arrest when microvoiding occurs. All the data quoted are at 0.5 cm/min and only slight rate effects are observed.

6.9 Polypropylene (PP)

This is a semi-crystalline polymer which forms spherulitic crystals of up to 70% of the material. It has a glass transition of about −30°C and is thus utilised above this value in contrast to the polymers so far mentioned.

The homopolymer is quite brittle at low temperatures and this is confirmed in

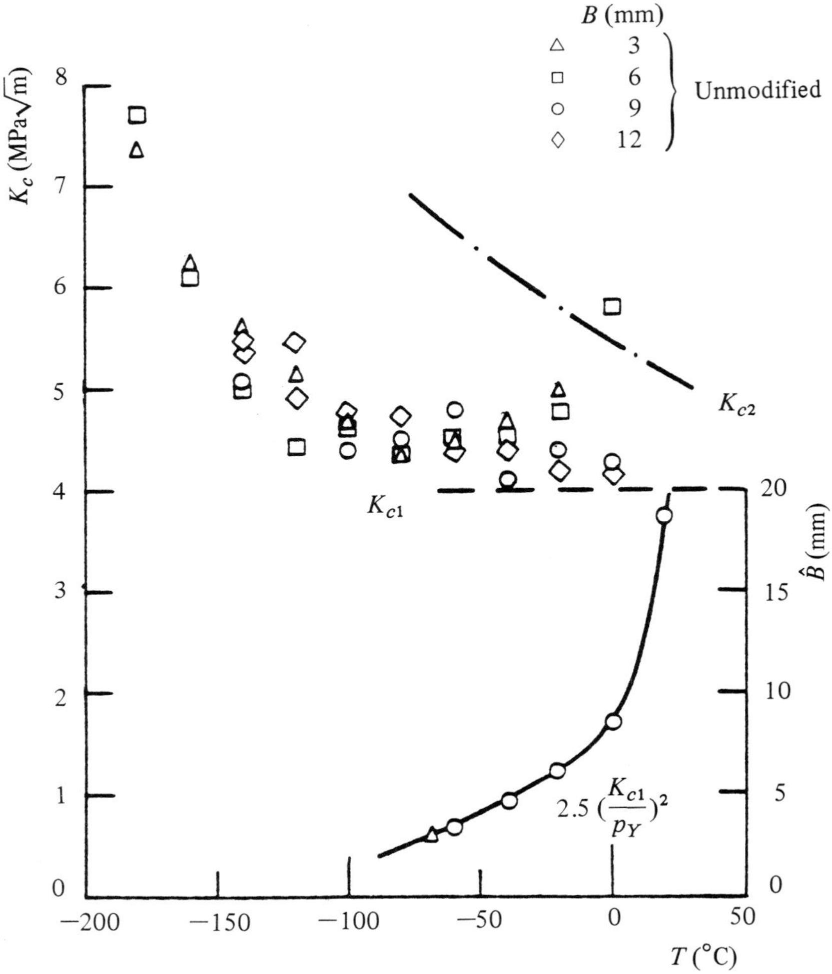

Fig. 6.32 Fracture data for PVC, SEN tension, $D = 50$ mm

Fig. 6.33 which shows $p\,Y$ versus $a^{-1/2}$ for several geometries at -60°C. The rather high value of K_c of 4.7 MPa$\sqrt{m}$ is typical and remains essentially constant between -100°C and $+20$°C. The transition to ductile tearing is abrupt with no evidence of thickness dependence or a transition region. There is evidence of a very small amount of whitening around the crack tip. At -60°C, the plane strain limiting thickness value, $2.5\,(K_{c1}/p_Y)^2 \simeq 9.3$ mm, and the tension specimens are within the criterion. At 20°C, this value rises to 61 mm and yet brittle failures are observed. The bending specimens violate the criterion even at -60°C, since $D = 96$ mm, but consistent values of K_{c1} are obtained.

Table 6.2

Temperature (°C)	−40	−20	0	+20
p_Y (MPa)	92.0	80.0	68.0	46.0
$(K_{c2}/p_Y)^2$ (mm)	4.4	5.2	6.3	12.3
$4r_0$ (mm)	3.5	4.4	5.0	9.8
$2.5\ (K_{c1}/p_Y)^2$ (mm)	4.7	6.2	8.6	18.9

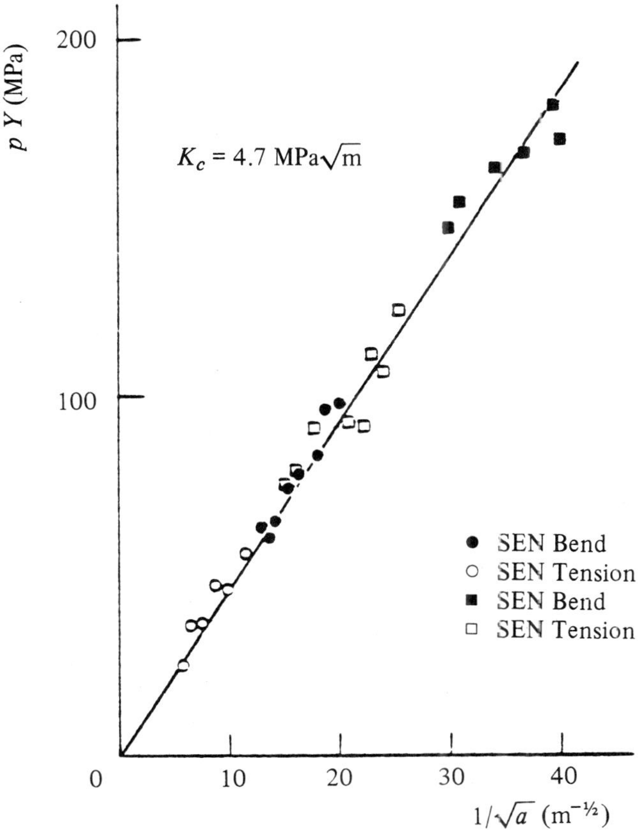

Fig. 6.33 Fracture toughness data from four test methods for the homopolymer PP at −60°C (from reference [13])

Modified polymers are produced by the addition of ethylene-propylene rubber or by polymerisation with additional ethylene. Again, this results in a dispersion of rubber particles with attendant increases in whitening around the crack tip. The yield stress is reduced as demonstrated in Fig. 6.34, where the

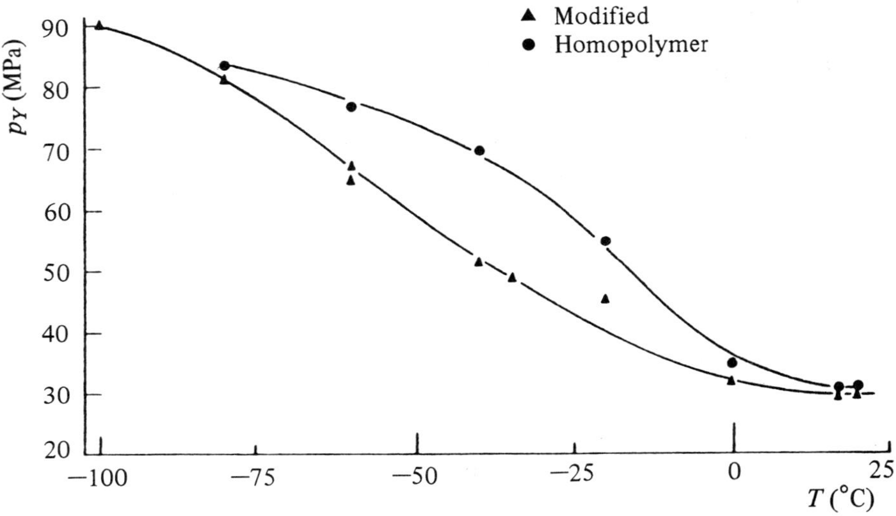

Fig. 6.34 The variation of yield stress with temperature for homopolymer PP and modified PP (from reference [13])

values for an 8% ethylene material are compared with those for the parent homopolymer. The largest difference is in the range −70°C to 0°C where the rubber phase has its glass transition. The fracture behaviour is changed profoundly, as is shown in Fig. 6.35 where SEN tension data with $B = 9.5$ mm are compared with bending data at $B = 50$ mm. Below −100°C, the results are coincident with $K_c = 4.1$ MPa$\sqrt{m}$, but between −100°C and −30°C there are brittle failures with substantial differences between the two loading modes. Data from several thicknesses at −60°C are compared in Fig. 6.36 and that from bending can be seen to be constant for $B > 10$ mm. There is a slight increase below this value and the ASTM thickness transition value shown at 9.7 mm confirms its validity for B in the bending mode. However, D is 6 mm and clearly in violation but still gives K_{c1} values. The tension data are higher and show a pronounced variation which is replotted in Fig. 6.37 against B^{-1} where it can be seen to extrapolate to the bending plane strain value, K_{c1}, in accordance with equation (5.11). Again, the brittle fractures in tension are sustained to very low thicknesses (5 mm in this case) and show no evidence of the transitions with thickness observed in PC.

K_{c2} may be computed in the usual way and is given in Table 6.3, together with p_Y and the computed values of 2.5 $(K_{c1}/p_Y)^2$ and $(K_{c2}/p_Y)^2$. K_{c2} is unusual in that

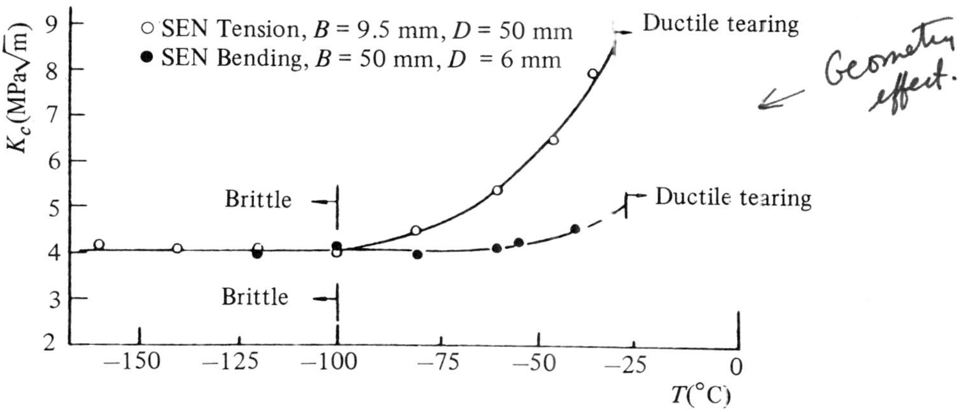

Fig. 6.35 The variation of fracture toughness with temperature for the modified PP (from reference [13])

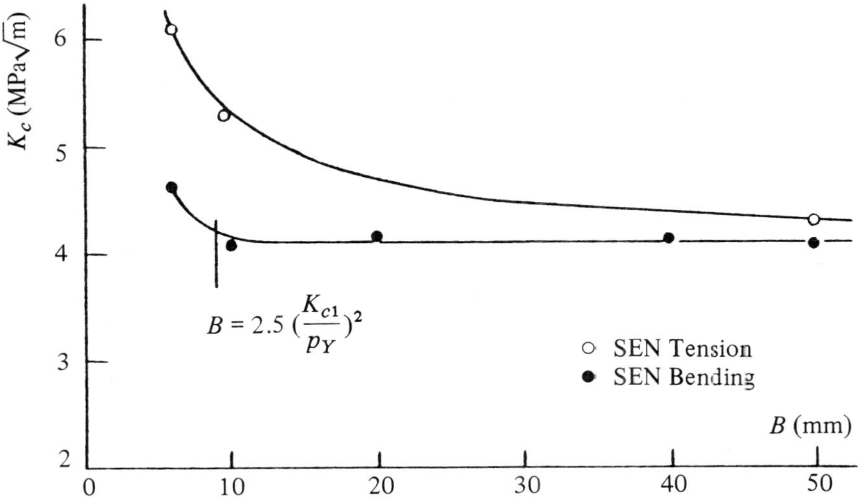

Fig. 6.36 The variation of fracture toughness with thickness for modified PP at −60°C (from reference [13])

it decreases with decreasing temperature and becomes equal to K_{c1} at -100°C.

The pattern of behaviour here is that K_{c1} remains approximately constant with temperature at about 4 MPa$\sqrt{m}$ and that for temperatures below about -60°C, the yield stress is sufficiently high that all the fractures are in plane strain with no mode of loading dependence. This type of behaviour is sustained

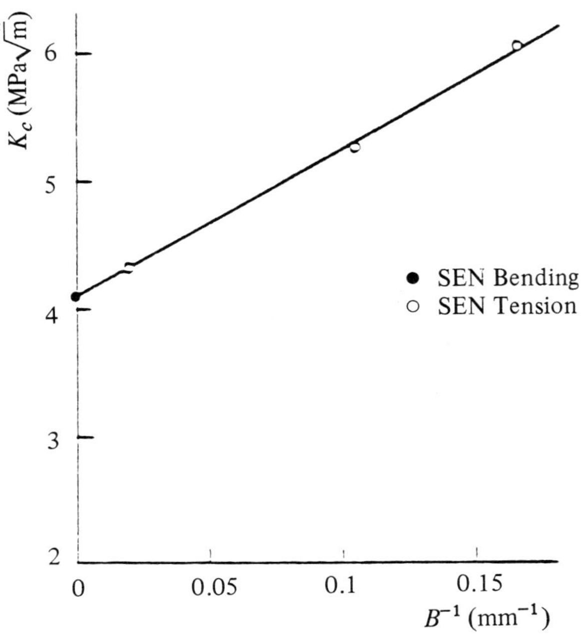

Fig. 6.37 Fracture toughness versus inverse thickness for modified PP at −60°C (from reference [13])

Table 6.3

Temperature (°C)	−35	−45	−60	−80	−100
p_Y (MPa)	49.0	55.0	67.1	82.0	91.4
K_{c2} (MPa$\sqrt{m}$)	8.3	7.7	7.4	6.5	4.1
$2.5\,(K_{c1}/p_Y)^2$ (mm)	17.5	13.7	9.2	6.2	5.0
r_0 (mm)	4.6	3.1	1.9	1.0	0.3
$(K_{c2}/p_Y)^2$ (mm)	28.7	19.6	12.2	6.3	2.0

up to 20°C for homopolymer, even though normal size parameters would suggest the onset of thickness or width effects. There is little evidence of shear yielding and it appears that this is a consequence of a microvoiding process. The copolymers give a modified version of this behaviour with enhanced voiding but also additional shear yielding since size and mode of loading effects become apparent, although, again, brittle failures can occur with extensive yielding. The very shallow but wide SEN bend specimens are remarkable in this regard in that they give brittle failures at true K_{c1} values when normal calculations would suggest fully yielded conditions. The modification of the homopolymer seems to make very little difference to the toughness of the material, as illustrated in Fig. 6.38, where K_{c1} and K_{c2} are plotted as a function of ethylene content. The variation in apparent K_c' shown is due entirely to changes in p_Y. This, again, is consistent with the observations in HIPS where the basic matrix property determines K_{c1} in the composite.

Initiation J_{c1} values are high here, being around 10 kJ/m^2, probably because of microvoiding. As for PVC, these high values would not be achieved in circumstances which restrict microvoiding, as, for example, in impact (see Section 8.3). J_{c2} is similar to other polymers, being about 25 kJ/m^2.

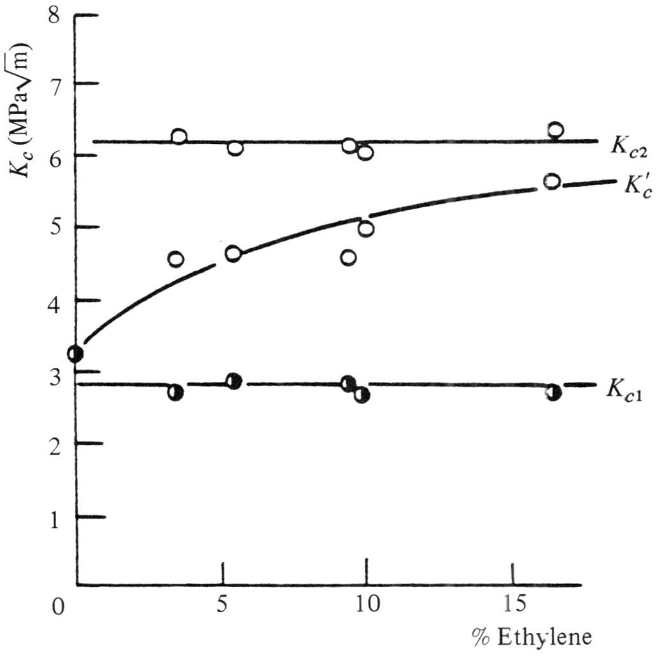

Fig. 6.38 Fracture toughness values for modified PP as a function of ethylene content at −60°C (from reference [14])

6.10 Polyethylene (PE)

This is a further example of a partially crystalline polymer and is produced with a range of densities from about 900 kg/m^3 to 960 kg/m^3. The molecular configuration is very simple which gives rise to the crystallinity and in the low density forms chain branching results in rather weakly bonded materials with low stiffnesses and yield stresses. The low density materials are used mainly for packaging applications and are extremely ductile but have insufficient stiffness for engineering applications. The higher density materials are produced with more linear molecules and are considerably stiffer but also less ductile. The toughness is thus strongly dependent on density and increases as the density decreases. The yielding process is shear stress controlled, giving constant volume deformation with lateral contraction and only slight evidence of whitening at high deformation.

In all but the very highest densities, it is not possible to produce brittle failures in tension specimens of normally available sizes above about $-150°C$ and recourse must be made to the higher constraint of bending. The size effects are very similar to those seen in metals in that the loss of plane strain conditions leads to pronounced non-linearity in the load-deflection curves and K_c is no longer useful. Fig. 6.39 illustrates this effect where K_c for a 955 kg/m^3 material at 20°C is shown as a function of thickness for $D = 50$ mm. The points are for various a/D values and show that above about 10 mm, K_c gives an essentially constant value independent of a/D and B, while below this value it changes with both. The usual size criterion is also shown and is a reasonable estimate of the transition. Results in which D was varied for $B = 20$ mm is shown in Fig. 6.40, and here K_c decreases at low D values below K_{c1} ($\simeq 1.55$ MPa$\sqrt{\text{m}}$), in contrast to the low B values which gave higher apparent values. This is because the low B condition leads to a loss of constraint and thus an elevation in K_c, while low D maintains constraint but tends to plastic collapse conditions.

The line drawn is for the simple plasticity correction given in equation (5.16), which may be written with apparent $K \equiv p\sqrt{\pi\, a}$ and $K_c{}^2 = E\, J$, so that:

$$K = K_{c1}\left[1 + \left(\frac{1}{2\pi}\left(\frac{K_{c1}}{M\, p_Y}\right)^2 \frac{D}{a}\right)\frac{1}{D}\right]^{-1/2} \tag{6.4}$$

For the data used here, $M = 1$ and an average $a/D = 0.3$ is used, and the trend is correct suggesting that a plane stress model works reasonably well. Also shown is $2.5\,(K_{c1}/p_Y)^2$ and it is apparent that this would result in low K_c values if used as a limit for D. About twice this value would appear to be necessary and will be discussed in more detail in Section 6.14.

K_{c1} measurements in SEN bending which give valid plane strain values have been made for materials of densities 960 kg/m^3, 955 kg/m^3 and 940 kg/m^3 and are shown as a function of temperature in Fig. 6.41. The yield stresses obtained in simple tension are shown in Fig. 6.42 and there is only a slight decrease with decreasing density. The K_{c1} values, however, show a marked increase as density decreases so that, while it is possible to achieve valid conditions for the 960 kg/m^3 and 955 kg/m^3 density materials, it was not so for temperatures of greater than $-100°C$ for the 940 kg/m^3 material. Those values shown at

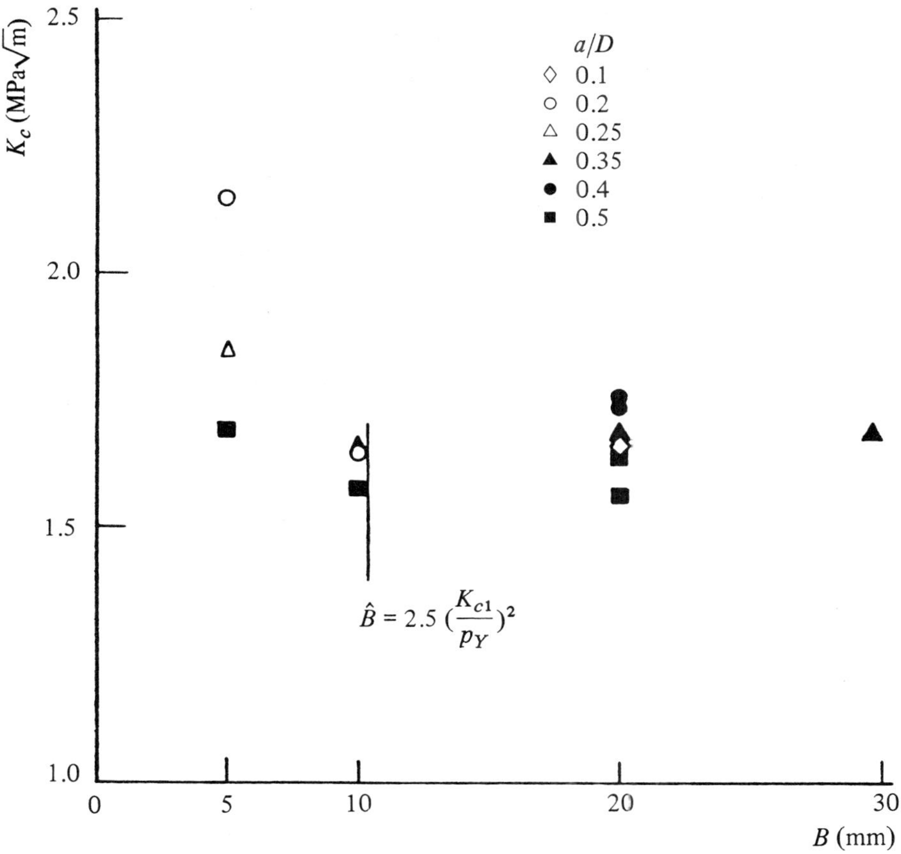

Fig. 6.39 Thickness effects for PE (density 955 kg/m^3) at 23°C (from reference [15])

-100°C, -80°C and -60°C are for invalid specimens. The rise at very low temperatures observed in PVC is also seen here and an influence of the liquid nitrogen used in cooling is suspected. The variation in the size parameter 2.5 $(K_{c1}/p_Y)^2$ versus temperature is shown in Fig. 6.43 and illustrates the special problem of the low density materials. It should be noted that this is not a homopolymer and is effectively a modified material, although no pronounced whitening was observed.

The problem of characterising crack initiation in the very ductile 940 kg/m^3 density material can be addressed using the J_c concepts outlined in Section 5.5. If bending is used with deep notches, then the high constraint of elastic deformation is approximately maintained into fully plastic collapse. Thus, failure is in plane strain fully plastic conditions and usually there will be

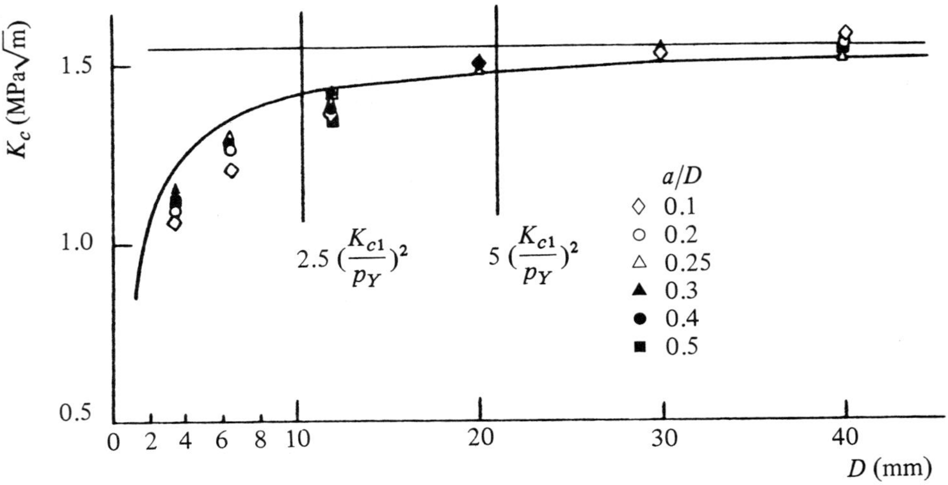

Fig. 6.40 Depth effects for PE (density 955 kg/m^3) at 23°C (from reference [15])

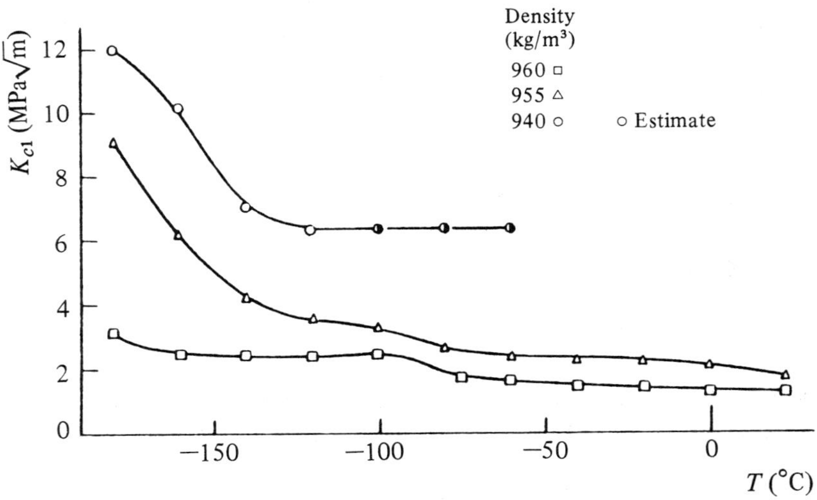

Fig. 6.41 Toughness as a function of temperature for PE at three densities (from reference [15])

pronounced non-linearity in the load-deflection diagram, as shown schematically in Fig. 6.44(a). J may be deduced from the total energy dissipated, since $\eta_p = 2$ for this case (see Section 5.5), so that:

$$J = \frac{2U}{B(D-a)} \tag{6.5}$$

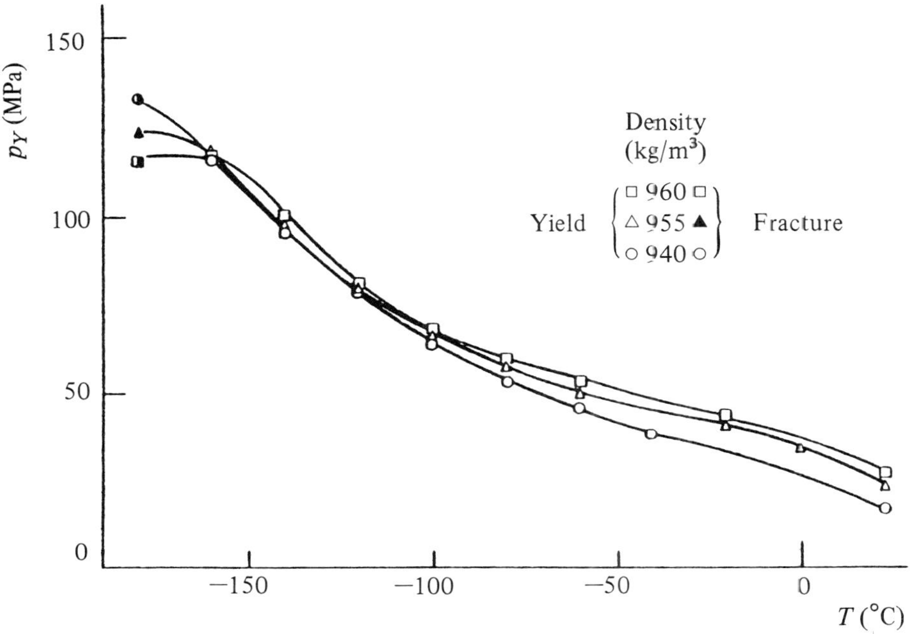

Fig. 6.42 Yield stress as a function of temperature for PE for three densities (from reference [15])

and J may be determined for the load-deflection diagram up to the point of crack initiation, as shown in Fig. 6.44(c). This point is difficult to observe in these opaque materials and has to be found by loading a series of identical specimens to loads less than final failure but above initiation. If these are then broken open at low temperatures, the slow growth, Δa, can be measured, as shown in Fig. 6.44(b). If the J for each specimen is then plotted versus Δa as shown in Fig. 6.44(d), the J–Δa curve is then extrapolated back to initiation, which must be corrected for crack tip blunting, which is given by the COD expression:

$$J = \delta \, p_Y$$

and if we assume smooth blunting, as sketched in Fig. 6.44(d), then $\Delta a \simeq \delta/2$ and:

$$J = 2p_Y \, \Delta a \tag{6.6}$$

The intercept of the growth and blunting lines is thus the initiation value for cracking under plane strain conditions, J_{c1}.

A series of these tests has been performed on the 940 kg/m³ density material over the temperature range −80°C to +23°C using $B = 20$ mm and $D = 40$ mm, and the J–Δa lines are shown in Fig. 6.45. The lines are reasonably linear and J_{c1} values can be found and are shown in Fig. 6.46 both as J_{c1} and as equivalent K_{c1}

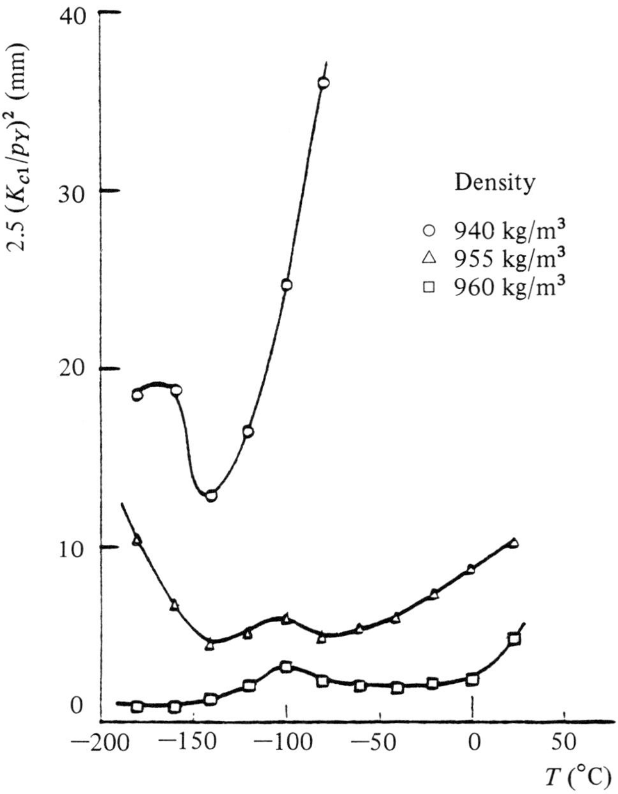

Fig. 6.43 Size parameter for PE

values to compare with the conventionally determined values. The coincidence of the data at $-80°C$ is simply a reflection of self-consistency of the method since the J method is an alternative route to K_{c1} in these now elastic specimens, since, for this geometry, equation (6.5) is valid for the case also, i.e.

$$G = \frac{(1-a/D)}{\phi} \frac{U}{B(D-a)} \simeq \frac{2U}{B(D-a)}$$

for $L/D \simeq 6$ (see Table 4.2, Charpy data).

6.11 Thermoset resins

Thermoset polymers are so called because they are produced by crosslinking (or setting) a basic polymer to produce a stiff, three-dimensional molecular network. There is very little thermal softening of this structure and over most of their operating range they behave as very stiff glasses. There is no reliable evidence of

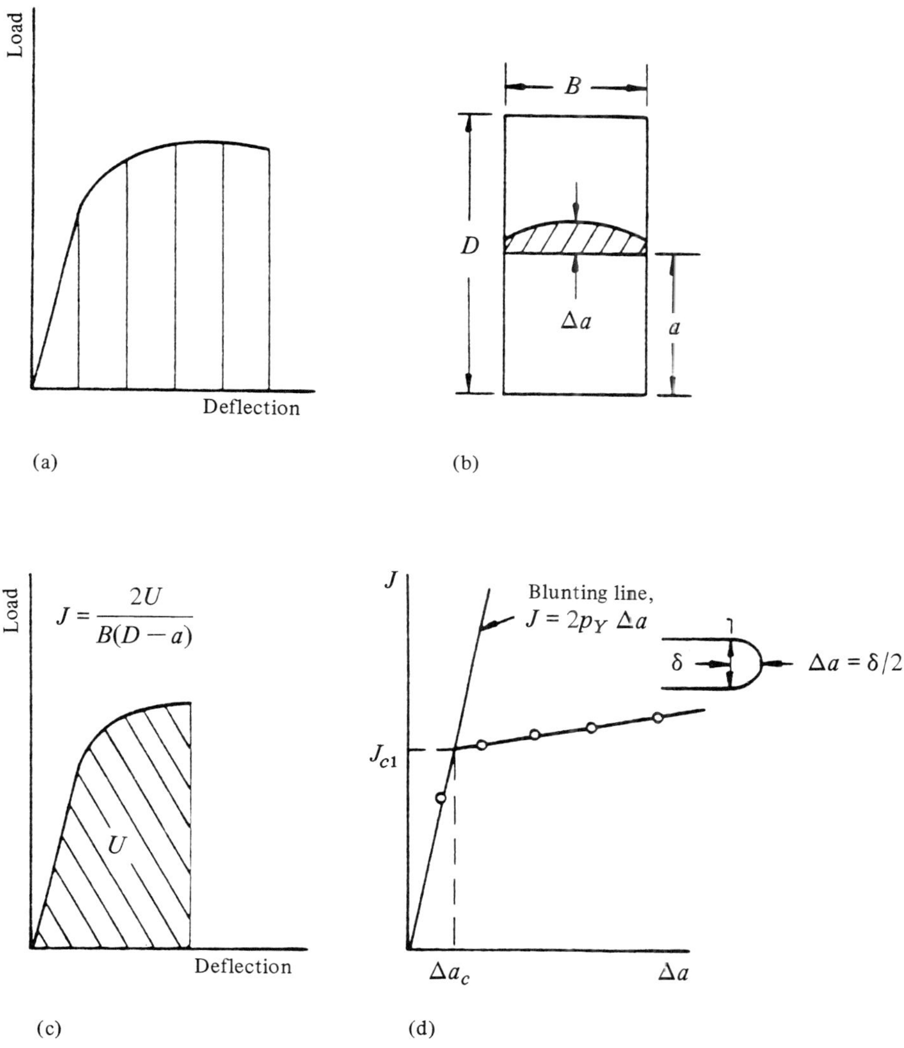

Fig. 6.44 J_{c1} measurement procedure (from reference [16])

crazing occurring and energy absorption is all by a shear yielding process. The group contains epoxies, polyesters and formaldehyde resins, and they are frequently used in conjunction with glass and other fibres for composites and also as adhesives. Rubber modified versions of several thermosets are produced.

In general, all these materials are brittle and have a low K_c value of about 0.6 MPa$\sqrt{m}$. Since the yield stress is approximately 100 MPa, the size parameter $2.5\,(K_{c1}/p_Y)^2 \simeq 0.1$ mm, so that most conventional test geometries are well

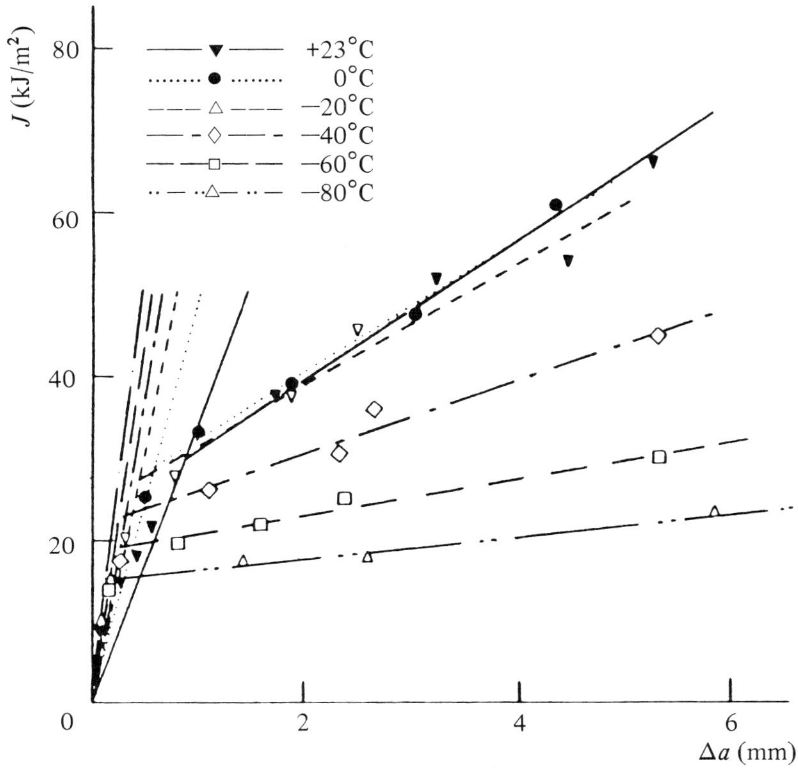

Fig. 6.45 J-crack extension lines – PE, 940 kg/m^2 density (from reference [16])

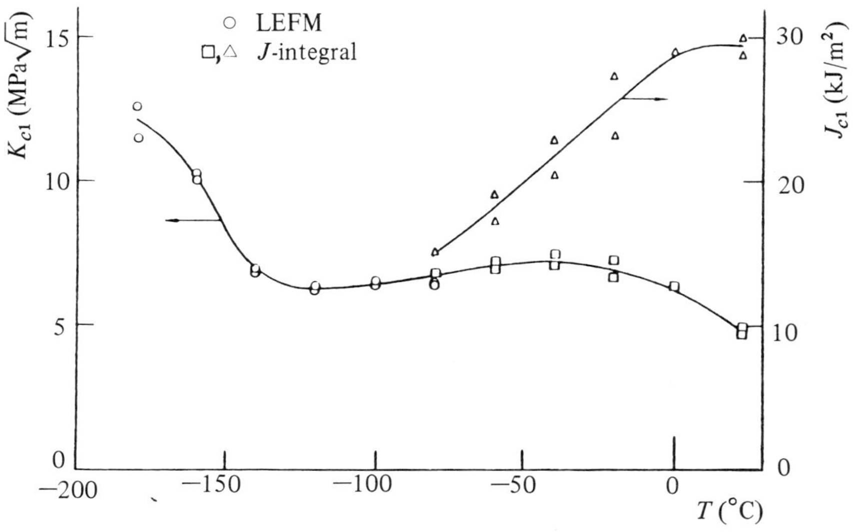

Fig. 6.46 J_{c1} and K_{c1} for 940 kg/m^3 density PE (from reference [16])

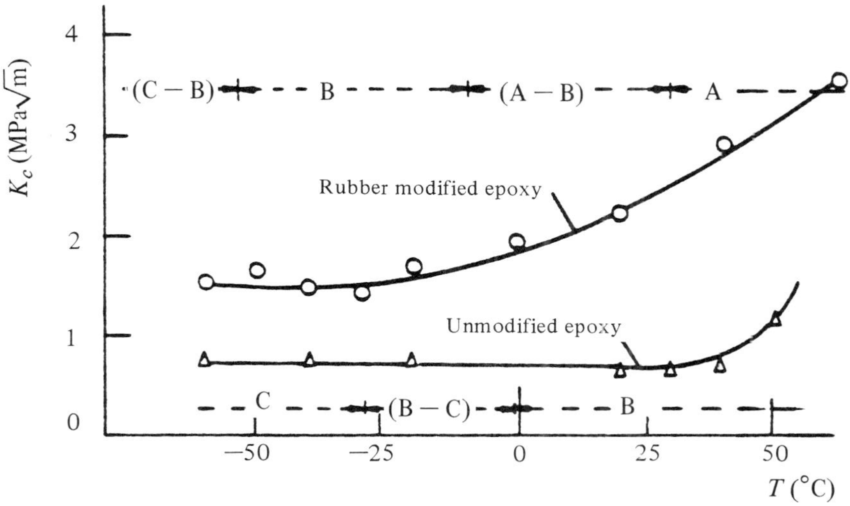

Fig. 6.47 Fracture of epoxy resins (from reference [17])

within the plane strain requirements. There is very little evidence of thickness effects and failures go abruptly to ductile tearing with increasing temperature. Fig. 6.47 shows some data for two epoxy resins obtained on 6 mm thick compact tension specimens. The rubber modified version shows the expected elevation in toughness and an enhanced temperature dependence. Three types of crack growth are identified:

A: ductile tearing,
C: stable brittle crack growth, and
B: unstable, stick-slip, brittle crack growth.

These materials are slightly viscoelastic so that the stable brittle growth is the same as that seen in PMMA. The unstable growth is a more unusual phenomenon in that cracks initiate at K_c values significantly above their stable growth values and subsequently arrest so that cracks progress in a series of jumps, i.e. 'stick–slip'. These initiation values correlate very closely with the yield stress of the material, as is illustrated in Fig. 6.48 where the ratio of the initiation value to that at stable growth is shown as a function of p_Y. The variations shown were produced by temperature, formulation, curing and rubber modification. The stable growth values in fact change very little and the critical size condition for plane strain for $B = 6$ mm used here is reached when $p_Y \simeq 40$ MPa. This corresponds closely to the onset of ductile tearing, as shown, so that all the unstable behaviour (B) is within the plane strain region.

The most likely explanation of this effect is that plastic crack tip blunting occurs, which results in an elevation of K_c at initiation [18]. This may be modelled using the elastic stresses around a crack tip of radius ρ given in

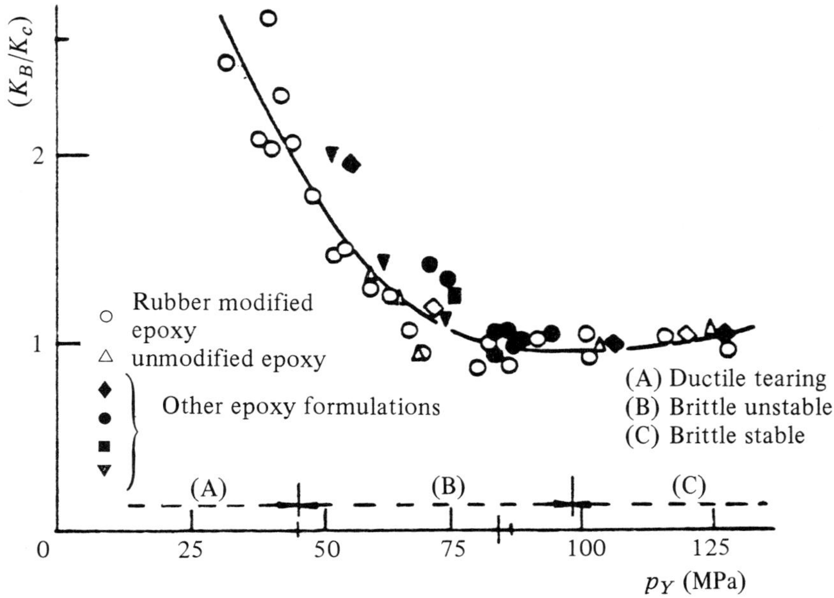

Fig. 6.48 Toughness as a function of yield stress for epoxy resins (from reference [17])

equation (4.9), i.e.

$$K_B = p\sqrt{\pi a} = K_c \frac{(1+\rho/2l_0)^{3/2}}{(1+\rho/l_0)} \qquad (6.7)$$

and assuming that ρ given by the COD such that:

$$\rho \simeq \delta = \frac{K_B{}^2}{E\,p_Y} \qquad (6.8)$$

(This assumption appears to work better in this case than the smooth blunting model $\rho = \delta/2$ used for PE.) In this model, failure is assumed to occur when a critical stress, p_0, is reached at a distance l_0 so that:

$$K_c{}^2 = p_0{}^2\, 2\pi\, l_0$$

and two parameters are necessary to characterise this type of behaviour. Fig. 6.49 shows the rubber modified data plotted according to this relationship as K_B/K_c versus $\delta^{1/2}$ and a fitted line of $p_0 = 200$ MPa and $l_0 = 10$ µm. The open points are δ values computed as COD values, while the filled points are from specimens made with drilled holes and confirm the basic blunting concept. The major point to note is that $K_B \simeq K_c$ for $\delta \leqslant 3.23\, l_0 = 32$ µm, i.e. $\delta^{1/2} = 5.6\sqrt{\mu m}$ and there is no self-blunting, while for $\delta > 32$ µm an elevation in K_c will occur. This is

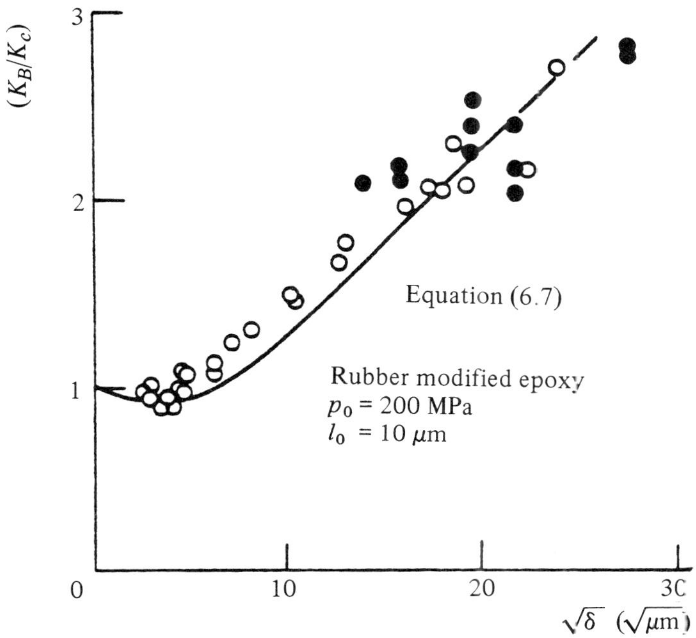

Fig. 6.49 Crack blunting model predictions for rubber modified epoxy resin (from reference [17])

equivalent to saying that the blunting is determined by a characteristic size in the material, l_0, under these plane strain conditions.

6.12 Fibre reinforced polymers

Fibre reinforcement of polymers with glass or other fibres is widely used in practice. Glass fibre thermoset composites are used in many structural applications and K_c (or G_c) values may be determined, providing that any damage zone around the crack tip satisfies the usual criterion. For randomly reinforced systems, there are no special problems, but in highly oriented materials it is often only possible to obtain colinear crack growth along the fibre direction. Typical values for a randomly reinforced polyester resin (33% glass) are $K_c \simeq 7$ MPa$\sqrt{m}$ ($G_c = 7$ kJ/m^2). Data for a randomly glass fibre reinforced polyester resin (a bulk moulding compound) are shown in Fig. 6.50 for SEN and compact tension methods, confirming the basic method. The very low stress values shown for small crack lengths often occur in inhomogeneous composites when the introduced crack length approaches the inherent flaw size in the material and failure occurs away from the notch.

Random fibre reinforcement is also used for thermoplastics; notably, PP and

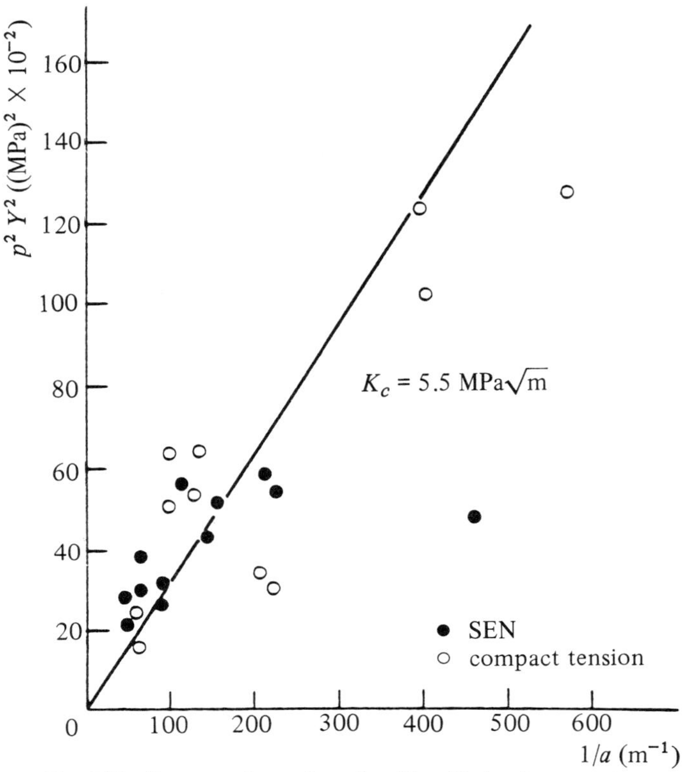

Fig. 6.50 Fracture of a random glass fibre filled polyester resin

Nylon. Again, K_c values may be obtained and are typically $K_c \simeq 4$ MPa$\sqrt{m}$ ($G_c = 2.5$ kJ/m^2).

6.13 Rubbers

A class of materials of special practical importance is that of rubbers which are highly flexible crosslinked (vulcanised) networks. In common with other crosslinked polymers, they fracture in a brittle manner in the sense that any plastic flow is very local but this is accompanied by large, non-linear, elastic deformations. The general G_c analysis can thus be employed and it is usual to use a directly calibrated specimen such as a parallel strip (see Section 2.4.3(vi)). Most rubbers show time dependence and a strong influence of method of preparation and G_c values in the range 3–10 kJ/m^2 are typical.

G_c may also be obtained from SEN tension tests, where the strain energy density form of relationship as in equation (2.32) may be used:

$$G = 2k\, W_0\, a \tag{6.9}$$

where W_0 is the strain energy density remote from the crack in a uniformly loaded sheet containing a small flaw of length a. k in this case is a function of strain which is conveniently expressed in terms of the extension ratio in the specimen, λ (extended length/original length). This function has been obtained both experimentally [19] and from computations [20], and is shown in Fig. 6.51(a). The major influence here is the shortening of the crack due to the

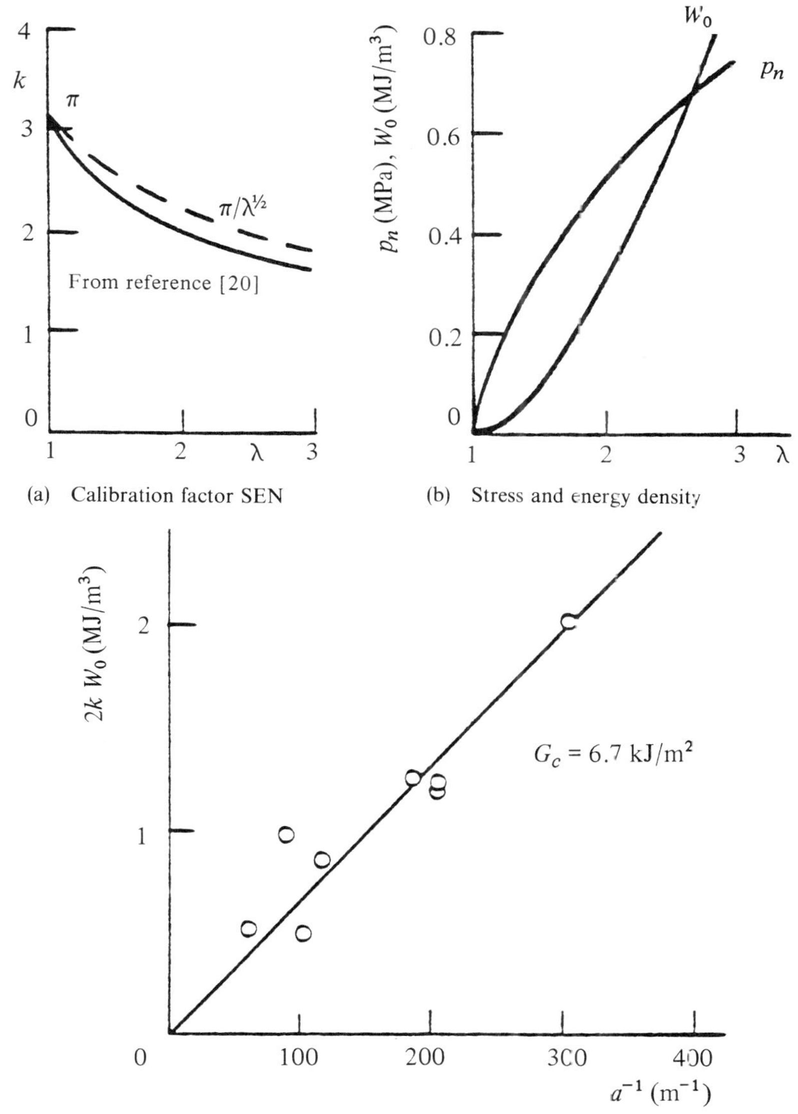

(a) Calibration factor SEN

(b) Stress and energy density

(c) Fracture data

Fig. 6.51 Fracture of rubber (from reference [21])

large strains, as is demonstrated by the function $\pi/\lambda^{1/2}$ which models this effect. Fig. 6.51(b) shows a nominal stress, p_n, versus λ curve for a particular vulcanised natural rubber derived from data given in reference [21], together with its integration W_0, and in Fig. 6.51(c) we have fracture results (from the same reference) plotted as $2k\ W_0$ versus a^{-1}, showing good linearity and a G_c value of 6.7 kJ/m^2.

6.14 Summary

A large volume of information is presented in this chapter and it is apparent that there are substantial differences between the various polymers. The basic concept of fracture mechanics of describing or characterising fracture by means of G_c, K_c or J_c is borne out in general, particularly in terms of being a crack length independent parameter. It is also apparent that size effects are important in determining the conditions under which the parameter is measured and thus its magnitude. Of major importance are the degree of constraint and the extent of plasticity at which crack growth occurs.

Interpretation of the data is often confused by the effects of instabilities in the observations. The criteria which are discussed here are for the onset of crack growth or initiation and it is an entirely separate matter to determine if such growth is stable. Stability here means growth which is controlled by the input of external work and can be stopped or started at will. Unstable growth is that in which the internal energy release rate of the system is in excess of that required to cause propagation, so that acceleration occurs and the growth is uncontrollable externally (see Section 8.7). Such conditions are described in equations (2.7) and (2.8) in that:

$$G = R \equiv J \tag{6.10}$$

and:

$$\frac{dG}{dA} > \frac{dR}{dA} \equiv \frac{dJ}{dA}$$

It should be noted that G describes the energy release rate of the whole system and can include energy stored in the loading system, while R (and J) describe the material behaviour. It should also be noted that brittle is not synonymous with unstable. Brittle cracks are those which maintain a sharp tip and have little gross distortion, while ductile implies the presence of such distortion. Both may be stable or unstable, although brittle cracks are more often unstable since dR/dA is generally low. The analysis of stability will be pursued in more detail in Chapters 7 and 8.

In the observations described here, initiations were sometimes, but not always, accompanied by instability and so when a final instability is recorded it is frequently difficult to know if it is the first crack initiation or some subsequent one. Plane strain initiation occurs first since it is the lowest toughness condition, and for materials with high yield stresses there are only very slight plane stress effects to promote stability so that it is common for such materials to become unstable immediately. It is also the case that some geometries, via high dG/dA values, tend to promote instability and those with low average stresses, such as bending, are more prone to give controlled elastic behaviour. If the plane strain fracture does not go unstable, then plane stress effects can become involved and the average toughness, K_c', increases. Again, when fracture occurs with the surface plane stress regions at K_{c2}, the result may be stable or unstable. If it is the latter, then this is what is recorded, but if not then stable growth will occur leading to larger scale plasticity and ductile tearing.

The criterion of $B > 2.5\,(K_{c1}/p_Y)^2$ does seem adequate on the limited evidence available for polymers to give conditions for plane strain initiation, but it is *not* a condition for ensuring instability at fracture. The criterion of $B < 2r_0 = 1/\pi\,(K_{c2}/p_Y)^2$ to give plane stress seems reasonable for those materials in which there is some microvoiding present so that lateral contraction is limited and crack sharpness maintained. When there is significant lateral contraction, then ductile tearing can ensue prior to reaching, full plane stress at $B < 4r_0 = 2/\pi\,(K_{c2}/p_Y)^2$.

The thickness is influential in controlling the degree of constraint present, while the crack length (a) and ligament size ($D-a$) control the stress level and hence the extent of plastic effects. The onset of ductile tearing which accompanies the more extensive plasticity in all degrees of constraint is thus determined by these two parameters. It is, of course, possible to lose constraint because of an inadequate ($D-a$) so there can be a coupling of the two effects. Indeed, the criterion $2.5\,(K_{c1}/p_Y)^2$ is usually taken as a lower limit on B, a and ($D-a$) when defining plane strain conditions. The criteria for yield controlled behaviour would certainly have an upper bound of plastic collapse, as in specimens for which $D > (K_c^2/p_Y^2)$, but experimental evidence would suggest a much more conservative criterion is necessary. One such can be derived on the basis of the nominal elastic stress at the crack tip, p_N, reaching some fraction of p_Y. We may write:

$$p = p_N f\left(\frac{a}{D}\right)$$

and since:

$$K_c^2 = Y^2 p^2 a$$

at fracture, we may rearrange the two equations as:

$$\frac{1}{D}\left(\frac{K_c}{p_Y}\right)^2 = \left[Y^2 f^2\left(\frac{a}{D}\right)\right]\left(\frac{p_N}{p_Y}\right)^2 \qquad (6.11)$$

The expression in square brackets is a function of (a/D) only and if the appropriate Y and f functions are used, can be shown to maximise in the region $0.2 < a/D < 0.5$. The appropriate expressions for geometries of interest are:

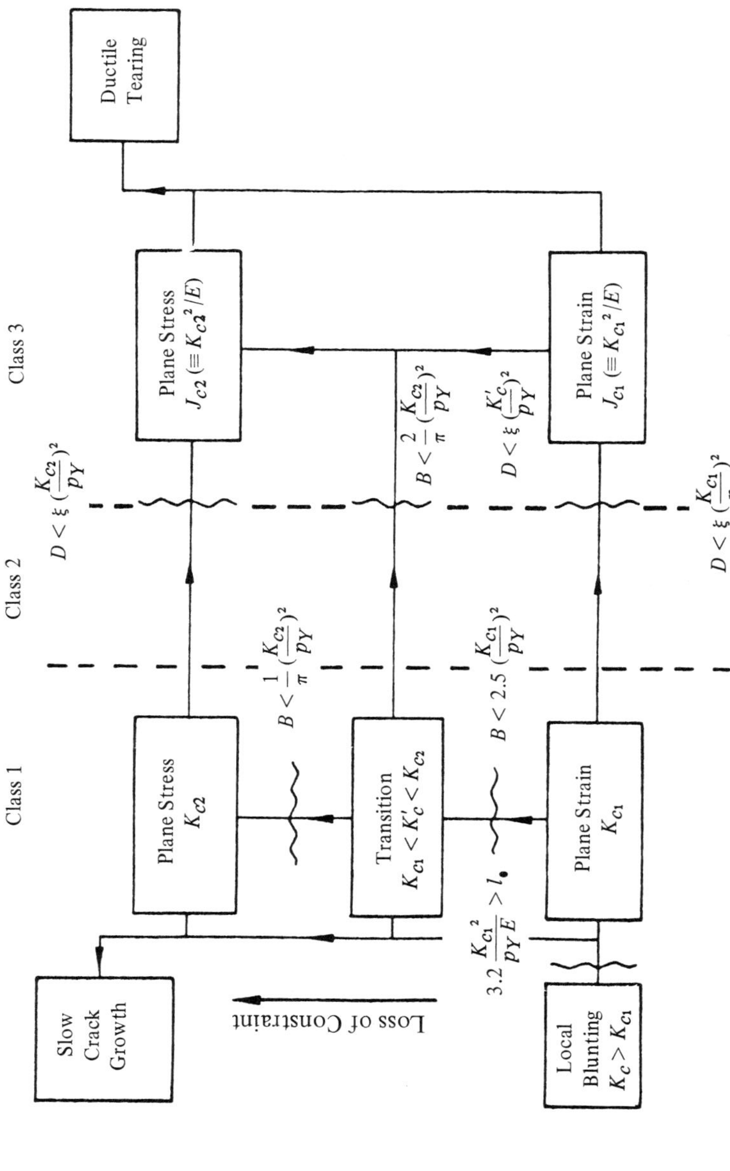

Fig. 6.52 Transitions in fracture of polymers

	f	$\xi = 1/D\,(K_c/p_Y)^2{}_{max}$	$(c/D)_{at\ max}$
CN Tension	$(1-a/D)$	$0.56\,(p_N/p_Y)^2$	0.45
SEN Bending	$(1-a/D)^2$	$0.30\,(p_N/p_Y)^2$	0.25
SEN Tension	$\dfrac{(1-a/D)^2}{1+2a/D}$	$0.25\,(p_N/p_Y)^2$	0.22

Most of the data reported here are for SEN tension, for which:

$$\frac{1}{D}\left(\frac{K_c}{p_Y}\right)^2 = 0.25$$

for $p_N = p_Y$, the upper limit, which agrees quite well with the transitions to ductile tearing observed.

A summary of the various types of behaviour observed in polymers is shown in Fig. 6.52 in diagrammatic form. The vertical direction represents decreasing constraint, while that from left to right is for increasing plasticity effects moving from class 1 through class 2 to class 3 (see Chapter 5). K_c increases from the plane strain value, K_{c1}, on the lower line to K_{c2} on the upper for plane stress.

Table 6.4

K_{c1} Values at 20°C

Material	K_{c1} (MPa$\sqrt{\text{m}}$)
PMMA	0.7–1.6
PS	0.7–1.1
PC	2.2
PES (Polyethersulphone)	1.2
HIPS	1.0–2.0
ABS	2.0
PVC	2.0–4.0
PP	3.0–4.5
PE	1.0–6.0
POM (Acetal)	$\simeq 4$
PA (Nylon)	2.5–3.0
Epoxies	0.6
Polyesters	0.6
PET (Polyethylene terephthalate)	$\simeq 5$

Also shown on the diagram is the transition to local crack tip blunting in plane strain based on a material size criterion. Two boxes are shown for stable growth after initiation; ductile tearing at high loads and brittle slow crack growth. These processes and their interaction with G give rise to instability and discussion of them will form the basis of Chapter 7.

In Table 6.4, a list of K_{c1} values for various polymers at 20°C are given as a guide in design. Detailed information can be obtained from the appropriate section for most of the materials listed.

References for Chapter 6

[1] Marshall, G. P., Coutts, L. H. & Williams, J. G., *J. Mater. Sci.*, **9** (1974) 1409–1419.
[2] Marshall, G. P., Culver, L. E. & Williams, J. G., *Int. J. Fracture*, **9** (1973) 295–309.
[3] Williams, J. G., *Polym. Engng. Sci.*, **17** (1977) 144–149.
[4] Morgan, G. P. & Ward, I. M., *Polymer*, **18** (1977) 87–91.
[5] Parvin, M. & Williams, J. G., *J. Mater. Sci.*, **11** (1976) 2045–2050.
[6] Nikpur, K. & Williams, J. G., *J. Mater. Sci.*, **14** (1979) 467–473.
[7] Parvin, M. & Williams, J. G., *Int. J. Fracture*, **11** (1975) 93–972.
[8] Irwin, G. R., Kies, J. A. & Smith, H. L., 'Fracture strengths relative to onset and arrest of crack propagation', NRL Report No. 5222, November (1958).
[9] Parvin, M. & Williams, J. G., *J. Mater. Sci.*, **10** (1975) 1883–1888.
[10] Fraser, R. A. W. & Ward, I. M., *Polymer*, **19** (1978) 220–224.
[11] Ferguson, R. J., Marshall, G. P. & Williams, J. G., *Polymer*, **14** (1973) 451–459.
[12] Newmann, L. V. & Williams, J. G., *J. Mater. Sci.*, **15** (1980) 773–780.
[13] Fernando, P. L. & Williams, J. G., *Polym. Engng. Sci.*, **20** (1980) 215–220.
[14] Fernando, P. L. & Williams, J. G., *Polym. Engng. Sci.*, **21** (1981) 1003–1009.
[15] Chan, M. K. V. & Williams, J. G., *Polym. Engng. Sci.*, **21** (1981) 1019–1026.
[16] Chan, M. K. V. & Williams, J. G., *Int. J. Fracture*, **23** (1983) 145–159.
[17] Kinloch, A. J., Shaws, S. J., Todd, D. A. & Hunston, D. L. *Polym.*, **24** (1983) 1341–1354, 1355–1363.
[18] Kinloch, A. J. & Williams, J. G., *J. Mater. Sci.*, **15** (1980) 987–996.
[19] Greensmith, H. W., *J. Appl. Polym. Sci.*, **7** (1963) 993–1002.
[20] Lindley, P. B., *J. Strain Analysis*, **7** (1972) 132–140.
[21] Rivlin, R. S. & Thomas, A. G., *J. Polym. Sci.*, **10** (1953) 291–318.

7

Slow crack growth and stability

7.1 Introduction

Chapter 6 dealt with the conditions under which a crack would initiate and included some limited discussion of subsequent behaviour. A simple brittle fracture has a single value of G_c (or K_c) and is either stationary or, after initiation, is usually of an undetermined velocity. Observations on several polymers demonstrate that this is not an adequate description of their behaviour. Cracks grow considerable distances at low speeds after initiation and prior to instability, and this can occur with plane strain failures as well as in ductile tearing. In general, therefore, we must examine the mechanisms which can lead to slow crack growth and in which the fracture criterion is not single valued. Such growth may be simply the amount, as in an R (or J)$-\Delta a$ curve (discussed in Section 2.3), the rate, $\dot{a}$, arising from time-dependent properties, or the growth per cycle for fatigue. The various sections of this chapter will be concerned with the description of these different forms of behaviour and their measurement.

7.2 Time-dependent fracture

The derivation of $J(t)$ for visco-elastic materials discussed in Section 5.7 showed that, as one would expect, both crack initiation and propagation are time-dependent parameters. For initiation, a rate dependence is predicted and is observed in several polymers, as described in Chapter 6. As cracks propagate, the analysis suggests that the time dependence is controlled by local events at the crack tip, and thus the crack speed is the controlling variable, so that $J_c(\dot{a})$ is an appropriate representation. We shall now consider the supporting evidence for this view.

7.2.1 *Craze growth and crack initiation times*

For a material in which both the modulus and the craze or yield stress are time-dependent, it is convenient to use power-law dependencies of the form:

$$c(t)=\left(\frac{t}{\tau_0}\right)^n \frac{1}{E_0} \text{ and } p_c=p_0\left(\frac{t}{\tau_1}\right)^{-m}$$

as given in Section 5.7. If a crack specimen is loaded, the crack tip craze zone, c, will grow and the crack opening displacement increase. This growth will occur, even for a constant load, because of the decrease in p_c and modulus and is sometimes referred to as 'relaxation-controlled' growth for this reason [1]. For specimens under constant load, we would thus expect craze growth given by the time-dependent line zone result (equation (5.61)):

$$c(t)=\frac{\pi}{8}\frac{K^2}{p_c^{\,2}(t)}=\frac{\pi}{8}\frac{K^2}{p_0^{\,2}}\left(\frac{t}{\tau_1}\right)^{2m} \tag{7.1}$$

Some data are shown in Fig. 7.1 for three polymers as log $c(t)$ versus log t and the slope is compared with $2m$, where m was measured from separate craze stress tests in tension to determine the rate dependence. The agreement is good and confirms that visco-elastic relaxation is the controlling mechanism. The data for HIPS is at several K levels and in Fig. 7.2, the results are presented as K^2 versus c/t^{2m} and give the linearity expected from equation (7.1) with $p_0=20$ MPa, which is in agreement with tensile measurements (see Fig. 6.22, for example).

As the craze grows, both the crack opening displacement and J increase and, by postulating either as a criterion, we have expressions for initiation times, t_i, as given in equations (5.66):

$$\lambda_{nm}{}^{1/n}\, t_i=\left(\frac{J_c E_0}{K^2}\right)^{1/n}\tau_0 \text{ for } J=J_c$$

or:

$$\left(\frac{\tau_0}{\tau_1}\right)^m \lambda_{nm}{}^{1/(m+n)}\, t_i=\left(\frac{p_0 E_0 \delta_c}{K^2}\right)^{1/(n+m)}\tau_0 \qquad \text{for } \delta=\delta_c$$

Thus, $t_i \propto K^{-2/n}$ for $J=J_c$ and $t_i \propto K^{-2/(m+n)}$ for $\delta=\delta_c$. Fig. 7.3 shows log K versus log time data for various PVC specimens, in which both the initiation and final failure is recorded [2]. The initiation time is a significant proportion of the life and lines are drawn for $m=0.045$ and $n=0.008$ with a slope of $(m+n)/2$ which describe the data quite well and suggest a δ_c criterion. This applies to both the 'good' and 'bad' samples and, in both, a rather more rapid decrease in properties is indicated over part of the life.

There is a good deal of evidence of crack initiation at various rates and a similar analysis gives:

$$K \propto \dot{x}^{n/2} \text{ for } J=J_c$$

and:

$$K \propto \dot{x}^{(m+n)/2} \text{ for } \delta=\delta_c$$

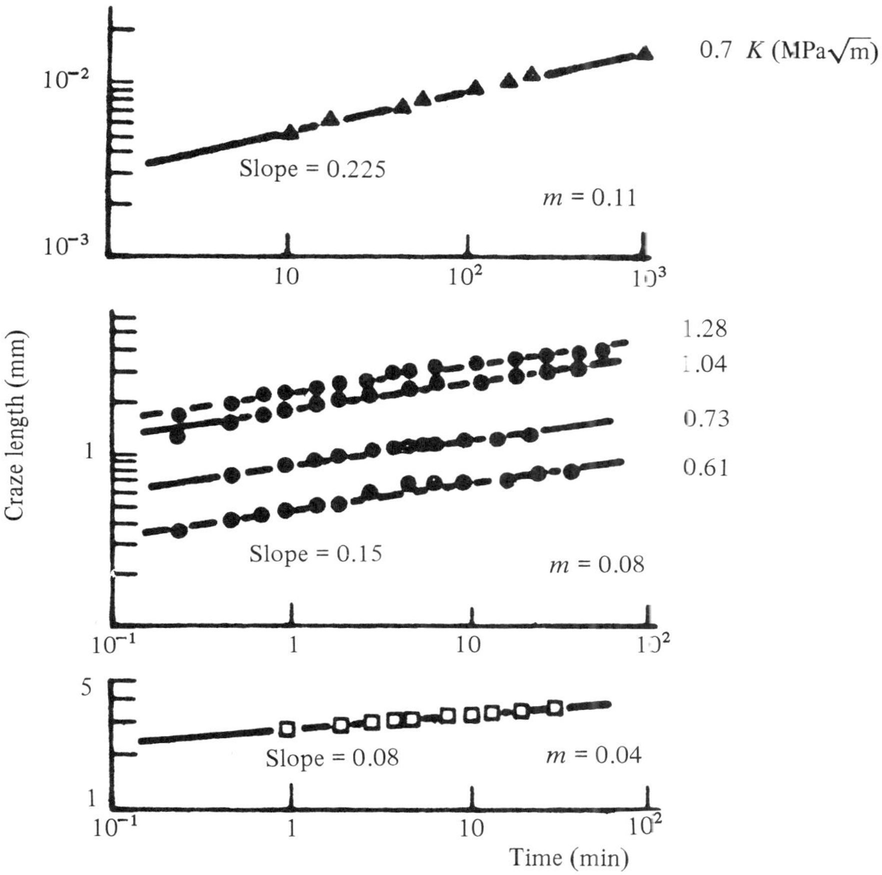

Fig. 7.1 Craze growth curves for various polymers tested in air at +20°C and 50% rh. △ PMMA, ● PC, □ HIPS (from reference [1])

The results shown in Fig. 6.3 for PMMA, for example, at 20°C give a power of 0.10 and since $m \sim n \sim 0.1$ for this material, it appears that $\delta = \delta_c$ is again the appropriate criterion. The data for HIPS shown in Fig. 6.23 shows rather a strong dependence (~ 0.1) and, although m is significant due to crazing, n is very small again, suggesting that a constant δ is more likely.

7.2.2 *Crack propagation*

For a crack propagating at a constant rate, it is possible to perform both a J or δ analysis, as outlined in Section 5.7, and it is usually assumed that this relationship will hold for cracks whose speeds are varying only slowly. Thus,

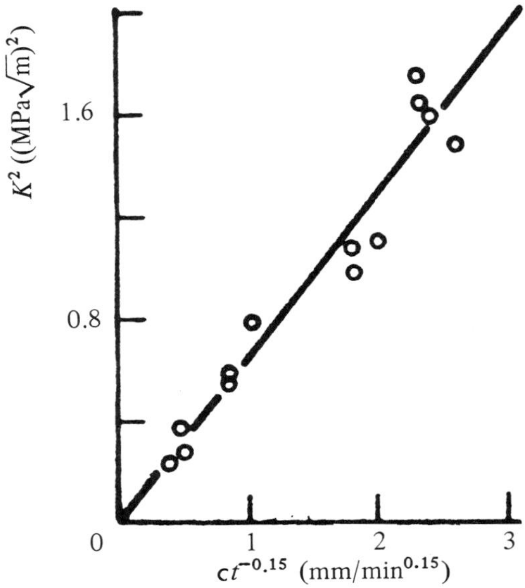

Fig. 7.2 Craze growth in HIPS (from reference [1])

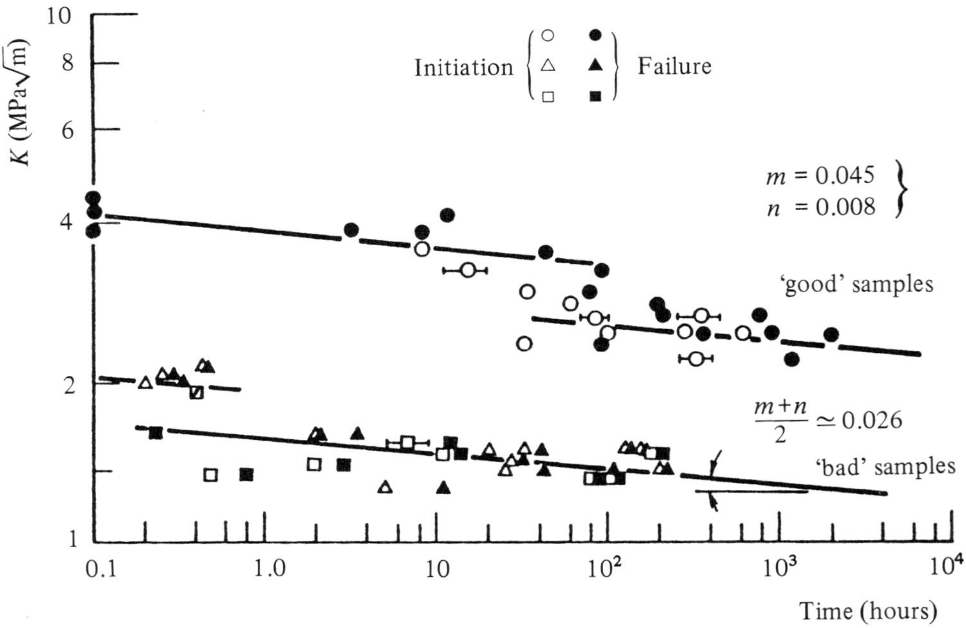

Fig. 7.3 Initiation and final fracture times for notched PVC samples (from reference [2]) – crack length 4 mm

from equation (5.65), we have:

$$K \propto \dot{a}^{n/2(1-2m)} \text{ for } J = J_c$$

and: $$K \propto \dot{a}^{(m+n)/2(1-2m)} \text{ with } J \propto \dot{a}^{n/2(1-2m)} \text{ for } \delta = \delta_c$$

The constant J criterion is, of course, the perfectly brittle case and does not exhibit the general effect to be discussed here, although it may appear to when plotted in terms of K because of modulus changes. The general problem of measuring these functions is most conveniently discussed in terms of $J_c(\dot{a})$ or $G_c(\dot{a})$ and it is clear from the analysis that a system in which $\dot{a}$ is a constant or slowly varying is required. Most testing is, therefore, carried out on specimens in which G is a constant for a fixed load, so that $C \propto a$ and dC/da is constant. The most commonly used for stiff materials are the contoured cantilever beam test or the double torsion test (see Section 2.4.3(ii) and (iii)), although other geometries are used for rubber. In general, we may write:

$$\frac{u}{P} = C = C_0 + C_1\left(\frac{a}{a_1}\right) \tag{7.2}$$

where C_0, C_1 and a_1 are constants, and C_0 may include the external stiffness of the loading system. The strain energy release rate is thus:

$$G = \frac{P^2}{2B}\left(\frac{C_1}{a_1}\right) \tag{7.3}$$

and if $G_c(\dot{a})$, then for a fixed load the crack will run at a speed to give a G_c corresponding to the G generated by the load, i.e.

$$G_c(\dot{a}) = \frac{P^2}{2B}\left(\frac{C_1}{a_1}\right) \tag{7.4}$$

Fig. 7.4 shows various situations which may arise in such tests for a typical form

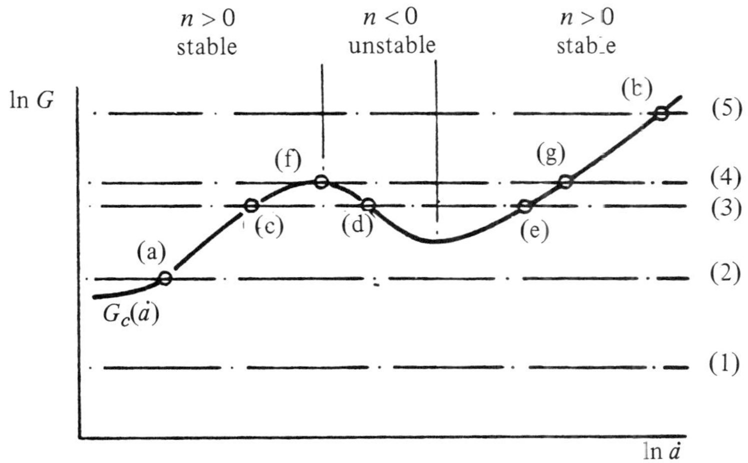

Fig. 7.4 $G_c(\dot{a})$ and constant load G curves

of $G_c(\dot{a})$. For $G < G_0$, a threshold value as in line (1), there will be no growth. For lines (2) and (5), $G > G_0$ and there are unique crack speeds for each load at which growth will occur as marked (a) and (b). For line (3), we have three possible speeds at (c), (d) and (e) which all satisfy equation (7.4). A load which gradually increases will clearly cause initiation at a low speed so that condition (c) is the most likely to be achieved. However, a decreasing load could, in principle, achieve equilibrium at any of the three values. Reference must now be made to the stability of such behaviour and for this we return to the criterion given in equations (2.8) and (2.9), i.e.

$$G = G_c \text{ and } dG > dG_c$$

Previously, we have considered changing a, but here a perturbation of equilibrium involves $\dot{a}$ so that the condition becomes:

$$\frac{dG}{d\dot{a}} > \frac{dG_c}{d\dot{a}} \tag{7.5}$$

for instability. For fixed load, $dG/d\dot{a} = 0$ and so the crack is stable for $dG_c/d\dot{a} > 0$ and unstable for $dG_c/d\dot{a} < 0$. If we assume that $G_c \propto \dot{a}^n$, then $dG_c/d\dot{a} \propto n\,\dot{a}^{n-1}$ and since $\dot{a} > 0$, then we conclude that growth is *stable for* $n > 0$ and *unstable for* $n < 0$. This may be confirmed by reference to Fig. 7.3, where an increase in $\dot{a}$ at (c) results in $G < G_c$ and a tendency to return, while a decrease in $\dot{a}$ gives $G > G_c$ and an accelerating trend, so that we have a condition of stable equilibrium for $n > 0$. The reverse is true at (d), so that a crack could not be grown in a stable fashion here under constant load, but both (c) and (d) could be achieved. In practice, a rising load test which proceeded at constant rate would give the G_c line up to (f) which would then jump to (g) and continue to rise with increasing load. On unloading, the speed would decrease to the minimum point (thus giving (e)) and then jump to the lower speed as the load decreases further.

The characteristics of the test may be changed by driving the displacement at a constant rate, $\dot{u}$, so that $\Delta = \dot{\Delta} t$ and from equation (7.2) we have:

$$P = \frac{\dot{u}\,t}{C_0 + C_1\,(a/a_1)}$$

A constant load condition can be achieved here for a constant crack speed if $C_1\,(a/a_1) \gg C_0$, which is true for very stiff loading systems, where C_0 is small. (Note that for a very soft loading system, $C_0 \gg C_1(a/a_1)$ and crack growth does not decrease P, thus being equivalent to the previous case). Under such circumstances, we may write:

$$G \simeq \frac{\dot{u}^2}{2B\,a_1\,C_1}\,\frac{1}{\dot{a}^2} \text{ with } P \simeq \frac{\dot{u}\,a_1}{C_1\,\dot{a}}$$

so that $G \propto \dot{a}^{-2}$ and in Fig. 7.3 the lines for G would have a slope of -2 for each rate and stable equilibrium would be obtained for $n > -2$ in this case. In practice, the method works well for $n > 0$ and does enable cracks to grow at any speed for $n = 0$, but the presence of a maximum in G_c makes it difficult to achieve stable conditions for $n < 0$. These occur when $\dot{a}$ is just sufficient to correct for $\dot{u}$, thus keeping P constant, and if $\dot{a}$ is less than this value then P will continue to rise. As the specimen is loaded, P will increase rapidly until initiation and, if this is below the desired state, as in Fig. 7.4, then the load will continue to rise but at a decreasing rate. If the required conditions correspond to points (a) or (c) in Fig. 7.4, then the load maximises to give stable conditions, as shown in Fig. 7.5(a). If we are seeking condition (d), however, at the appropriate load, $\dot{a}$ will be too low and the load will rise until (f) is reached, whereupon, if loading continues, the crack jumps to (g). If the condition sought is at a higher speed than (g), then it will subsequently stabilise, but if not $\dot{a}$ will be too large and P will fall as shown in Fig. 7.5(b). This will occur for all crack speeds greater than that at (d) and it can only restabilise at some speed less than at (c), after which P increases again. This leads to a 'stick–slip' mechanism which can occur when there are maxima and dependent upon the particular loading conditions used. The loading curves have a distinct saw-tooth form when this is occurring, as shown in Fig. 7.5(b).

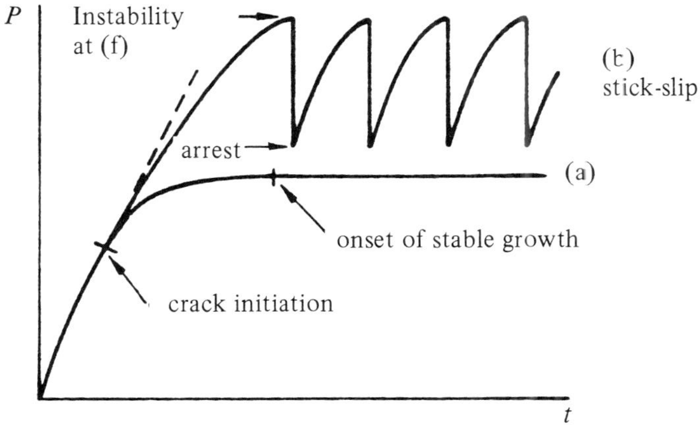

Fig. 7.5 Stable crack growth and stick–slip at $\dot{u}$ constant

This phenomenon is beautifully demonstrated in some data on natural rubber vulcanisates shown in Fig. 7.6 and taken from reference [3]. For vulcanisate A (on the left) at $T > 25°C$, there are rising $G_c - \dot{a}$ curves and stable growth was observed in a tearing test. At the lower temperatures, the slope becomes low and the broken line indicates stick–slip and only the upper value is shown. This should, of course, be the maximum and thus remain constant, as it does. The right-hand curve shows data with a pronounced minimum with a stable, rising, part after the minimum.

Stick–slip can arise from a number of sources since it is a consequence of G_c

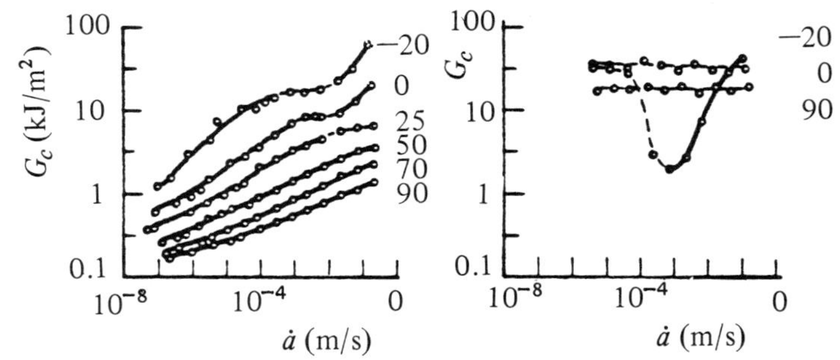

Fig. 7.6 Crack growth data on natural rubber (from reference [3]); — stable growth, --- unstable growth; log scales

being dependent on $\dot{a}$. An extreme example is found in epoxy resins where local crack tip blunting gives rise to high initiation values, while an essentially constant, and a much lower value, exists for a propagating crack. A blunt crack will often produce a similar effect and it may be expressed in terms of G_c as:

$$G_c = G_c', \ \dot{a} = 0$$
$$G_c = G_c'' < G_c', \ \dot{a} > 0 \tag{7.6}$$

When $G = G_c'$, all specimens will be unstable and the crack will grow rapidly before arresting at $G = G_c''$ if unloading can occur. Continued loading results in re-initiation at G_c', thus giving stick–slip.

Stable crack growth can be produced in glassy polymers in both the contoured cantilever beam or double torsion, as mentioned previously. The beam specimens are preferable because they give an almost straight crack front but are difficult to control because of crack wandering, even with side grooves. There is also a tendency for the crack surface to become curved, leading to arrest and subsequent instability. These effects are much less prevalent in double torsion but the crack front shows marked curvature as sketched in Fig. 7.7.

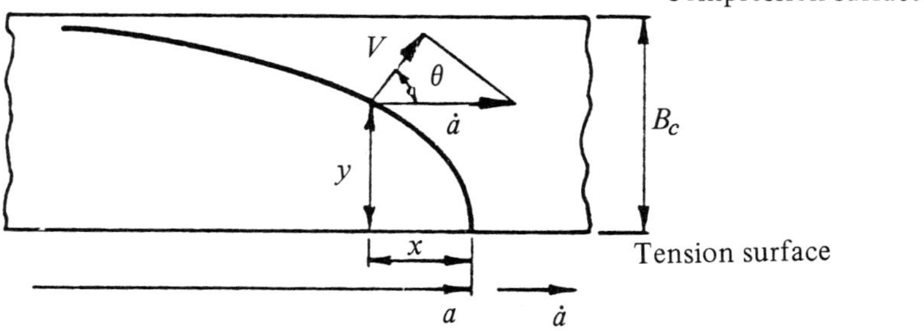

Fig. 7.7 Crack front profile in the double torsion test

Stability is achieved with this profile propagating at a constant rate along the specimen. The controlling or true crack growth rate is that normal to the front and shown as V in Fig. 7.7 and is given by:

$$V=\dot{a}\cos\theta=\dot{a}\frac{dy/dx}{\sqrt{1+(dy/dx)^2}}$$

Assuming that there is a growth law of the form $G_c \propto V^n$, then the apparent, or measured, G_c' may be expressed in terms of the true value at $V=\dot{a}$ as:

$$\frac{G_c'}{G_c}=\frac{1}{B_c}\int_0^{B_c}\left[\frac{dy/dx}{\sqrt{1+(dy/dx)^2}}\right]^n dy \tag{7.7}$$

The profile is determined by the elastic deformation required to give the essentially constant K along the crack front and it has been shown that this can be approximated by assuming a fixed displacement on the front with the beam arms rotating about some point a distance SB_c from the crack tip [4, 5]. This leads to a form:

$$\frac{y}{B_c}=\frac{x}{SB_c+x} \tag{7.8}$$

and finally:

$$\frac{G_c'}{G_c}=\int_0^1\left(\frac{\xi^2}{\sqrt{S^2+\xi^4}}\right)^n d\xi \text{ for } \frac{a}{B_c}\gg S \tag{7.9}$$

S is partially determined by the nature of the material, but there is also a strong geometrical influence and a simple analysis would suggest that $S \propto \sqrt{(B^3\,D)/B_c^{\,5}}$. Fig. 7.8 shows G_c'/G_c as a function of S for $0.02<n<0.2$ from equation (7.9) and clearly it is desirable to keep S as low as possible. This may be achieved by using low B and D values but the strongest influence is B_c and to achieve low S values, the side grooves should be as shallow as possible. A photograph of a typical surface for a specimen in PMMA is shown in Fig. 7.9, for which $S=1.04$ and the theoretical curve for equation (7.8) is also shown fitted at $x/B_c=1$. In general, it is difficult to predetermine S and the best technique is to fracture a trial specimen and from the surface profile determine S. Since $0<n<0.1$, a value of $S<1$ is desirable to give errors in G_c of about 10% which correspond to variations in K_c of 5%.

Fig. 7.10 shows a comparison of K_c data for tapered DCB and double torsion for PMMA at 20°C taken from reference [6] and in which $n\simeq 0.08$ and $S\simeq 1$ so that the double torsion data would be expected to be about 7% less than that for tapered cleavage. There is very little difference between the two sets of points and the lines are drawn 7% apart and are a reasonable description of the scatter observed. Double torsion data for PMMA for a range of temperatures are shown in Fig. 7.11 as log K_c versus log $\dot{a}$. Stable growth could be achieved in the temperature range -60°C to $+80$°C and, in general, the slope was $\simeq 0.08$ on the log–log plot. Again, this is consistent with a constant δ_c criterion since

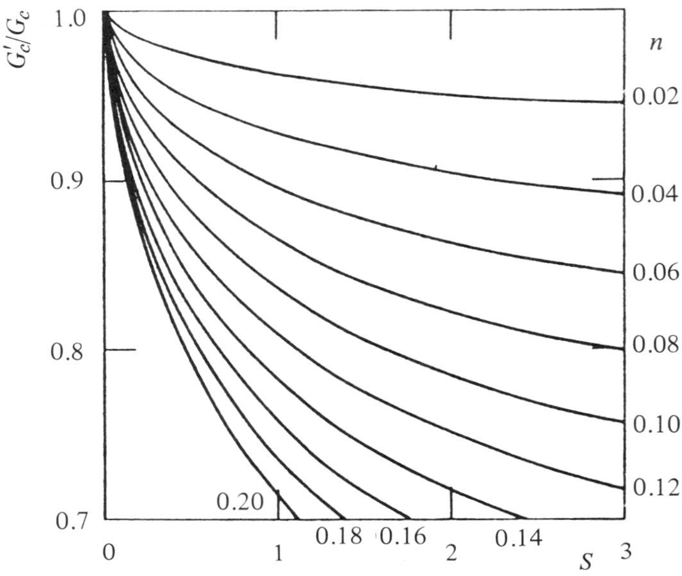

Fig. 7.8 Correction factor for G_c (equation (7.9))

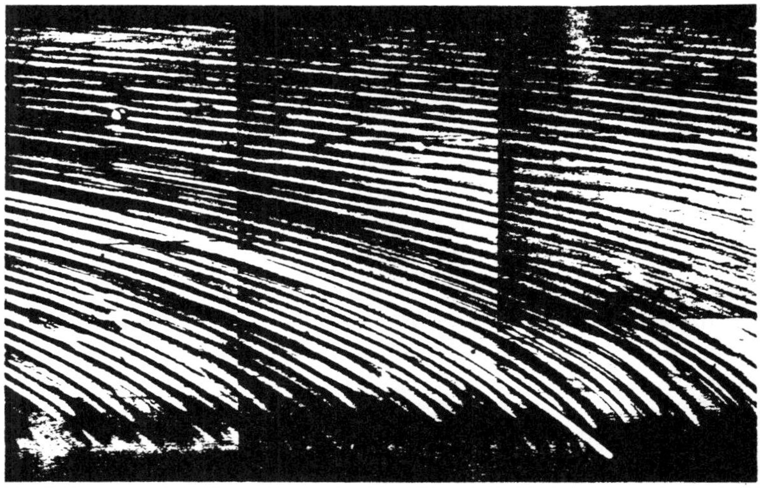

Fig. 7.9 Fracture surface for PMMA in the double torsion test; line is from equation (7.8) with $S = 1.04$

$m = n = 0.08$ in this case and, as shown in equations (5.65), we would expect:

$$K_c \propto \dot{a}^{\frac{m+n}{2(1-2m)}} \simeq \dot{a}^n \tag{7.10}$$

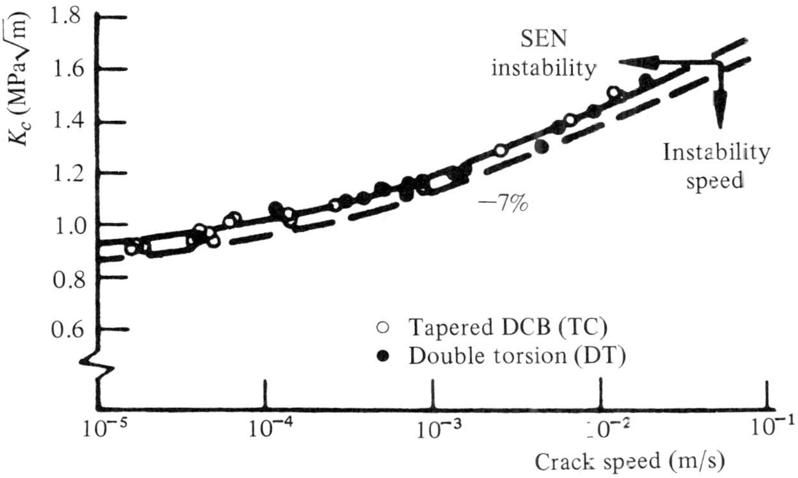

Fig. 7.10 A comparison of tapered DCB and double torsion data for PMMA at 20°C (from reference [6])

An examination of the data shows that $\delta_c \simeq 2$ μm provides a good criterion over the whole temperature range for this material and this has been confirmed by direct optical measurements [7]. Similar data can be obtained when a material is sufficiently brittle and has sufficient time-dependence in terms of m and n. The combination is not particularly common and difficulties are frequently encountered arising from ductility effects and low time-dependence. When such curves can be obtained, there are important results which may be deduced from them.

7.2.3 *Instability conditions*

The occurrence of maxima in the G_c or K_c versus $\dot{a}$ curves are of considerable importance in determining behaviour in practice. In general, they lead to rapid crack acceleration or jumping which can precipitate fracture. The presence of time-dependence, and consequent loss peaks, leads to rising curves but, in general, does not lead to maxima in the $K_c - \dot{a}$ curves. It is certainly possible that rate-dependent processes can cease to operate, as for example in crazing, thus leading to a decrease in energy absorption at high rates, but there is little direct evidence of this in the literature.

Instabilities are observed in PMMA as shown in Fig. 7.11, where the single edge notch K_c values are used to define accurately the instability condition in double torsion. Because the crack is controlled by visco-elasticity, the $K_c - \dot{a}$ curves in Fig. 7.11 can be shifted along the rate axis to give quite a good master curve with the appropriate energy for the β process in PMMA [6]. The instability speeds clearly do not follow this relationship and do not occur at any loss peaks.

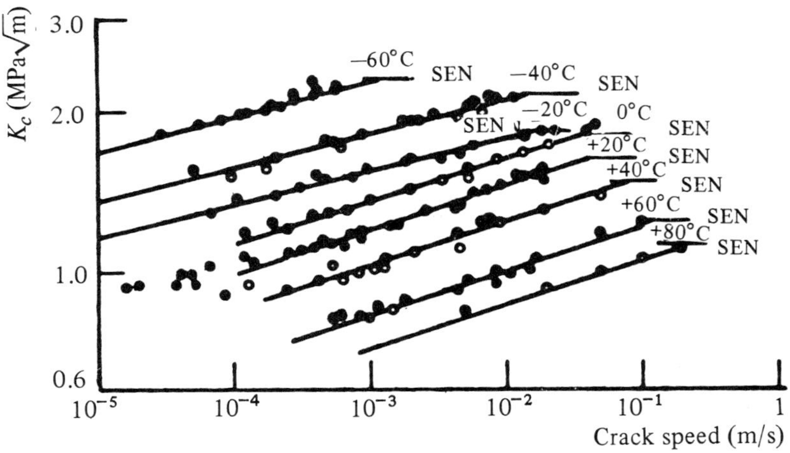

Fig. 7.11 K_c as a function of crack speed at various temperatures – stable crack growth in DT tests (from reference [6]) – log scales

A possible explanation [6] is that the heat generated in the crack tip zone is not dissipated at the higher speeds so that, for a fixed δ_c criterion, the thermal softening will result in a falling K_c with increasing $\dot{a}$ and thus a maximum will occur. The maximum temperature generated in a line zone of thickness b by an energy dissipation of G_c is:

$$\Delta\tilde{T} = \frac{G_c}{\rho\, C\, b}$$

where ρ is the density, and C the specific heat. For a strip with heat generated at a fixed rate, we may determine the temperature rise at a time t from [8]:

$$\Delta T = \Delta\tilde{T}\left(1 - 4i^2\, erfc\sqrt{\frac{b^2\,\rho\, C}{16\, k\, t}}\right) \tag{7.11}$$

where k is the thermal conductivity, and i^2 *erfc* is the double integral of $(1 - erf)$, where *erf* is the error function. For crazes, b is of the order of δ_c and thus small, and we may use the limit of this equation as $b \to 0$, i.e.

$$\Delta T \to \frac{G_c}{\sqrt{\pi\,\rho\, C\, k}}\frac{1}{\sqrt{t}} \tag{7.12}$$

Since $\dot{a} \propto c/t \propto K_c^2/t$ and $G_c \propto K_c^2$, then $\Delta T \propto K_c\sqrt{\dot{a}}$ and on differentiation, we have:

$$\frac{dT}{d\dot{a}} = (T - T_0)\left(\frac{1}{K_c}\frac{dK_c}{d\dot{a}} + \frac{1}{2\dot{a}}\right) \tag{7.13}$$

where T_0 is the original temperature. The temperature dependence of K_c may be expressed via an activation energy, H, so that:

$$K_c \propto \dot{a}^n e^{n\frac{H}{R}\left(\frac{1}{T}-\frac{1}{T_0}\right)}$$

and hence:
$$\frac{1}{K_c}\frac{dK_c}{d\dot{a}} = n\left(\frac{1}{\dot{a}} - \frac{H}{R\,T^2}\frac{dT}{d\dot{a}}\right) \tag{7.14}$$

where R is the gas constant. The condition of $dK_c/d\dot{a}=0$ in equation (7.14) can be achieved with $n=0$ but also when:

$$\frac{dT}{da} = \frac{R\,T^2}{H\,\dot{a}}$$

which, on substitution into equation (7.13), gives the condition that:

$$\Delta T \simeq \frac{2R}{H}\,T_0^{\,2}$$

so that we have finally:
$$K_c^{\,2}\,\dot{a}_c \propto T_0^{\,4} \tag{7.15}$$

at instability. A graph of the instability data taken from Fig. 7.11 plotted in this form is shown in Fig. 7.12 and the linearity confirms the nature of the mechanism.

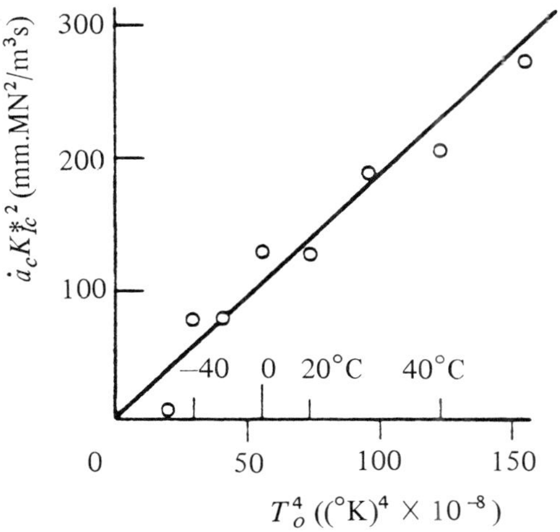

Fig. 7.12 Instability crack speeds for PMMA as a function of temperature

A critical time for these isothermal-adiabatic transitions may be deduced from equation (7.12), i.e.

$$t_c = \left(\frac{H}{2R}\right)^2 \frac{1}{\pi\,\rho\,C\,k}\frac{G_c^{\,2}}{T_0^{\,4}}$$

For glassy polymers: $\left(\frac{H}{2R}\right)^2 \frac{1}{\pi \rho C k} \simeq 10 \left(\frac{m^2\,{}^\circ C^2}{J}\right)^2.s$

and if $G_c \simeq 10^3$ J/m^2, then: $t_c \simeq \frac{1}{10}\left(\frac{100}{T_0}\right)^4.s$

so that at 20°C, $t_c \simeq 1$ ms, while at -60°C this has increased to 5 ms.

7.2.4 *Crack propagation times*

As outlined in Section 5.7, the time, t_p, taken in propagating a crack from some initial value a_0 under constant load to a length $a_0 + \Delta a$ may be found simply by integrating the inverse crack speed:

$$t_p = \int_{a_0}^{a_0+\Delta a} \frac{da}{\dot{a}} \tag{7.16}$$

The various growth laws all lead to relationships of the form:

$$\dot{a} \propto a^{\xi}, \text{ where } \xi \text{ is a constant and } \gg 0$$

which, on integration, gives:

$$t_p \propto \frac{1}{(1-\xi)} [(a_0 + \Delta a)^{1-\xi} - a_0{}^{1-\xi}]$$

If we now take t_p to be the time to achieve $\Delta a \gg a_0$, then, since $\xi \gg 0$:

$$t_p \propto \frac{1}{(\xi - 1)\, a_0{}^{\xi-1}}$$

giving: $$t_p = \frac{1}{(\xi-1)}\left(\frac{a_0}{\dot{a}_i}\right)$$

where $\dot{a}_i$ is the crack velocity at initiation. For small values of m and n and using a constant δ_c criterion, it was shown in Section 5.7 that:

$$t_i \simeq \left(\frac{\delta_c}{\delta_0}\right)^{\frac{1}{m+n}} \tau_0 \text{ and } \dot{a} \simeq \frac{1}{\sqrt{5}}\left(\frac{\delta_0}{\delta_c}\right)^{\frac{1}{m+n}} \frac{c_0}{\tau_0}$$

so that at initiation:

$$\dot{a}_i \simeq \frac{1}{\sqrt{5}} \frac{c_0}{t_i} \text{ where } c_0 = \frac{\pi}{8} \frac{K_0{}^2}{p_0{}^2}$$

and since $\xi = (1-2m)/(m+n)$, in this case we have:

$$t_p = \left(\frac{m+n}{1-3m-n}\right)\sqrt{5}\left(\frac{a_0}{c_0}\right) t_i \tag{7.17}$$

Thus, the propagation time is a factor times the initiation time and this is determined by the ratio of original crack size to the characteristic zone size, c_0. When log K_c versus log t data is used, as in Fig. 7.3, it would be expected that the initiation and final failure points would be displaced a fixed distance along the log time axis and this is essentially what is observed in the data. The displacement on this scale will be:

$$\log\left(1+\left(\frac{m+n}{1-3m-n}\right)\sqrt{5}\,\frac{a_0}{c_0}\right)$$

$$\simeq \log\left(1+0.14\frac{a_0}{c_0}\right) \text{ in this case}$$

For the higher points, the displacement $\simeq 0.5$, while for the lower ones it is much less and $\simeq 0.2$, so that:

$$\frac{a_0}{c_0} \simeq 15 \quad \text{for the upper points}$$

and

$$\frac{a_0}{c_0} \simeq 4 \quad \text{for the lower points}$$

Since $a_0 = 4$ mm, we have zone sizes of 0.3 mm and 1 mm, respectively, which, with the appropriate K_c values, suggest p_c values of approximately 110 MPa and 30 MPa.

For typical glassy materials, inherent flaws appear to be around 50 μm and if we take $c_0 \simeq 200$ μm with $m = n = 0.05$, we have:

$$\frac{t_i}{t_p} \simeq 14$$

i.e. the initiation phase would be expected to be about an order of magnitude greater than that for propagation. The stress-time dependence in such a test would, of course, take the form $p \propto t^{-(m+n)/2}$, as in the initiation phase.

7.3 Environmental crack and zone growth

The term environmental stress cracking (or crazing) is usually applied to conditions in which crack or craze growth is promoted by the presence of an environment and a stress. In most cases, there is no chemical interaction between the environment and the polymer, and the diffusion of the environment into the bulk polymer is usually very slow. Stressed components, however, show rapid crack growth in circumstances when it does not occur without the environment and there has been much debate about the mechanism involved. The most likely mechanism appears to be that the presence of the stress and a flaw or crack results in a local microvoided zone or craze at the crack tip and that it is the interaction within this zone which gives the enhanced crack growth. The porous zone has a very high area to volume ratio and bulk diffusion over very short periods can completely plasticise the small ligaments. The craze stress, p_c, is thus reduced by some factor, say α, so that the new G_c becomes:

$$G_c = \delta_c \, \alpha \, p_c \tag{7.18}$$

If δ_c remains the same, then G_c also reduces by α and crack growth occurs at a reduced G_c (or K_c) level. The visco-elastic nature of p_c may change but that of E, if K_c is used, will not. It should be noted that this mechanism promotes an already formed dry craze and hence results in growth in conditions where it would not otherwise occur. The effect is shown schematically in Fig. 7.13.

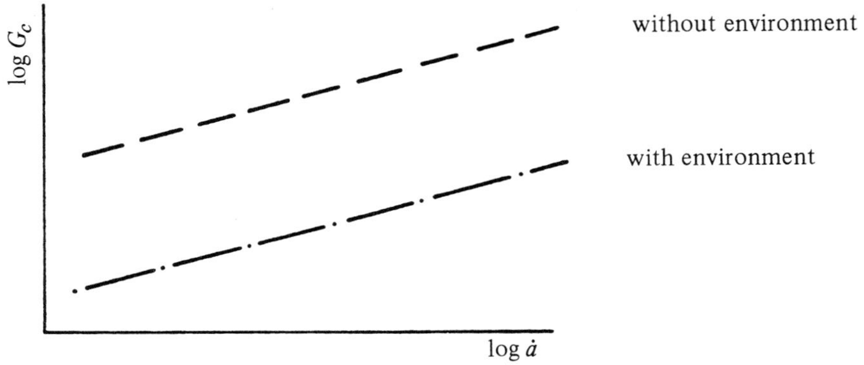

Fig. 7.13 Environmental effects on a G_c–$\dot{a}$ curve

If this were the only mechanism involved, there would be little to add since the description is entirely in terms of a lower p_c value. However, the environment must diffuse into the porous zone in order to effect the plasticisation. It is helpful to consider this as being described in terms of Darcy's law for diffusion in porous media:

$$\dot{u} = \frac{A}{12\eta}\left[-\frac{dp}{dx}\right] \tag{7.19}$$

where $\dot{u}$ is the flow velocity, A is the pore area, η is the viscosity, and $[-d\bar{p}/dx]$ is the driving pressure gradient. [Equation (7.19) may, in fact, be derived from the usual equations for viscous flow].

A crack or zone will grow at the rate determined by the slowest controlling process and when this is the flow, we have flow-controlled growth.

7.3.1 *Zone growth in environments*

For a craze in which the environment has penetrated a distance x, as shown in Fig. 7.14, equilibrium requires that there is an unplasticised zone of length r_0 as shown. The total stress intensity may be written using equation (4.20) for the two

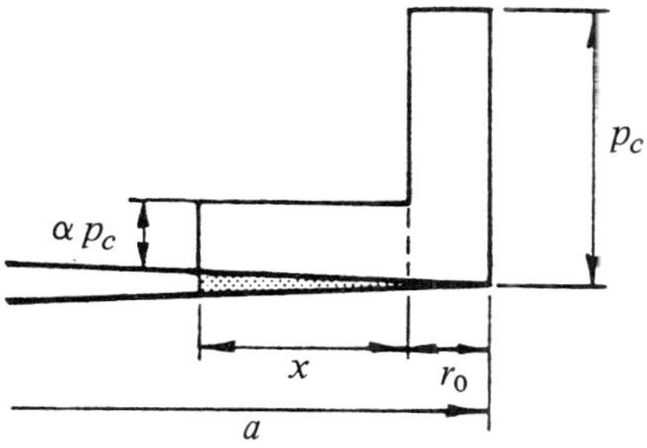

Fig. 7.14 Stresses on a plasticised zone

zones:

$$K=\sqrt{\pi\, a}\,\frac{2}{\pi}\left[\int_{1-(x+r_0)/a}^{1-r_0/a}\frac{\alpha\, p_c\, d\xi}{\sqrt{1-\xi^2}}+\int_{1-r_0/a}^{1}\frac{p_c\, d\xi}{\sqrt{1-\xi^2}}\right]$$

i.e.

$$K=\frac{2}{\pi}\,p_c\sqrt{\pi\,(a_0+x+r_0)}\left[\alpha\cos^{-1}\left(\frac{a_0}{a_0+x+r_0}\right)\right.$$
$$\left.+(1-\alpha)\cos^{-1}\left(\frac{a_0+x}{a_0+x+r_0}\right)\right] \quad (7.20)$$

where $a_0=a-x-r_0$, the true crack length. Thus, for a given value of K, the zone size can take up any value between the extremes of $x=0$ and $r_0=0$. For the usual case of $x, r_0 \ll a_0$, equation (7.20) reduces to:

$$K=2\frac{\overline{2}}{\pi}\,p_c\,[\alpha\sqrt{x+r_0}+(1-\alpha)\sqrt{r_0}] \quad (7.21)$$

and the limiting values are:

$$\frac{\pi}{8}\frac{K^2}{p_c^{\,2}}<x+r_0<\frac{\pi}{8}\frac{K^2}{\alpha^2\, p_c^{\,2}} \quad (7.22)$$

Within this range, x is flow-controlled according to equation (7.19) and the growth laws may be deduced by considering the generalisation of this relationship to two dimensions:

$$\dot{u}=\frac{A}{12\eta}\left(-\frac{\partial\bar{p}}{\partial x}\right) \text{ and } \dot{v}=\frac{A}{12\eta}\left(-\frac{\partial\bar{p}}{\partial y}\right) \quad (7.23)$$

Continuity of fluid flow requires that:

$$\frac{\partial\dot{u}}{\partial x}+\frac{\partial\dot{v}}{\partial y}=0$$

so that the pressure distribution must satisfy the Laplace equation:

$$\nabla^2 \bar{p} = \frac{\partial^2 \bar{p}}{\partial x^2} + \frac{\partial^2 \bar{p}}{\partial y^2} = 0 \tag{7.24}$$

The flow-controlled growth of a craze may then be predicted by first determining the appropriate $\bar{p}$ distribution from equation (7.24) and then using this with equations (7.23) to determine the growth rates. Some special cases are of practical importance.

(i) *One-dimensional flow*

This is the most important since it models a small zone at a wide crack tip and we have:

$$\frac{d^2 \bar{p}}{dx^2} = 0$$

and $\bar{p} = A\,x + B$. For a craze of current length x_0, $\bar{p} = \bar{p}_0$ at $x=0$ and $\bar{p}=0$ at $x = x_0$, so that:

$$\bar{p} = \bar{p}_0 \left(1 - \frac{x}{x_0}\right) \text{ and } \frac{d\bar{p}}{dx} = -\frac{\bar{p}_0}{x_0}$$

The velocity $\dot{u} = dx_0/dt$ and we have:

$$\frac{dx_0}{dt} = \frac{A\,\bar{p}_0}{12\eta\, x_0}$$

i.e.

$$x_0 = \sqrt{\frac{A\,\bar{p}_0}{6\eta}} \sqrt{t} \tag{7.25}$$

Thus, flow-controlled craze growth would be expected to have a $t^{1/2}$ dependence and this has been observed in several polymer-liquid systems. Fig. 7.15 shows a $\log x_0$ versus $\log t$ graph for PMMA in methanol [1], in which the slope of 0.5 is well-defined. The arrest condition is also shown for different K values and from these it is possible to compute the fully plasticised craze stress, p_c. The craze

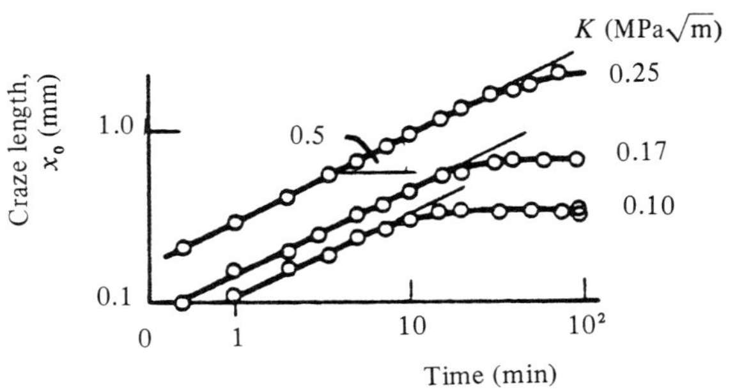

Fig. 7.15 Craze growth in an environment – PMMA in methanol (from reference [1])

length is frequently substantial compared with the crack length so that the complete solution from equation (7.20) for $r_0 = 0$ is required, i.e.

$$\frac{\pi}{2}\frac{p}{\alpha p_c} = \cos^{-1}\left(\frac{a_0}{a_0 + x}\right) \tag{7.26}$$

where $K = p\sqrt{\pi\,(a_0 + x)}$. Fig. 7.16 shows craze arrest data for PMMA in ethanol plotted in this way from which αp_c may be found.

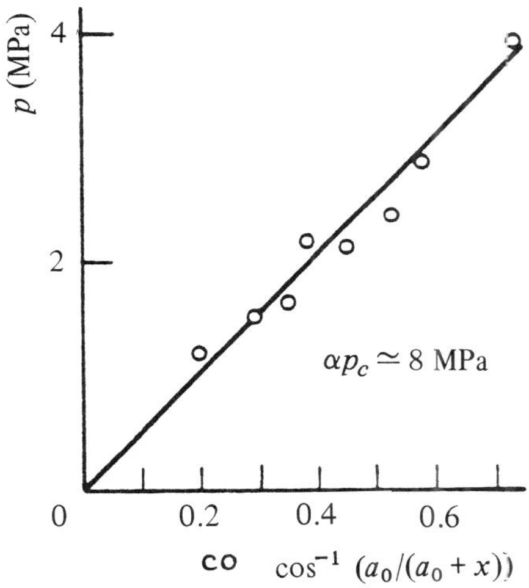

Fig. 7.16 Craze arrest lengths as a function of applied stress according to equation (7.26) – PMMA in ethanol

(ii) *Axial symmetry*

For crazes growing either inwards from circumferential notches or outwards from internal penny-shaped cracks, the pressure distribution must be axisymmetric so that the Laplace equation becomes:

$$\frac{d^2\bar{p}}{dr^2} + \frac{1}{r}\frac{d\bar{p}}{dr} = 0 \text{ and } \bar{p} = A \ln r + B$$

For the inward growing case, we have:

$$\bar{p} = \bar{p}_0 \text{ at } r = a \text{ and } \bar{p} = 0 \text{ at } r = a - x_0$$

so that the pressure gradient at the craze tip is given by:

$$\frac{dp}{dr} = -\frac{\bar{p}_0}{(a - x_0)\ln(1 - x_0/a)}$$

and the growth law is:

$$\frac{dx_0}{dt} = -\frac{A\,\bar{p}_0}{12\eta\,(a-x_0)\ln(1-x_0/a)}$$

which may be integrated to give:

$$\phi\, x_0 = \sqrt{\frac{A\,\bar{p}_0}{6\eta}}\sqrt{t}$$

$$(7.27)$$

where: $$\phi^2 = \frac{(1-x_0/a)^2\,[\ln(1-x_0/a)^2-1]+1}{2(x_0/a)^2}$$

is the necessary correction factor for circular geometry. Fig. 7.17 shows data for PMMA in methanol plotted as x_0/a versus $\sqrt{t}$, both with and without ϕ, and clearly the correction is effective. The correction for outward growing crazes is given by replacing $(1-x_0/a)$ with $(1+x_0/a)$ in equations (7.27).

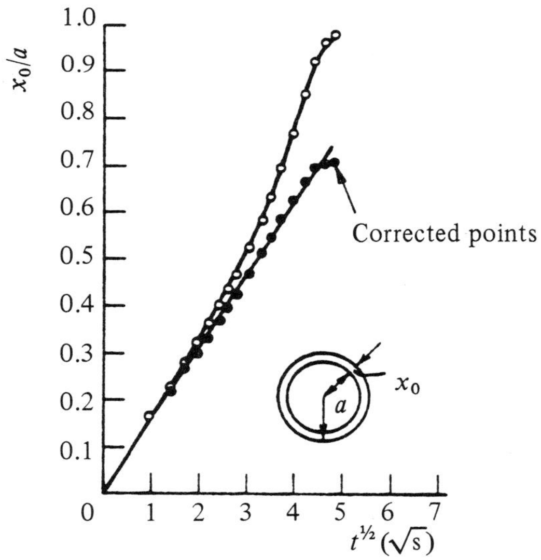

Fig. 7.17 Axisymmetric craze growth – PMMA in methanol (from reference [8])

(iii) *Growth at a free surface*

When crazes grow near a free surface, the flow can be augmented by flow inwards directly through the surface. At the junction of the craze front and the surface, there are the conflicting boundary conditions of $\bar{p}=\bar{p}_0$ and $\bar{p}=0$ which would lead to infinite rates. These do not occur and, in general, there is a threshold in terms of loading before this type of flow occurs, probably when a surface skin breaks, and, in addition, a limiting and equal flow condition occurs along and normal to the surface.

Fig. 7.18 shows a diagram of a craze growing along a surface from an initial

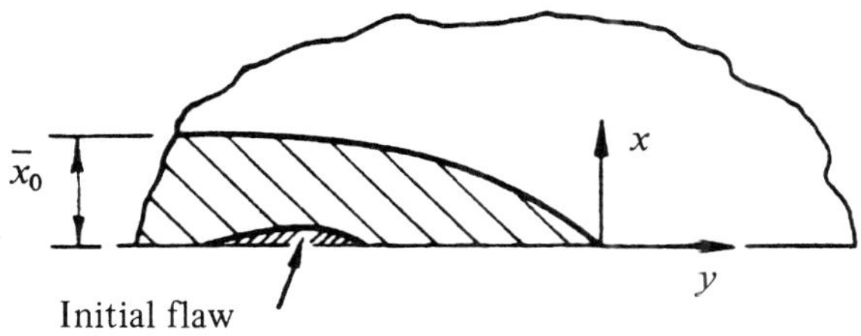

Fig. 7.18 Craze growth at a free surface

flaw which has a depth $\bar{x}_0$ remote from the front. A pressure distribution of the form:

$$\bar{p}=\bar{p}_0\left(1-\frac{x}{\bar{x}_0}-e^{y/\bar{x}_0}\cos\frac{x}{\bar{x}_0}\right) \tag{7.28}$$

satisfies both the Laplace equation and the boundary conditions. For large negative y, the one-dimensional result is recovered and the craze front profile is given when $\bar{p}=0$, i.e.

$$\frac{x_0}{\bar{x}_0}=1-e^{y_0/\bar{x}_0}\cos\frac{x_0}{\bar{x}_0}$$

and the velocities are:

$$\frac{dx_0}{dt}=\frac{A\,\bar{p}}{12\eta}\frac{1}{\bar{x}_0}\left[1-\tan\frac{x_0}{\bar{x}_0}\left(1-\frac{x_0}{\bar{x}_0}\right)\right]$$

$$\frac{dy_0}{dt}=\frac{A\,\bar{p}}{12\eta}\frac{1}{\bar{x}_0}\left[1-\frac{x_0}{\bar{x}_0}\right] \tag{7.29}$$

$\bar{x}_0$ is the arrest length and is given by:

$$\bar{x}_0=\frac{\pi}{8}\frac{K^2}{\alpha\,p_c^{\,2}}$$

and at the surface, we have:

$$\frac{dx_0}{dt}=\frac{dy_0}{dt}=\frac{A\,\bar{p}}{12\eta}\frac{1}{\bar{x}_0} \tag{7.30}$$

i.e. the pressure gradient is determined by the arrest craze length.

(iv) *Growth from an edge notch*

A special case of some interest is growth from an edge notch, as shown in Fig. 7.19. If $B/2 < \bar{x}_0$, then the two surface flows meet and the pressure distribution becomes:

$$\bar{p} = \bar{p}_0 \left[1 - e^{\frac{\pi}{2}\frac{y}{B}} \cos\frac{\pi}{2}\frac{x}{B} \right]$$

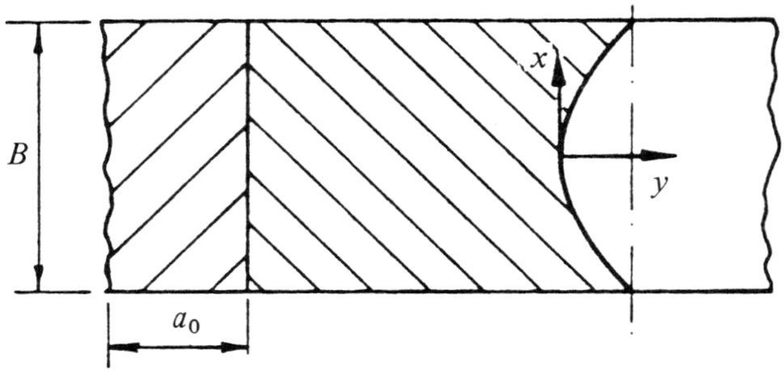

Fig. 7.19 Side flow in an edge notch specimen

and the profile has the form:

$$e^{-\pi y_0/2B} = \cos\frac{\pi x_0}{2B}$$

i.e. concave, as shown in Fig. 7.19. The velocity in the y direction then becomes:

$$\frac{dy_0}{dt} = \frac{A\,\bar{p}_0}{12\eta}\frac{\pi}{2B} \tag{7.31}$$

Fig. 7.20 shows data for PMMA in methanol in which higher loadings have been

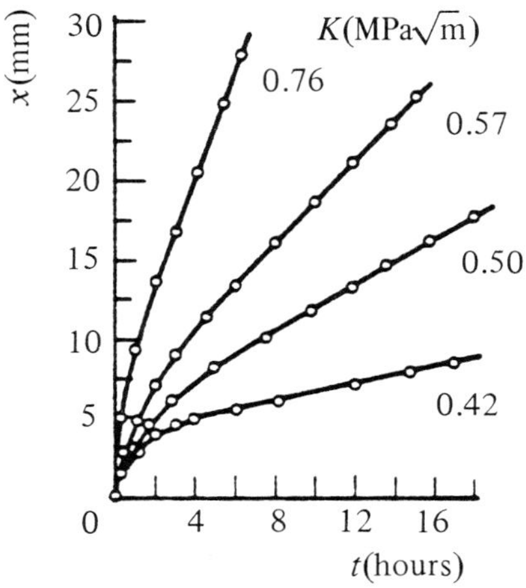

Fig. 7.20 Constant rate craze growth – PMMA in methanol (from reference [9])

applied and it is clear that constant speed growth occurred and suggests augmented side flow. Observations on the crazes confirm this by showing concave fronts for the constant speed cases and convex for $t^{1/2}$ growth. In addition, shielding the sides of the specimens with glass plates [9] prevents constant speed growth.

7.3.2 *K-dependence and driving forces*

It is clear from Figs. 7.15 and 7.20 that, although the time-dependence of flow-controlled growth may be predicted from the flow model, the parameter $A\,\bar{p}_0$, the driving force, is dependent on loading conditions and requires separate consideration. It has been found [9] that the initial stress intensity factor, K, is the controlling variable and this is illustrated in Fig. 7.21, where $x_0/\sqrt{t}$ is shown to be a linear function of K for end flow growth, while in Fig. 7.22 it is demonstrated that constant speed growth gives $dy_0/dt \propto K^2$, both for PMMA in

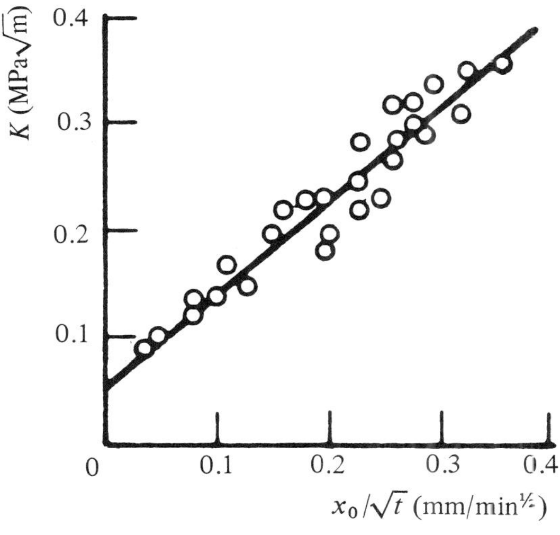

Fig. 7.21 The dependence of craze growth on stress intensity factor, K (from reference [1])

methanol. This has been shown to be so for several polymer–liquid combinations [10] and implies that:

$$A\,\bar{p}_0 \propto K^2$$

for low values of K. It does appear, however, that above some limiting values of K, $A\bar{p}_0$ becomes constant [9] and this is consistent with direct observations of craze void contents [10] which give constant values of about 50% for fully developed crazes.

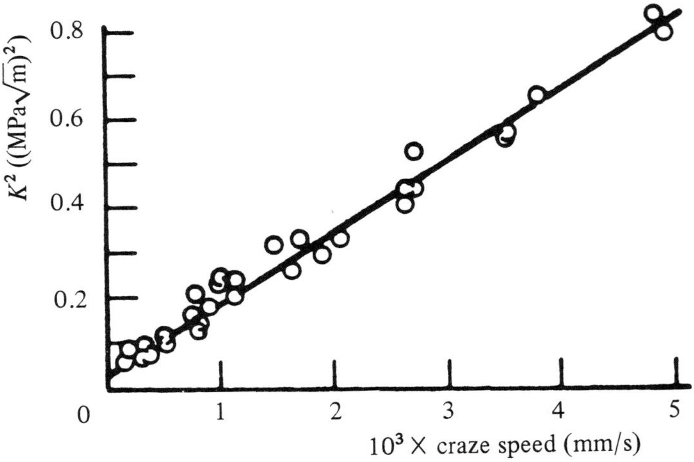

Fig. 7.22 Dependence of constant speed craze growth on K – PMMA in methanol (from reference [1])

A simple model of the initial phase of void growth is illustrated in Fig. 7.23, in which void initiation sites as a spacing of l_0 are envisaged. The strain in the zone is given by δ/d_0 where δ is the local displacement and d_0 the original zone thickness, and if constant volume deformation is assumed the ligament thickness d' becomes:

$$d' = \frac{l_0}{\sqrt{1+\delta/d_0}}$$

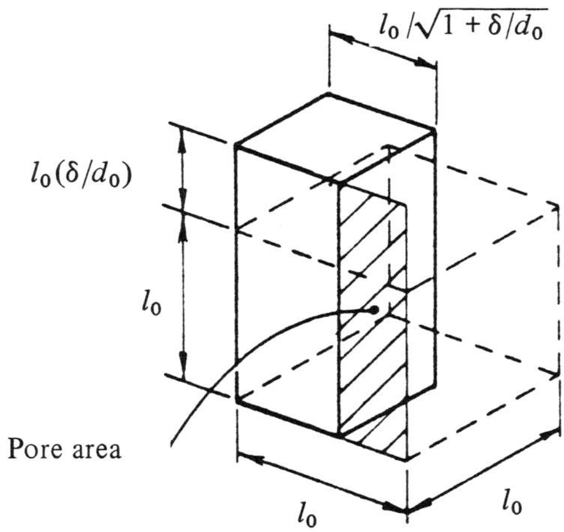

Fig. 7.23 The generation of a craze ligament

and the pore area: $$A = l_0^2\left(1+\frac{\delta}{d_0}\right)\left(1-\frac{1}{\sqrt{1+\delta/d_0}}\right)$$

The strain may be expressed in terms of the volume fraction of fibres, v_f:

$$v_f = \left(1+\frac{\delta}{d_0}\right)^{-1}$$

and since v_f becomes about 0.5, then δ/d_0 increased to about unity before new material is drawn into the craze to maintain v_f constant, presumably as a consequence of ligament work-hardening. Within this range of δ/d_0, A may be approximated to within 20% by:

$$A \simeq \frac{l_0^2}{2}\frac{\delta}{d_0}$$

δ here will be that of the initial craze so that we have finally:

$$A \simeq \frac{l_0^2}{2d_0}\frac{K^2}{p_c E} \tag{7.32}$$

The pressure term $\bar{p}_0$ is made up of a contribution from the external fluid pressure (atmospheric, usually) because of the vacuum in the newly formed voids, say p_a, and some contribution from the surface tension of the fluid. This latter could be expressed as a capillary pressure arising from the meniscus generated at the fluid surface and we have the Laplace equation for this:

$$p = \gamma_L\left(\frac{1}{R_1}+\frac{1}{R_2}\right) \tag{7.33}$$

where γ_L is the fluid surface tension, and R_1 and R_2 are the principal radii of curvature of the meniscus. The radii are determined by the pore geometry and the contact angle, θ, at the fluid–solid interface. For a circular pore of radius r, for example, we have $R_1 = R_2 = r/\cos\theta$ and $p = 2\gamma_L \cos\theta/r$, and for a narrow slit of depth d, $p = 2\gamma_L \cos\theta/d$ also.

The concept of a contact angle implies an equilibrium of forces at the interface, as shown in Fig. 7.24, such that:

$$\gamma_S = \gamma_{SL} + \gamma_L \cos\theta$$

where γ_L is the surface tension of the solid, and γ_{SL} is that for the solid–fluid interface. It has been shown [11] that γ_{SL} can be estimated using a simple energy averaging concept such that:

$$\gamma_{SL} \simeq \gamma_S + \gamma_L - 2\sqrt{\gamma_S}\sqrt{\gamma_L}$$

giving an approximation for $\cos\theta$ of:

$$\cos\theta \simeq 2\sqrt{\gamma_S/\gamma_L} - 1 \tag{7.34}$$

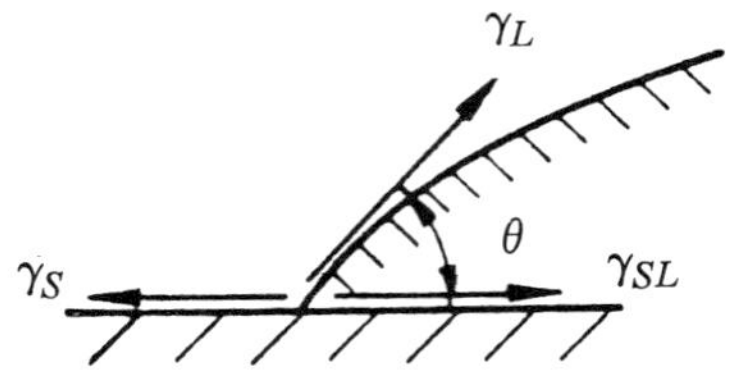

Fig. 7.24 The solid–fluid interface with a contact angle

For $\frac{1}{4} \leqslant \gamma_S/\gamma_L \leqslant 1$, $\cos\theta$ is positive $(0 \leqslant \theta \leqslant \pi/2)$ and so is p, while for $0 < \gamma_S/\gamma_L < \frac{1}{4}$, $\cos\theta$ and p are negative. Thus, for contact angles between 0 and $\pi/2$, the capillary pressure would assist flow, while for those greater than $\pi/2$ it would act as a resistance.

An additional effect can also be envisaged for $\gamma_S/\gamma_L > 1$ since $\theta = 0$ and it is not possible to establish surface equilibrium. In this case, the capillary pressure is augmented by a surface spreading force, S, given by:

$$S = \gamma_S - \gamma_{SL} - \gamma_L \simeq 2(\sqrt{\gamma_S}\sqrt{\gamma_L} - \gamma_L) \tag{7.35}$$

We may now write the following expression:

$$\bar{p}_0 = p_a + \frac{2\gamma_L}{d}(2\sqrt{\gamma_S/\gamma_L} - 1)$$

which includes both the positive contact angle cases and those for surface wetting when $\theta = 0$. For most polymers, $\gamma_S \simeq 0.035$ N/m and typical fluids used vary from those which wet, such as methanol with $\gamma_L = 0.02$ N/m and water which does not as $\gamma_L = 0.07$ N/m. The parameter $\gamma_L \cos\theta$ varies very little, in fact only from 0.033 N/m to 0.029 N/m, while S varies from 0.013 N/m to -0.041 N/m.

The choice of the appropriate value of d for the craze presents some difficulties. Direct observations suggest that the fibril diameters are about 10^{-8} m [12] and, since γ_L is about 0.02 N/m for most of the liquids studied, we have a capillary pressure of approximately 40 bar. Taking 10^{-8} m as the characteristic dimension of the craze, we may compute $A\,\bar{p}_0$ approximately and this becomes 2×10^{-10} N. It is possible to estimate $A\,\bar{p}_0$ from $x_0/\sqrt{t}$ and dy_0/dt data and using $\eta = 0.6 \times 10^{-3}$ Ns/m^2, both types of data give 5×10^{-12} N for PMMA in methanol, i.e. two orders of magnitude lower. It would appear that these very high capillary pressures do not arise in practice, probably because the craze structure does not form continuous tubes of the pore size but that the fibrils act more as inhibitors to flow. The dimension more appropriate to the meniscus is the craze height, d_0, which is of the order of 10^{-6} m giving capillary pressures of about 0.4 bar and thus of a similar magnitude to atmospheric pressure. Indeed, whether a meniscus forms at all is open to question and if it did not, then the driving force from surface tension would arise solely from the spreading force S as suggested by Shanahan & Schultz [13] and is consistent with a pronounced difference between wetting and non-wetting liquids. The form of the driving force expression using S alone would be:

$$A\,\bar{p}_0 = \frac{l_0^{\,2}}{2d_0}\frac{K^2}{p_c\,E}\left[p_a + \frac{2S}{d_0}\right] \tag{7.36}$$

where S can be estimated from equation (7.35), and equation (7.25) may be written as:

$$\frac{x_0^{\,2}}{t} = \frac{l_0^{\,2}}{12d_0\eta}\frac{K^2}{p_c\,E}\left(p_a + \frac{2S}{d_0}\right) \tag{7.37}$$

From Fig. 7.21 for PMMA in methanol, we have that $x_0^{\,2}/K^2t \simeq 1.7\times10^{-8}$ m/(MPa)^{2}s and using $\eta = 0.6\times10^{-3}$ Ns/m^2, $p_c = 100$ MPa and $E = 3$ GPa gives:

$$\frac{l_0^{\,2}}{d_0}\left(p_a + \frac{2S}{d_0}\right) = 3.7\times10^{-5}\ \text{N/m}$$

The upper limit for K is 0.6 MPa$\sqrt{\text{m}}$ when $\delta \simeq d_0$ so that $d_0 \simeq 1.6$ μm and if $S = 0.02$ N/m and $p_a = 10^5$ N/m^2, we have that $l_0 \simeq 0.7\times10^{-8}$ m which is in accord with direct observations.

It is of interest to note that degassed methanol gives increased craze speeds by about a factor of 2.5 [14]. Gas bubbles are observed with normal methanol and it appears that the low pressure at the craze tip results in dissolved air coming out of solution and providing a back pressure. An estimate of this effect from the vapour pressure would be 0.25 bar and, since removing this increases the effective pressure by a factor of 2.5, a pressure of 0.4 bar is implied, confirming that the pressure is of the order of 1 bar.

The validity of equation (7.36) is further demonstrated in Fig. 7.25, where $x_0^{\,2}/K^2t$ is shown as a function of η^{-1} for PMMA in various liquids and mixtures. All those given have rather similar S values and the change in growth rate can be seen to be largely controlled by the change in η. In these systems, there was a well defined initiation value of K, K_m, and this is shown as a function of αp_c in Fig. 7.26. The linear correlation suggests a constant zone size of 10 μm and the criterion for initiation is probably the plasticisation of the original craze which would be of this magnitude.

Also shown in Fig. 7.26 are data for water, paraffin, ethylene glycol and white spirit which follow the same craze growth dependence. Their growth rates were much slower, however, and all have negative S values so that such an effect would be expected. Water, for example, gave $x_0^{\,2}/K^2t \simeq 10^{-9}$ m/(MPa)^{2}s and, since $S \simeq -0.04$ N/m, we may use equation (7.36) and $\eta = 10^{-3}$ Ns/m^2 with the previous values of l_0 and d_0. The result is 0.4×10^{-9} m/(MPa)^{2}s, which agrees quite well with the measured value.

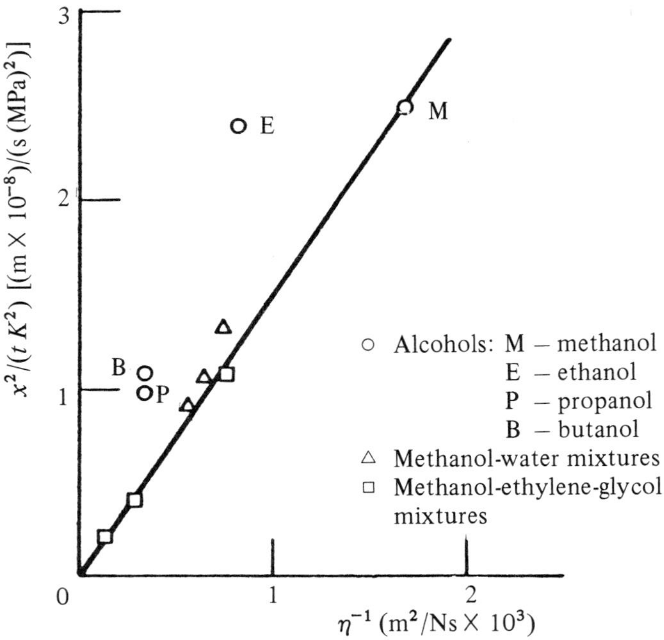

Fig. 7.25 Flow parameter as a function of inverse viscosity

7.3.3 *Flow-controlled crack growth*

Flow-controlled zone growth will, as in the visco-elastic case, result in an increase in δ and thus J at the crack tip until fracture occurs. If J increases rapidly with a, then immediate failure can occur, but for slowly increasing J it is possible to envisage a stable situation where the crack speed is the same as the zone growth rate which, in turn, is that of the fluid flow. This flow-controlled crack growth will be given by:

$$\dot{a} = \frac{l_0^2}{2d_0} \frac{\delta}{x} \frac{1}{12\eta} \left(p_a + \frac{2S}{d_0} \right) \tag{7.38}$$

For a fully developed crack tip zone, $\delta \simeq K^2/p_c E$ and $x \simeq \pi/8\,(K/\alpha p_c)^2$ so that the crack speed is independent of K and given by:

$$\dot{a} = \frac{\alpha^2}{3\pi} \frac{l_0^2}{d_0} \frac{p_c}{\eta E} \left(p_a + \frac{2S}{d_0} \right) \tag{7.39}$$

Under some circumstances, the pressure drop does not occur over the zone length and x may be replaced by some fraction of the thickness as in the craze growth case given in equation (7.31) or perhaps some other characteristic length. In either case, x becomes a constant, say L_0, and $\dot{a}$ now depends on K in the

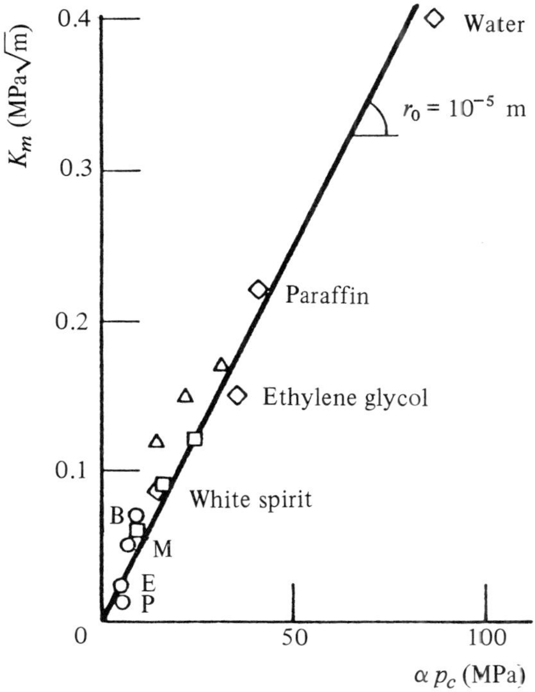

Fig. 7.26 Initiation K for craze growth as a function of craze stress

form:

$$\dot{a} = \frac{1}{24} \frac{l_0^2}{d_0 L_0} \frac{K^2}{p_c E \eta}\left(p_a + \frac{2S}{d_0}\right) \tag{7.40}$$

so that $K \propto \dot{a}^{1/2}$. The type of behaviour expected in this region is shown in Fig. 7.27 where a log K versus log $\dot{a}$ diagram is given. The upper and lower lines are for non-flow-controlled growth without and with the environment present, respectively. The lines for each type of flow-controlled growth will be at constant $\dot{a}$ or of slope $\frac{1}{2}$ as shown which, in general, will be greater than that of non-flow controlled behaviour.

An example of a constant speed transition is given in Fig. 7.28 for ABS in butyl oleate [15]. A series of other liquids with viscosities in the range 10^{-4} to 5 $\mathrm{Ns/m^2}$ were also used, for which S varies very little, and the transition speeds were found to correlate very well with η^{-1} and a craze dimension of about 4×10^{-8} m is predicted from equation (7.39).

$\dot{a}^{1/2}$ transitions have been observed in polyethylenes, as shown in Fig. 7.29 for crack growth in a detergent solution [16]. The low speed growth has an $\dot{a}^{1/4}$ form, to be discussed in Section 7.4, but a transition to $\dot{a}^{1/2}$ is observed, particularly at the higher temperatures. The transition, shown by the broken line, moves to higher speeds at increased temperatures which would be expected

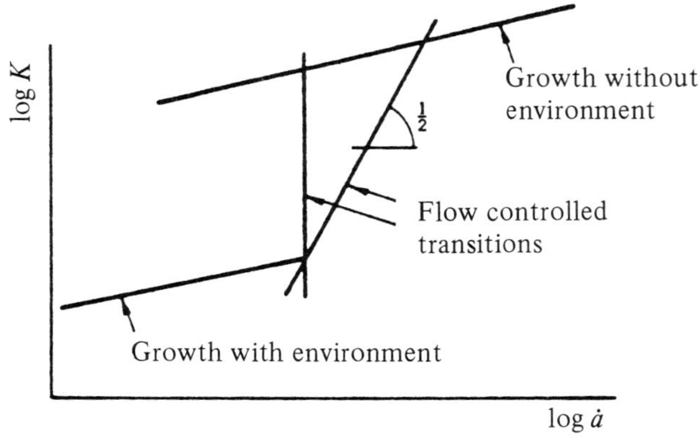

Fig. 7.27 Flow-controlled crack growth

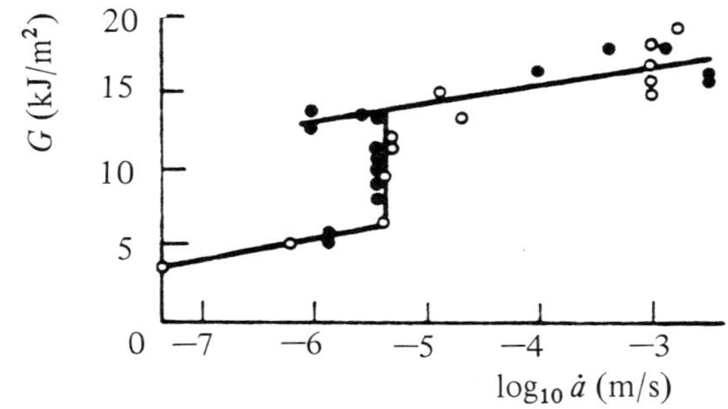

Fig. 7.28 Flow-controlled transition for ABS in butyl oleate (different points are two specimens) (from reference [15])

from the viscosity dependence. Fig. 7.30 shows the transition speeds plotted on a log-log basis with η and the expected slope of -1 is evident. The observed speeds are consistent with $l_0 \simeq 10^{-8}$ m with $\delta \simeq 30$ μm and $L_0 \simeq 1$ mm, although there is little supporting evidence for these fibril dimensions in polyethylenes. δ and L_0 are consistent with observed crack tip zone sizes but surface fibrils are around 1 μm in diameter. It seems unlikely that features on this scale play a part in flow control.

Some interesting support for the general ideas given here, and particularly the rôle of S, has been given by Shanahan & Schultz [13, 17, 18]. Fig. 7.31 shows data obtained on unnotched low density polyethylene samples tested to failure. They identify three zones of behaviour, of which the transition, zone two, is flow-controlled. Making the usual integration for an $\dot{a} \propto K^m$ relationship gives $p \propto t^{-1/m}$ and a line for $m=2$ has been added to Fig. 7.31 in zone two.

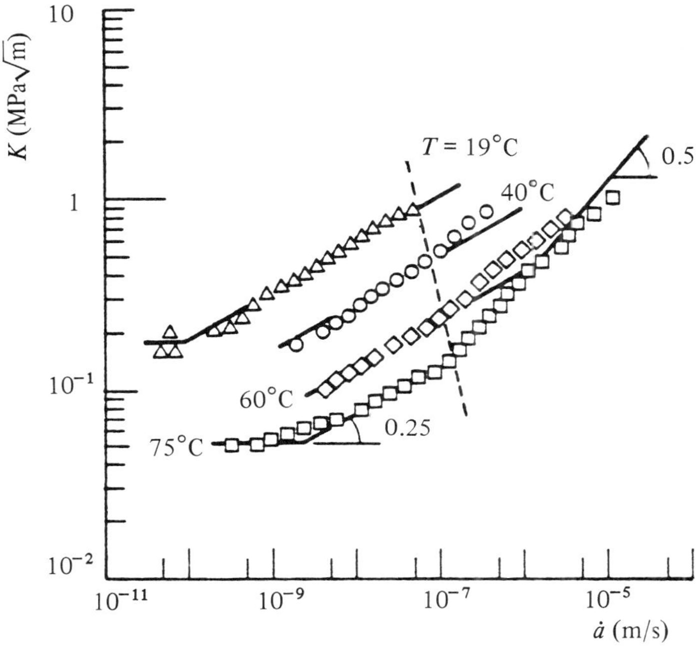

Fig. 7.29 Slow crack growth for PE in detergent solution at different temperatures (from reference [16])

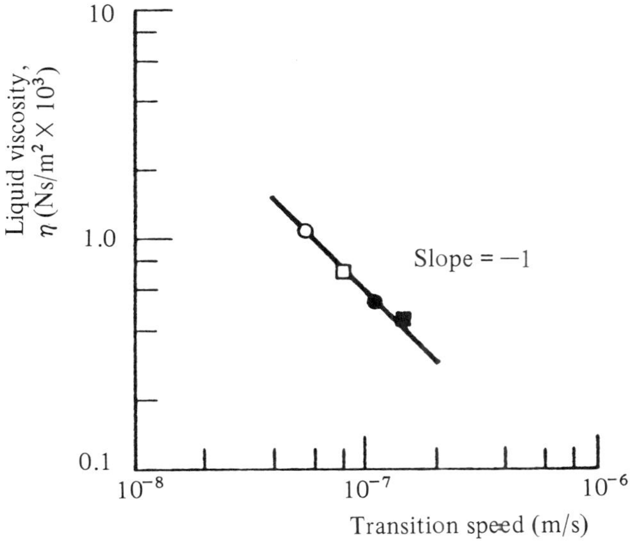

Fig. 7.30 Transition speed for $\dot{a}^{1/4} \rightarrow \dot{a}^{1/2}$ behaviour as a function of fluid viscosity (from reference [16])

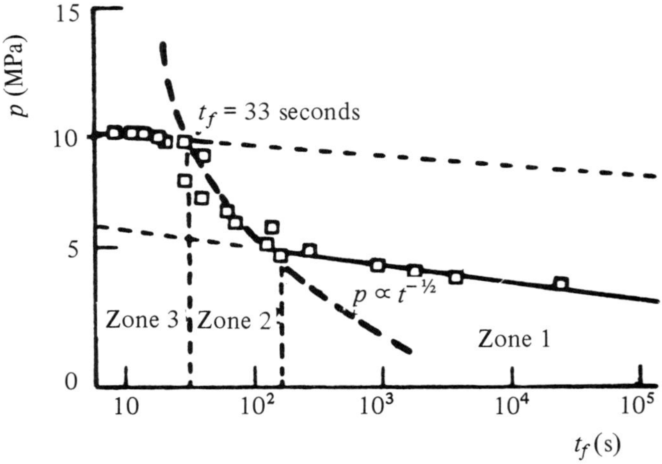

Fig. 7.31 Stress as a function of failure time – low density polyethylene in a silicon oil (from reference [18])

Equation (7.40) may be integrated to give:

$$\frac{\eta}{t_f} = \frac{\pi}{24} \frac{l_0{}^2 \, p^2}{d_0 \, L_0 \, p_c \, E \ln (a_f/a_0)} \left(p_a + \frac{2S}{d_0} \right) \tag{7.41}$$

where t_f is the failure time, a_0 is the initial flaw size, and a_f is the flaw size at failure. Fig. 7.32 shows η/t_f for the transition plotted versus S for a range of liquids and it is important to note that not only is the predicted linearity confirmed, but that $t_f \to \infty$ as $S \to \simeq -0.02$ N/m. From equation (7.41), we have that $d_0 \simeq 0.4$ μm, confirming that the order of the surface tension term is 1 bar.

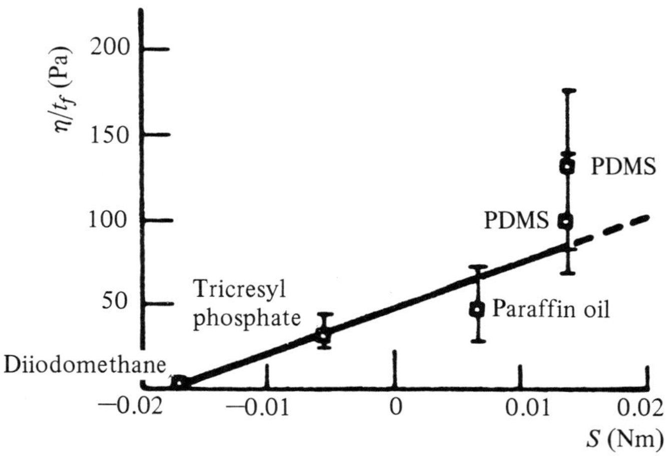

Fig. 7.32 Failure time as a function of viscosity and spreading factor (from reference [18])

7.4 Other controlling mechanisms

It was mentioned in Section 7.3.3 that slow crack growth in polyethylenes sometimes follows a $K \propto \dot{a}^{1/4}$ relationship. Such measurements are not easy to make since there is always the problem of ductility effects in the form of shear lips interacting and giving spurious time dependencies. It has been found, however, [16] that for specimens of sufficient size, three-point bending can overcome these problems and geometry-independent K–$\dot{a}$ lines may be obtained. This is demonstrated in Fig. 7.33 for polyethylene in detergent solution where the K–$\dot{a}$ lines are shown for tests in which the thickness, crack length,

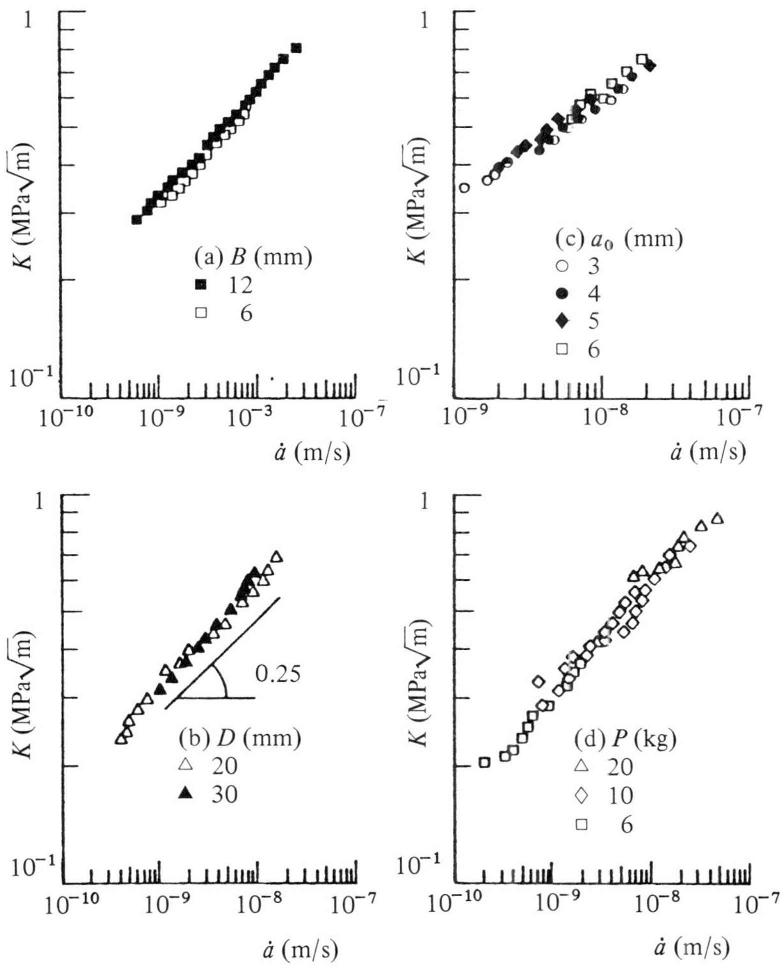

Fig. 7.33 Slow crack growth for PE in detergent solution – SEN, three-point bending – effect of test variables; span = 180 mm (from reference [16])

specimen depth and the load level were all changed and gave a single $K \propto \dot{a}^{1/4}$ line. Such dependencies are seen in practice and this is shown in Fig. 7.34 for polyethylene pipes tested in water at four temperatures. The open points are ductile failures but in all cases there is a sharp transition to brittle crack growth at long times. The slope is $-\frac{1}{4}$ on the log p–log t basis shown here confirming that it is of the same type as the $\dot{a}^{1/4}$ data.

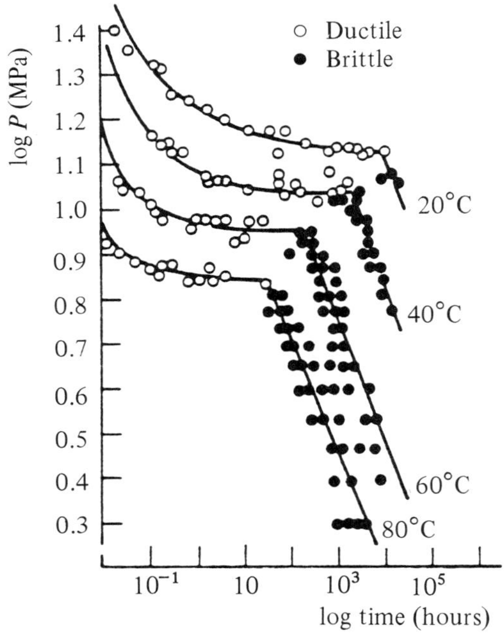

Fig. 7.34 Ductile and brittle failure in HDPE pipe (data kindly supplied by BP Chemicals Limited)

Such data suggest a transition to a separate controlling mechanism. The ductile failures do seem to be visco-elasticity controlled, although the effect is sometimes difficult to establish because of pronounced non-linear effects, particularly at short times. It has not been possible to rationalise the brittle failures on this basis, since the degree of time-dependence is too great for visco-elasticity to be the controlling mechanism. In addition, it is difficult to envisage why the form $\dot{a}^{1/4}$ should occur on this basis. Flow control is also unlikely since very similar lines can be produced in air and in oils of widely differing viscosities with no attendant shifts in crack speeds.

Part of the answer may be provided by the fracture surfaces which show a very distinctive structure. Although the surfaces are flat and macroscopically brittle, the failure is ductile on a small scale with the surface being made up of a series of cusps with highly deformed fibrils formed between them, as shown in Fig. 7.35.

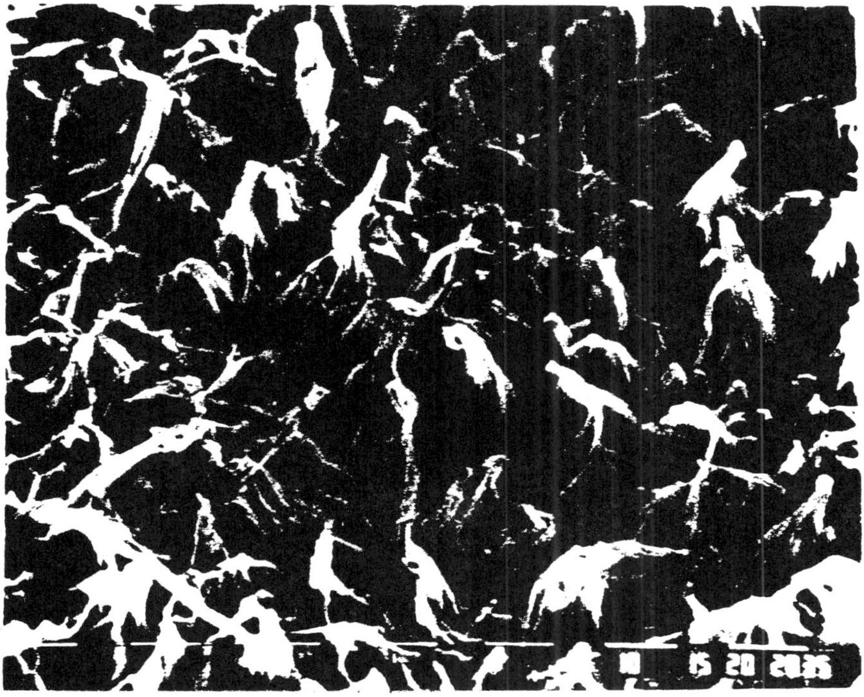

Fig. 7.35 Fracture surface for polyethylene in detergent solution showing micro-ductility

The cusps form from microvoids which appear to nucleate from sites within the material. The spacing between these site changes with crack speed and for low values (10^{-9} m/s) is about 1 μm and increases to about 50 μm at 10^{-7} m/s. From both direct observation and calculation, these values are close to the crack opening displacements which are not constant in this case. Polyethylenes do contain considerable amounts of particulate matter, such as catalyst residue, colouring agents and stabilisers, and it is reasonable to assume that if failure occurs by microvoiding, then the size of the voids is controlled by the spacing, and thus number, of such particles. This in turn determines the crack opening displacement and a simple spherical void model would suggest that $\delta \simeq l_0$, the spacing. If the particles diffuse within the polymer, then it is possible that the spacing would decrease with time according to:

$$l_0 = \bar{l}_0 \left(\frac{\tau_0}{t}\right)^{1/2}$$

where $\bar{l}_0$ and τ_0 are constants. Thus, we have finally:

$$K_c^{\,2} = E\, G_c = \delta\, E\, p_c = \bar{l}_0 \left(\frac{\tau_0}{t}\right)^{1/2} E\, p_c \tag{7.42}$$

i.e. $K_c \propto t^{-1/4} \propto \dot{a}^{1/4}$, as observed.

7.5 Fatigue

Fatigue crack growth is that which occurs under cyclic loading at load levels which would not result in growth for static loads. In one sense, the cyclic loading may be regarded as an environment which promotes cracking just as, say, a detergent acts on polyethylene. Fatigue appears to occur in all materials and, since almost all structures can be subjected to some sort of vibration, fatigue crack growth is a potential source of problems in design. The realisation of this fact is reflected in the large literature on the subject, but even so there are regular reports of practical failures attributed to this cause. Prior to the advent of fracture mechanics, testing consisted of determining the effects of many variables on the fatigue life of unnotched specimens which was presented in the form of the familiar stress-number of cycles to failure, or S–N, curves. Since failure arose from inherent flaws, there was always considerable scatter in the data so that large numbers of tests were necessary. The use of pre-cracked specimens and the measurement of crack growth rates to characterise fatigue behaviour greatly reduced the labour involved and this type of test is now the most widely used. The data are usually presented in the form of the range of stress intensity factor applied, ΔK, and the resulting crack growth rate per cycle, da/dN, plotted on a log–log basis. This form arose because it had been found for many metals that it is the range of load variations which controls growth rates and results in a unique crack growth for each cycle. It was further observed that large sections of such graphs were approximately linear and that a useful representation of the data is the Paris law:

$$\frac{da}{dN} = A\,\Delta K^m \tag{7.43}$$

where A and m are constants. In general, $2<n<6$ and frequently $m \simeq 4$, which has prompted much speculation on the existence of a general fatigue law.

It is well established that fatigue occurs in polymers and that it may be reasonably represented by the Paris law. Indeed, there has been sufficient study of this field to produce a book on the topic by Hertzberg & Manson [19] which provides a comprehensive review of the literature up to 1980. With this in mind, no attempt will be made here to review the literature in detail but rather to set some examples in the broader context of slow crack growth in general and to demonstrate how fatigue may be regarded not as a unique special case but as part of a spectrum of behaviour. It is also proposed to show that the line zone model can give useful insights into observed behaviour and link fatigue to conventional environmental stress cracking.

7.5.1 *Visco-elastic crack growth under cyclic loads*

It is useful to note that cyclic loading can result in a crack growth rate per cycle which arises solely from the visco-elastic response of the material. For a loading system of amplitude K_0, there will be some fraction which will be effectively constant, giving rise to an average growth rate $\dot{a}$ where $\dot{a}=(da/dN)\,\dot{N}$ in this case. By analogy with the constant load steady-state as discussed in Section 7.2.2, we may write, for a constant δ criterion:

$$K \propto a^{\frac{m+n}{2(1-2m)}} \propto \dot{N}^{\frac{m+n}{2(1-2m)}} \left(\frac{da}{dN}\right)^{\frac{m+n}{2(1-2m)}} \tag{7.44}$$

The constant of proportionality will be determined by the waveform of the loading. For a case in which, for example, $K \propto \dot{a}^n$, the waveform effect of a loading of $K=K_0\phi(t)$ may be approximated by:

$$\dot{a}=(\dot{a})_{K_0}\frac{1}{\dot{N}}\int_0^1 [\phi(t\dot{N})]^{1/n}\, d(t\dot{N}) \tag{7.45}$$

where $(\dot{a})_{K_0}$ is the crack speed evaluated at the maximum K value when under constant load. This is based on a pseudo-elastic analysis but is likely to be sensible for these steady-state conditions. For a cycle with a constant load for a fraction x of the period $\dot{N}^{-1}$, as shown in Fig. 7.36(a), we have:

$$\frac{da}{dN}=x\left[\frac{(\dot{a})_{K_0}}{\dot{N}}\right] \tag{7.46}$$

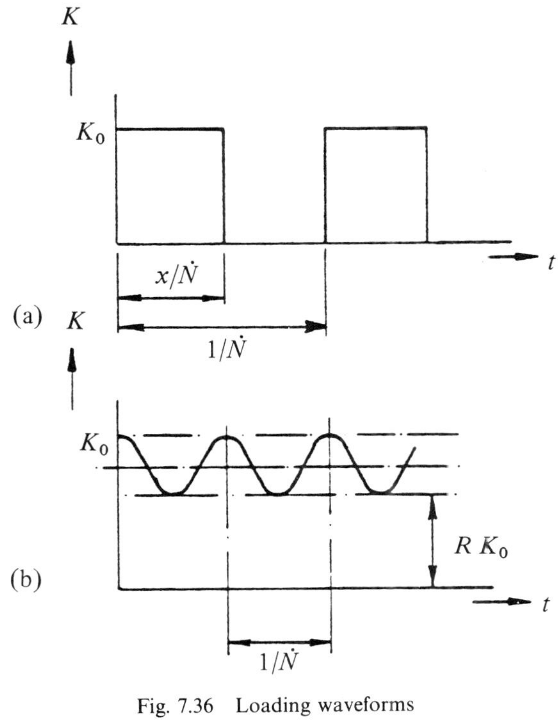

Fig. 7.36 Loading waveforms

while for the sinusoidal loading shown in Fig. 7.36(b), we have:

$$\frac{da}{dN} = \int_0^1 \left[\frac{(1+R)+(1-R)\sin 2\pi\,(t\dot{N})}{2} \right]^{1/n} d(t\dot{N}) \left[\frac{(\dot{a})_{K_0}}{\dot{N}} \right] \tag{7.47}$$

where R is the ratio of minimum to maximum K ratios.

In all cases, the form of the da/dN-K relationship is the same as the visco-elastic data giving, in general, rather high values for the power-law index in the Paris law, typically $\simeq 15$ to 20. In addition, da/dN has a frequency-dependence of $da/dN \propto \dot{N}^{-1}$ since $\dot{a}$ is a single function of K. It is also apparent from equation (7.47) that the dependence of da/dN on $K=(1-R)\,K_0$, noted in the Paris law, would not be expected from this type of behaviour.

7.5.2 *Visco-elastic heating effects*

A complicating factor in fatigue testing is visco-elastic heat generation during cycling. If sufficient heat is generated and not dissipated, specimens become hotter and under certain conditions they can fail because the temperature exceeds the softening value. The energy generation rate per unit volume generated during stress cycling is given by:

$$\dot{U} = \dot{N}\,\pi \frac{p_0{}^2}{E'} \tan\delta \tag{7.48}$$

where p_0 is the stress amplitude, E' the in-phase modulus, and $\tan\delta$ the loss factor. In effect, $p_0{}^2/2E'$ is the elastic energy per cycle, and $\tan\delta$ the fraction of it which is dissipated. In test specimens, some of this energy is lost from the surface via a heat transfer coefficient h, and we may write an expression for the nett temperature rise in the form:

$$\dot{T} = \frac{1}{\rho\,C} \left[\dot{N}\,\pi \frac{q^2\,p_0{}^2 \tan\delta}{E'} - \frac{h}{L}(T-T_0) \right] \tag{7.49}$$

where L is a specimen characteristic length given by the volume to surface area ratio, and q is a stress distribution factor ranging from $\frac{1}{2}$ for bending to unity for uniform stressing.

The material properties can all be functions of T, particularly $\tan\delta$ and E', but a useful guide to the behaviour may be obtained by assuming uniform values for which equation (7.49) may be solved to give:

$$T = T_0 + \dot{N}\,\pi\,L \frac{q^2\,p_0{}^2}{h} \frac{\tan\delta}{E'} (1 - e^{-ht/L\rho C}) \tag{7.50}$$

The temperature rise increases to a maximum value of $\Delta\bar{T} = \dot{N}\,\pi\,(q^2\,p_0{}^2)/h\,(\tan\delta)/E'$ with a characteristic time of $\tau = (L\,\rho\,C)/h$. In a test, if $T_0 + \Delta\bar{T} = T_s$, the softening temperature, then thermal failures occur and a limiting condition may be defined for $\dot{N}$ and p_0 in the form:

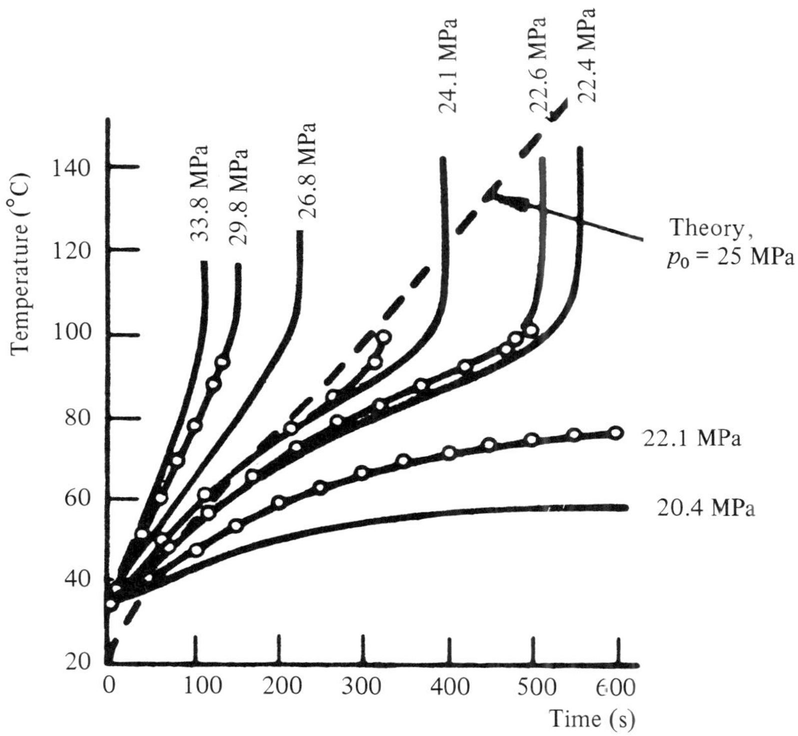

Fig. 7.37 Experimental temperature rise curves for PMMA – 6 mm round rod in rotating bending (from reference [20])

$$\dot{N}\,{p_0}^2 = \frac{h\,E'\,(T_s - T_0)}{\pi \tan\delta\, L\, q^2} \tag{7.51}$$

Measurements have been made of temperature rises and thermal failures observed [20] and Fig. 7.37 shows T versus t data for 6 mm diameter PMMA rod tested in rotating bending. The temperature rise is as expected and also $T_s \simeq 100$°C, after which T rises rapidly giving thermal failure. At this rotational speed (3000 rpm), $\dot{N} = 50$ Hz and it can be seen that the limiting stress is in the region of 22 MPa. Typical values for PMMA are $h \simeq 10$ J/m^2°Cs, $\tan\delta \simeq 0.06$, $E' \simeq 3$ GPa and, since L is the diameter for a cylinder, we have:

$$\dot{N}\,{p_0}^2 \simeq 2.5 \times 10^4\ (\text{MPa})^2\text{s}^{-1}$$

so that for $\dot{N} = 50$ Hz, $p_0 \simeq 22$ MPa, as observed. For the same values, $\tau \simeq 10^3$ s and equation (7.50) becomes:

$$T = T_0 + 295\,(1 - e^{-t/10^3})$$

for $p_0 = 25$ MPa and this line is shown in Fig. 7.37, confirming the general trend. p_0–N curves for thermal failure may be computed from equation (7.50) with $t = N/\dot{N}$ and, in general, the predictions are good, although variations in h can be large in practice.

Thermal effects are really a testing problem and there is little evidence of their occurring in practice but, as such, they can cause serious difficulties. As equation (7.51) demonstrates, at normal stress levels it is frequently unwise to go much above 10 Hz in testing and care must always be exercised.

As a final point, it should be noted that cyclic heating has little effect on the local crack tip other than through the bulk temperature change. This is because it is the crack speed which controls the time-scale and $\dot{N} \simeq 10^3$ Hz are necessary to give a sufficient effect.

7.5.3 *Fatigue crack propagation*

Some polymers conform well to the basic assumptions of the Paris law and Fig. 7.38 shows data for a PVC pipe grade plotted in the appropriate form. There are a wide range of R ratios and it is clear that ΔK is the major controlling

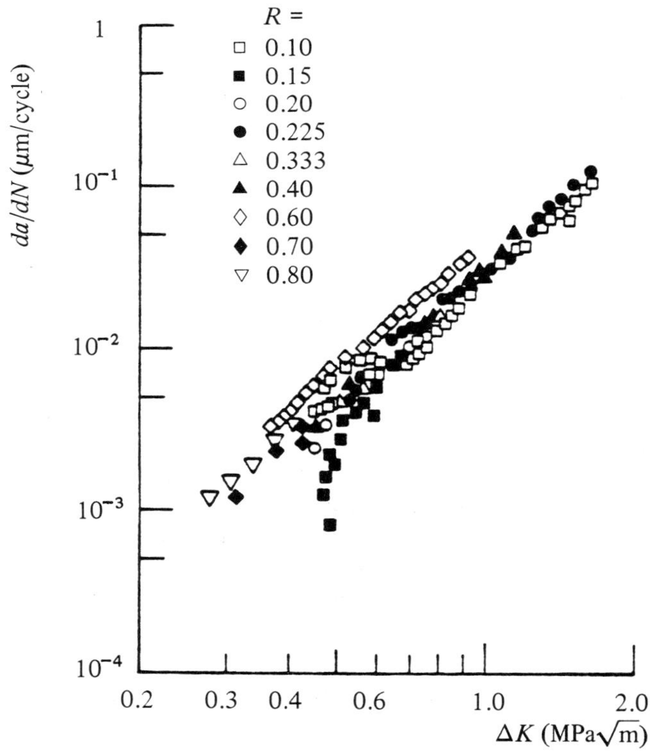

Fig. 7.38 Fatigue crack growth data for a pipe grade PVC at 10 Hz and 20°C

variable. There is also good linearity, except at very low growth rates, where the curves tend to have a much larger slope, again as with metals. The slope is about 2.4 in this case. The static K_{c1} value is 3.2 MPa$\sqrt{m}$ for this material with little evidence of slow crack growth so that the crack growth observed here is genuinely due to fatigue. The position is far less clear for more brittle behaviour and this is demonstrated in Fig. 7.39 for PMMA tested at 1 Hz. Here, there is a strong R ratio effect and the slope $\simeq 6$. The reason becomes apparent when the lines for visco-elastic crack growth using the results of Section 7.5.1 and Fig. 7.11 are added and it is clear that these constitute a significant portion of the growth and the high R-dependence. The growth curves are thus a combination of two effects and analysis must be treated with caution.

Curves of this form may be produced for any set of conditions and then used in design, but little is added to an understanding of the mechanisms involved and it is useful to consider now the use of a line zone model in fatigue [21]. When a crack is loaded under normal circumstances, a craze will form when some critical condition is reached and this will eventually fail after subsequent

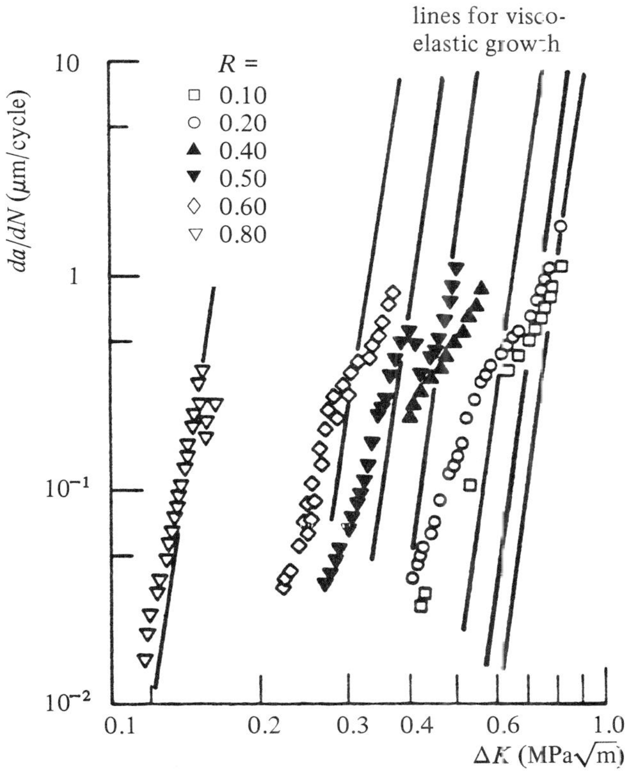

Fig. 7.39 Fatigue crack growth data for PMMA at 1 Hz and 20°C

growth at, say, a critical crack opening displacement. It is clear that under fatigue conditions, this craze formation and growth process is promoted by the unloading and reloading cycle. If we suppose that cycling of the very high stresses induced at the crack tip on loading results in the formation of a craze at below the normal initiation condition, then we have a mechanism for developing a craze at low stresses. Let us suppose that low applied K values induce local stresses of p_0 on a small zone of length r_0 at the crack tip and that these reduce to the craze stress p_c on unloading and reloading. This is a very drastic and oversimplifying assumption since subsequent cycles will almost certainly reduce p_c, but it is a useful first step. On reloading the crack, the original length r_0 now has a stress p_c so that a new zone carrying p_0 is formed which, to maintain equilibrium, must be less than the original r_0. Repeated unloading and reloading then results in the growth of a two-stage zone, as shown in Fig. 7.40, after, say, N cycles, of total length r_N with a high stress region of length r_0. This is an identical model to that used in Section 7.3.1 for environmental crack growth and equation (7.21) may now be used for the equilibrium condition:

$$K = 2\sqrt{2/\pi}\,[p_c\sqrt{r_N} + (p_0 - p_c)\sqrt{r_0}]$$

and for any K value, r_N must lie between:

$$\frac{\pi}{8}\frac{K^2}{p_0{}^2} < r_N < \frac{\pi}{8}\frac{K^2}{p_c{}^2}$$

It is convenient to express these relationships in terms of $r = (\pi/8)\,(K^2/p_0{}^2)$ and $\alpha = p_c/p_0$, so that:

$$\sqrt{r} = \alpha\sqrt{r_N} + (1-\alpha)\sqrt{r_0} \tag{7.52}$$

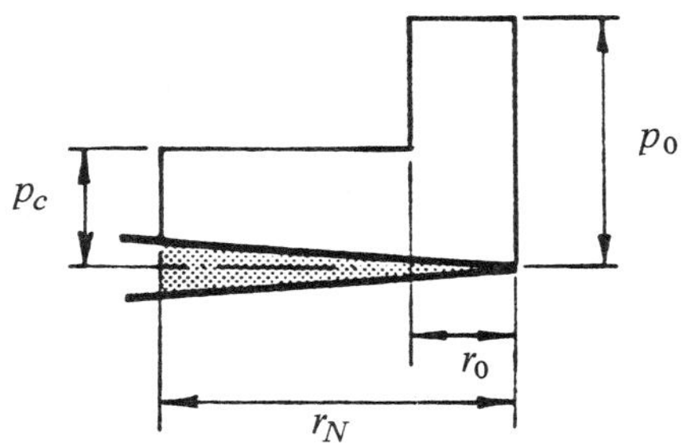

Fig. 7.40 Two-stage line zone in fatigue

The growth process may be expressed exactly as:

$$\sqrt{r} = \alpha\sqrt{r_{N+1}} + (1-\alpha)\sqrt{r_{N+1} - r_N} \tag{7.53}$$

starting at $r_1 = r$ and an r_N versus N expression may be deduced numerically.

For larger numbers of cycles, this may be written as a differential equation:

$$\sqrt{r}=\alpha\sqrt{r_N}+(1-\alpha)\sqrt{dr_N/dN}$$

which gives:

$$N=1+2\left(\frac{1-\alpha}{\alpha}\right)^2\left\{\frac{1}{(1-\alpha)}\left(\frac{1-\alpha}{1-\alpha\sqrt{r_N/r}}-1\right)-\ln\left(\frac{1-\alpha}{1-\alpha\sqrt{r_N/r}}\right)\right\} \tag{7.54}$$

In addition, as the zone grows, so the crack opening displacement increases and this may be derived by superposition from equation (4.23) to be:

$$\delta=\frac{8}{\pi}\frac{p_0}{E}r_N\left\{\alpha+(1-\alpha)\left[\sqrt{r_0/r_N}+\frac{1}{2}\left(1-\frac{r_0}{r_N}\right)\ln\left(\frac{1+\sqrt{r_0/r_N}}{1-\sqrt{r_0/r_N}}\right)\right]\right\} \tag{7.55}$$

which, for $r_0/r_N \ll 1$, may be approximated to:

$$\frac{\delta}{\delta_0}=2\sqrt{r_N/r}-\alpha\left(\frac{r_N}{r}\right) \tag{7.56}$$

by using equation (7.52), where $\delta_0=K^2/(p_0\,E)$, the initial value of δ. Fig. 7.41 shows both r_N/r and δ/δ_0 as functions of N for $\alpha=0.1$ as deduced from equations (7.54) and (7.56), respectively. r_N/r rises steeply as N increases and is very close to the limiting value of α^{-2} (100) by 10^4 cycles. δ reaches the limiting value at about a decade less.

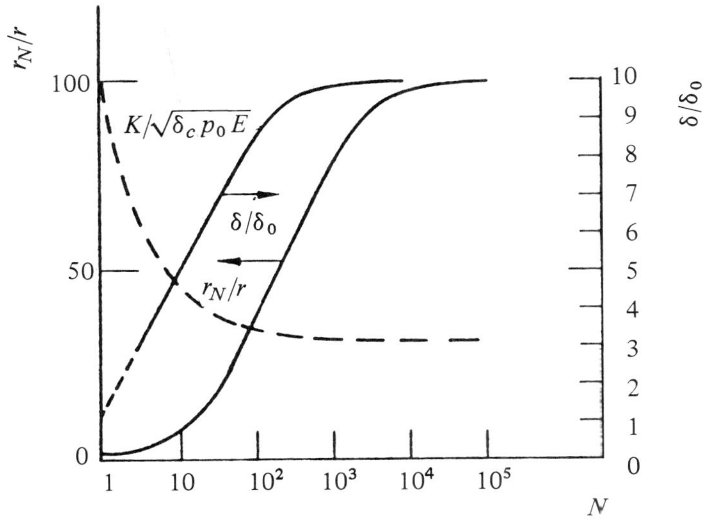

Fig. 7.41 Zone length and crack opening displacement as a function of number of cycles for the line zone model

A number of cycles to crack initiation may now be computed on the assumption that $\delta = \delta_c$ at this condition. This may be deduced directly from the δ/δ_0 line since $K/\sqrt{\delta_c\, p_0\, E} = \sqrt{\delta_0/\delta}$, and this is shown as the broken line in Fig. 7.41. We may thus model a crack incubation stage and the K–N line is, in effect, an S–N curve for initiation. The number of cycles is not large and there is little available experimental evidence to test these predictions.

The propagation stage may now be considered and this may proceed in two ways. Firstly, the crack may run through the entire craze region $(r_N - r_0)$ once it is initiated and then arrest with the formation of a new r_0 zone, from which craze growth would proceed as before. This type of propagation would proceed in a series of jumps and is most likely to occur in materials of low time-dependence since it requires unstable crack growth. This type of crack growth has been described in some detail in polymers, particularly by Hertzberg, *et al.* [e.g. 19], and is termed discontinuous crack growth. In general, the behaviour described is consistent with the model used here. Since fracture will usually occur when the craze is well developed, its length will be close to the limiting value with $r_0 \ll r_N$ so that observed band widths will be approximately the expected craze size and $\propto \Delta K^2$ as observed by Hertzberg [19].

A stable form of crack growth may also be envisaged in which crack propagation results in the crack growing only r_0 during each cycle. On reloading, this reproduces exactly the same two-stage zone (providing K does not change) so that a constant growth per cycle will be observed. If we again use the criterion of $\delta = \delta_c$, then we may eliminate r_N between equations (7.52) and (7.56) to give:

$$\frac{da}{dN} = r_0 = \frac{\pi}{8} \frac{1}{(1-\alpha)^2\, p_0{}^2} \left[K^2 - \alpha\,(\delta_c\, p_0\, E)\right] \tag{7.57}$$

This gives a linear dependence of da/dN on K^2 but with the important addition of a threshold condition of $K^2 = \alpha\,(\delta_c\, p_0\, E)$. In the conventional Paris law log–log graph, the slope m is given by:

$$m = 2\left(1 - \frac{\alpha\,\delta_c\, p_0\, E}{K^2}\right)^{-1}$$

so that $m \to 2$ when K is much larger than the threshold condition or if there is a very low threshold. As K approaches the threshold, m increases as observed. Fig. 7.42 shows the pipe grade PVC data of Fig. 7.38 plotted in accordance with equation (7.57) and good linearity is indicated with a strong R-dependence since it is plotted in terms of K_{max} here. There is a very low threshold so that $(\alpha\,\delta_c\, p_0\, E) \simeq 0.2$ $(\mathrm{MPa}\sqrt{\mathrm{m}})^2$ which accounts for $m \simeq 2.6$, as previously noted. Similar data for PMMA at 10 Hz and 20°C are shown in Fig. 7.43 and, here, there is linearity over a much smaller range. This is because of the presence of visco-elastic growth which increases da/dN so that only the lower K régime can be described by the fatigue analysis.

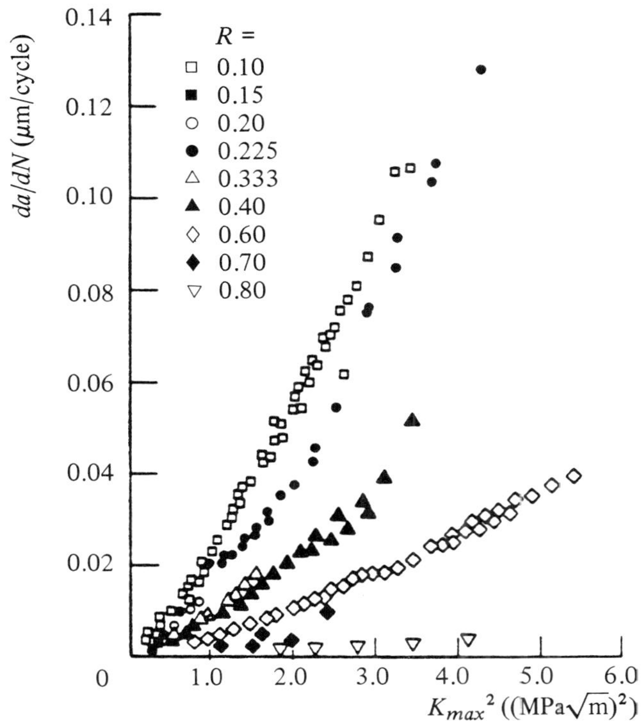

Fig. 7.42 Fatigue crack growth data for a pipe grade PVC at 10 Hz and 20°C plotted in accordance with equation (7.57)

The description of the R ratio effects requires the addition of two factors to the basic equation. Firstly, we must assume that the constant part of K, RK, gives a constant length zone R^2r which progresses with the fatigue fracture but does not change. As shown in Fig. 7.44, two situations may be visualised. In the first, case (i), the constant zone remains at the zone tip as shown in Fig. 7.44. Equation (7.52) may now be modified to:

$$\sqrt{r} = \alpha \sqrt{r_N} + (1-\alpha)\sqrt{r_0 + R^2 r}$$

and the effective K^2 value in equation (7.57) is given by $F_{(i)} K^2$ where:

$$F_{(i)} = 1 - (1-\alpha)^2 R^2 \tag{7.58}$$

In case (ii), shown in Fig. 7.44, the constant zone exists at the crack tip and requires a separate deformation mechanism, such as shear yielding, in order to form. In this case:

$$F_{(ii)} = (1-R)^2 + \alpha R^2 \tag{7.59}$$

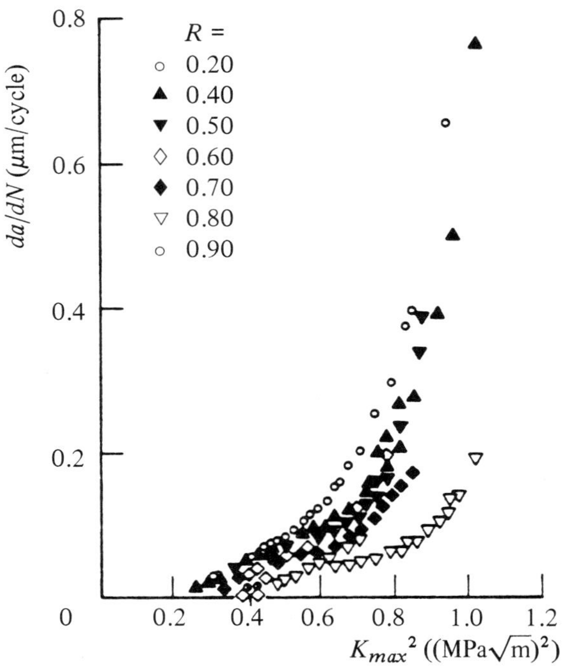

Fig. 7.43 Fatigue crack growth data for PMMA at 10 Hz and 20°C plotted in accordance with equation (7.57)

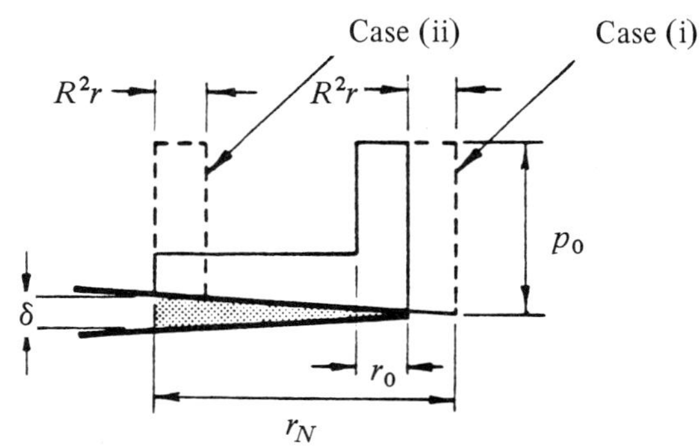

Fig. 7.44 Two-stage line zone with a mean stress

For $\alpha \ll 1$, $F_{(ii)} \rightarrow (1-R)^2$ so that the effective K is $(1-R)\,K = \Delta K$ as observed for PVC, while $F_{(i)} \rightarrow 1-R^2$, the strong dependence seen in PMMA.

An additional factor is that the loading and unloading results in residual compressive stresses around the crack tip and these may be modelled by a constant factor K_0^2 so that the final expression for growth rate is:

$$\frac{da}{dN} = \frac{\pi}{8} \frac{1}{(1-\alpha)^2 \, p_0^{\,2}} \left[F(R) \, (K^2 - K_0^{\,2}) - \alpha \, (\delta_c \, p_0 \, E) \right] \tag{7.60}$$

The various parameters may be obtained from the da/dN versus K^2 graphs since the slope is given by:

$$S = \frac{\pi}{8(1-\alpha)^2 \, p_0^{\,2}} F(R) \tag{7.61}$$

and the threshold:

$$\bar{K}^2 = \frac{\alpha \, (\delta_c \, p_0 \, E)}{F(R)} + K_0^{\,2} \tag{7.62}$$

K_0^2 may be determined independently of $F(R)$ since:

$$S \, \bar{K}^2 = \frac{\pi}{8} \left(\frac{\delta_c \, E}{p_0} \right) \frac{\alpha}{(1-\alpha)^2} + K_0^{\,2} \, S \tag{7.63}$$

Fig. 7.45 shows the slopes of the lines in Figs. 7.42 and 7.43 as functions of R with

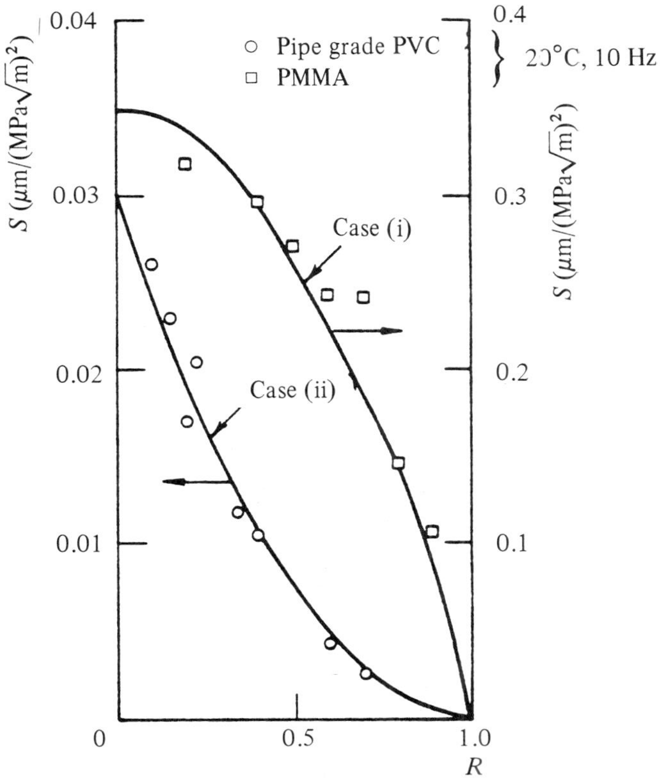

Fig. 7.45 Slope of fatigue crack growth lines as a function of R ratio (from reference [22])

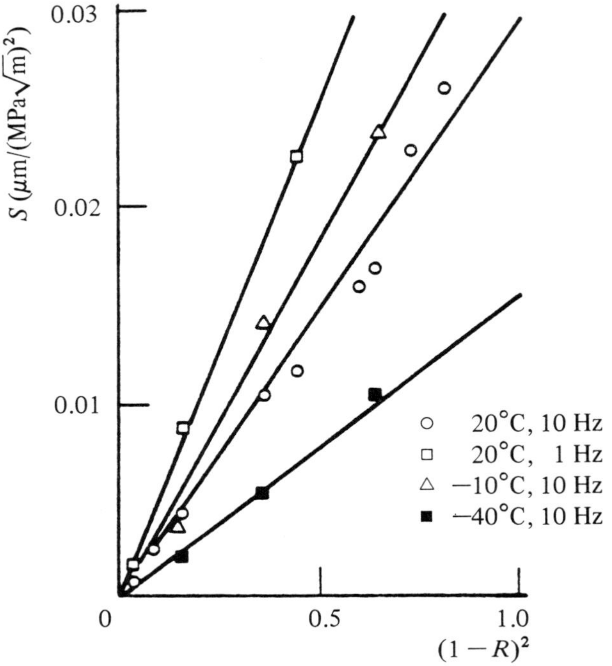

Fig. 7.46 Slopes of fatigue crack growth lines for pipe grade PVC plotted in accordance with case (ii) ($\alpha \ll 1$) (from reference [22])

the cases (i) and (ii) F functions shown for $\alpha \ll 1$. That for the PVC is very close to a $(1-R)^2$ dependence as expected and Fig. 7.46 shows data for a range of test conditions plotted on this basis. There is some evidence of intermediate types of behaviour in other materials [22] where both cases exist together. The existence of K_0 is demonstrated in Fig. 7.47 for PMMA where equation (7.63) is used for slope data at a range of R values and two frequencies.

Table 7.1 gives some values for the various parameters obtained on several polymers over a range of conditions. p_0 is very high and probably represents the elastic stress concentration prior to craze formation and is of the order of carbon–carbon bond strengths [23]. The true threshold value is remarkably constant at about 0.05 $(\text{MPa}\sqrt{\text{m}})^2$ which gives $G_c \simeq 20$ J/m^2, which is approximately the true surface energy of polymers. α may be estimated from the known craze stress values and at 20°C for PMMA is approximately 0.1 and for PVC 0.02. It is also possible to estimate δ_c from the threshold values so that, for example, for PMMA since $\alpha\, p_0 \simeq 100$ MPa and $E \simeq 3$ GPa, we have $\delta_c \simeq 0.2$ μm. The situation is somewhat complicated by the presence of K_0 and direct measurements in fatigue tests by Döll, *et al.* [24] have given values of 0.6 μm which is consistent with K_0 being included. Döll's data also show that δ_c is

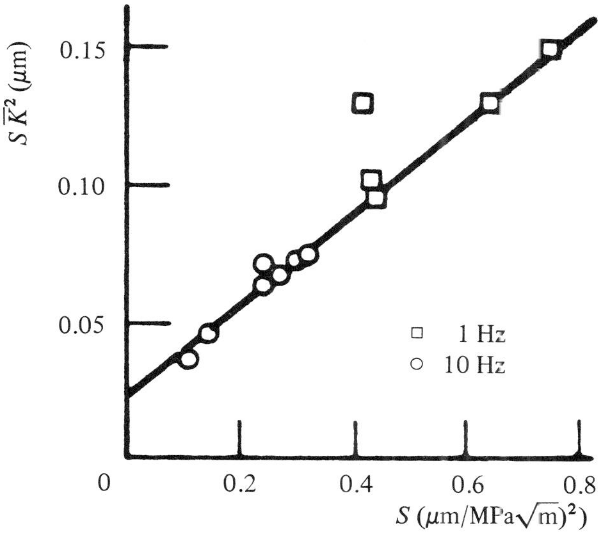

Fig. 7.47 Threshold data for PMMA at 20°C and two frequencies – plotted in accordance with equation (7.63) (from reference [22])

Table 7.1

Summary of measured data used in the model

Material	Temperature (°C)	Frequency (Hz)	p_0 (GPa)	$K_C{}^2$ ((MPa√m)^2)	$\alpha(\delta_c\, p_0\, E)$ ((MPa√m)^2)
PMMA	20	1	0.8	0.18	0.03
PMMA	20	10	1.0	0.18	0.05
Epoxy	20	10	2.5	0.05	0.06
Pipe Grade PVC	20	1	2.7	0.06	0.07
Pipe Grade PVC	20	10	3.6	0.10	0.03
Pipe Grade PVC	−10	10	3.2	0.18	0.06
Pipe Grade PVC	−40	10	4.9	0.49	0.17

constant only for $da/dN < 0.02$ μm/cycle and increases to the static load value by about 10 μm/cycle. The data in Fig. 7.43 would suggest a limit of about 0.1 μm/cycle before visco-elastic effects become effective.

Frequency effects are a somewhat contentious issue in the literature. Hertzberg [19] noted a strong correlation between frequency sensitivity and the proximity of the β-relaxation peak and shows a particularly high value for PMMA. This is consistent with the presence of visco-elastic crack growth along with fatigue as noted here, since this type of growth has a strong frequency dependence. The PVC tested here was believed to be giving almost completely

Table 7.2

Fatigue data for PS in air and environments at 0.15 Hz

Temperature (°C)	In Air		In Corn Oil		In Detergent	
	$K_0^2+\alpha(\delta_c p_0 E)$ $((\text{MPa}\sqrt{\text{m}})^2)$	p_0 (MPa)	$K_0^2+\alpha(\delta_c p_0 E)$ $((\text{MPa}\sqrt{\text{m}})^2)$	p_0 (MPa)	$K_0^2+\alpha(\delta_c p_0 E)$ $((\text{MPa}\sqrt{\text{m}})^2)$	p_0 (MPa)
60	0.071	337	0.026	35	0.022	37
40	0.15	336	0.062	46	0.058	30
30	0.25	283	—	—	—	—
20	0.19	620	0.106	85	—	—
0	0.26	622	—	—	—	—
−20						
	0.29	861	—	—	—	—
−60						

fatigue crack growth and there was no appreciable frequency dependence in the range 0.1 to 60 Hz.

The model may also be used for a combination of fatigue and environment, and Table 7.2 shows data for polystyrene over a range of temperatures in air and environments taken from [25]. The values are much as in other materials, although p_0 is somewhat lower, but in the environments, which promote crazing, p_0 becomes very close to a dry craze value and fatigue results in a further reduction of this value. Flow-controlled crack growth is also possible since a limiting $\dot{a}$ condition may be reached, resulting in the equivalent da/dN condition using $da/dN = \dot{a}/\dot{N}$ and the appropriate equation for $\dot{a}$ from Section 7.3.3.

An interesting example of the use of crack growth data for the prediction of fatigue life is shown in Fig. 7.48 which gives stress-number of cycles to failure data for a PVC 50 mm diameter pipe subjected to cyclic pressure at 1 Hz and 20°C [22]. For the stress range shown, the fracture surfaces showed very little ductility and semi-circular dark regions of about 0.5 mm were observed at the fracture initiation point. At the lower stress levels, there was evidence of initiation from several sites. A life prediction for this case can be obtained by integrating equation (7.60) ($R=0$) using $K^2 = p^2\,\pi\,a$ between an initial flaw of a_0 and the wall thickness B (3.7 mm in this case). The expression becomes:

$$N = \frac{8}{\pi^2}\left(\frac{p_0}{p}\right)^2 \ln\left[\frac{\pi\,p^2\,B - (K_0{}^2 + \alpha\,\delta_c\,p_0\,E)}{\pi\,p^2\,a_0 - (K_0{}^2 + \alpha\,\delta_c\,p_0\,E)}\right] \qquad (7.64)$$

and Fig. 7.48 shows lines computed from this expression for various values of a_0 and material parameters obtained from fatigue crack growth tests. The points

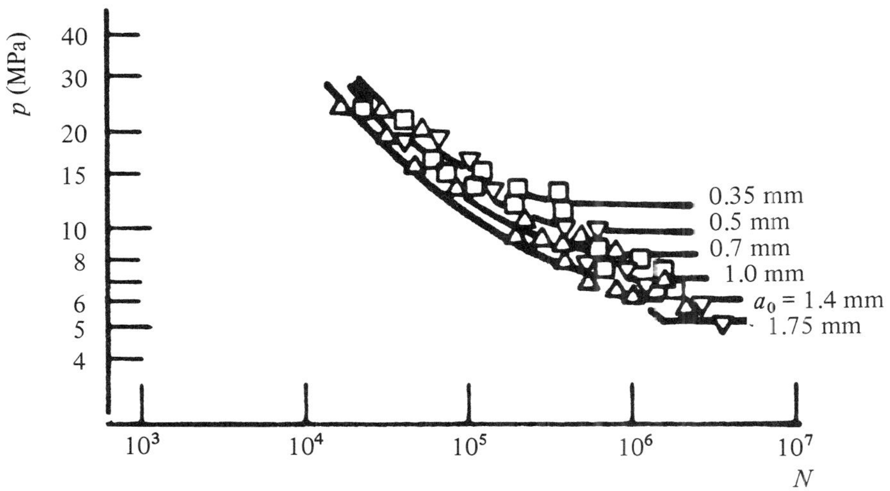

Fig. 7.48 Fatigue life data for PVC pipe under cyclic pressure, 20°C and 1 Hz (from reference [22])

group quite well about the lines and since the a_0 values increase approximately in integer values, it suggests that there is a basic flaw of about 0.35 mm and that the larger values are probably amalgamations of several of these. Calculations of incubation cycles suggested very low values (<500).

7.6 Ductile tearing

The presence of plastic flow which is more extensive than a truly local zone at the crack tip can lead to the resistance to crack propagation increasing as the crack grows. This results in the rising R curve, as discussed in Chapter 2, and for most loading systems, stable growth occurs prior to a possible instability. There has been much interest in describing this process in metals (see [26]) because of the need to design highly critical structures into the ductile failure range but relatively little effort has been devoted to this type of behaviour in polymers. Fully plastic failure under plane strain conditions, the usual case for metals, is uncommon in polymers because the necessary sizes to achieve this condition are generally far in excess of normal dimensions. When ductile tearing does occur, it is usually under plane stress conditions, although initiation is frequently in plane strain with a rather rapid transition to plane stress with gross lateral contraction. There is little experience to draw on for the use of fracture mechanics in this sort of situation, but, in principle, it is possible to use a J (or R) type of analysis and an outline of the procedure will now be given.

7.6.1 *J analysis of ductile tearing*

Let us suppose that ductile tears may be propagated across a specimen in a stable manner. In such an experiment, it would be possible to record a full load-deflection diagram and, by marking the specimen, record the crack growth throughout the test. Fig. 7.49 is an example of such data and was obtained on a low density polyethylene, using a single edge notch geometry [27]. Since the original load-deflection lines for a range of original crack lengths are marked with the crack growth, it is possible to interpolate the load-deflection lines for constant crack lengths. J may now be determined from these curves in the usual way by calculating the energy U as a function of crack length a at various fixed displacements. Such lines, derived from Fig. 7.49, are shown in Fig. 7.50 and show linearity. This would be expected for non-work hardening full yielding of the ligament as discussed in Section 4.5, since:

$$P = B\,(D-a)\,p_Y \tag{7.65}$$

i.e. constant, as observed in Fig. 7.49, for constant a. Thus:

$$U = P\,u = B\,(D-a)\,p_Y\,u \tag{7.66}$$

and U is linear in a at fixed u as observed in Fig. 7.50. Since $J = -(1/B)\,(\partial U/\partial a)$, we have:

$$J = p_Y\,u \tag{7.67}$$

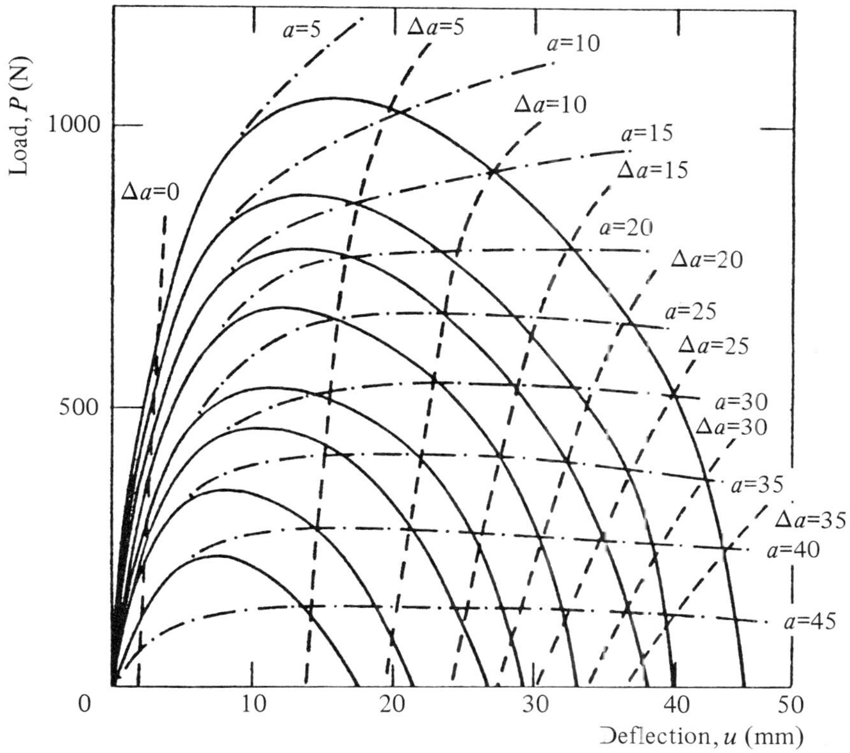

Fig. 7.49 Load-deflection diagram. — Experimental lines, —·—· interpolated constant crack length lines, - - - constant Δa lines (crack lengths in mm) (from reference [27])

and $J(u)$ may be determined from the slopes in Fig. 7.50 and is given in Fig. 7.51. The slope of this line is p_Y once the initial elastic region is passed and this remains approximately constant at 8 MPa, increasing slightly at large displacements. We may now use this result, together with the original P–u lines to determine J–Δa (crack growth) lines, as shown in Fig. 7.52. There is clearly no unique curve and the termination points indicate a linear dependence on ligament length. In addition, there is a $\Delta a = 0$ value of $J_1 = 24$ kJ/m^2, suggesting that J may be partitioned into a constant, crack tip specific part, and a deformation determined part which would be expected to be ligament size controlled. This is confirmed in Fig. 7.53 where $(J_c - J_1)/(1 - a_0/D)$ is plotted versus $\Delta a/(D - a_0)$, giving a single curve suggesting a form for J of:

$$J_c = J_1 + \left(1 - \frac{a_0}{D}\right) F\left(\frac{\Delta a}{D - a_0}\right) \tag{7.68}$$

The dependence of $(1 - a_0/D)$ probably defines an energy dependent on ligament volume, $\alpha(1 - a_0/D)^2$, while F represents the strain. Since $J_c(a)$ is now known, it is

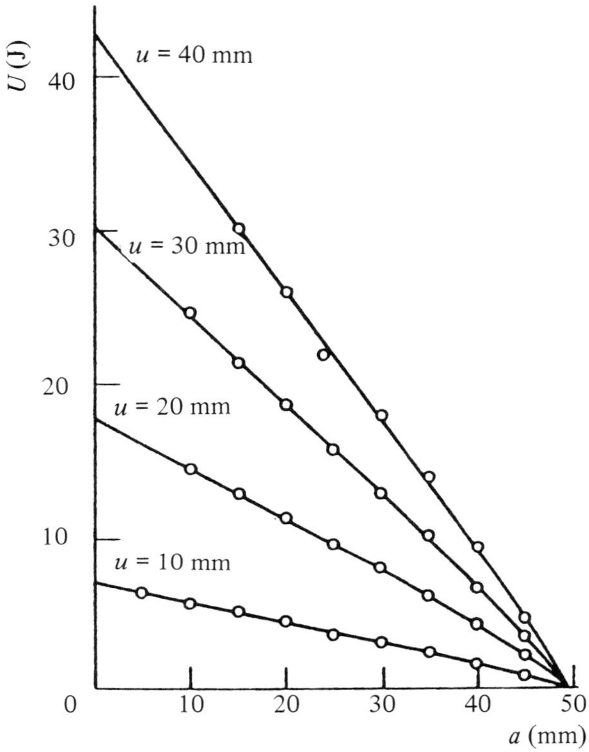

Fig. 7.50 Energy as a function of crack length at constant deflection – J is given by the slope (from reference [27])

possible to predict any P–u curve from the relationship:

$$P = B\, p_Y\, (D - a)$$

and:

$$J \simeq p_Y\, u$$

An interesting confirmation of the concept that J_1 is supplied by the energy release rate as given by an analysis suggested by Turner [26]. He points out that in fully ductile failure, there is not only the normal G arising from dC/da, but also an extra amount from the plastic load decrease, $B\, p_Y\, da$. This may be written as a total G:

$$G_{TOT} = -\left(\frac{P^2}{2B}\frac{dC}{da} + \frac{P\,C}{B}\frac{dP}{da}\right)$$

i.e.

$$G_{TOT} = -G\left(1 + \frac{C}{dC/da}\,\frac{2}{P}\,\frac{dP}{da}\right) \tag{7.69}$$

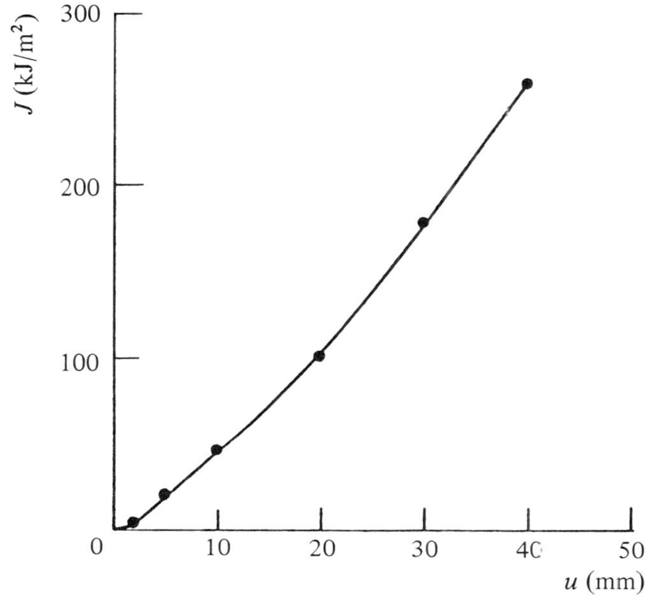

Fig. 7.51 J as a function of displacement (from reference [7])

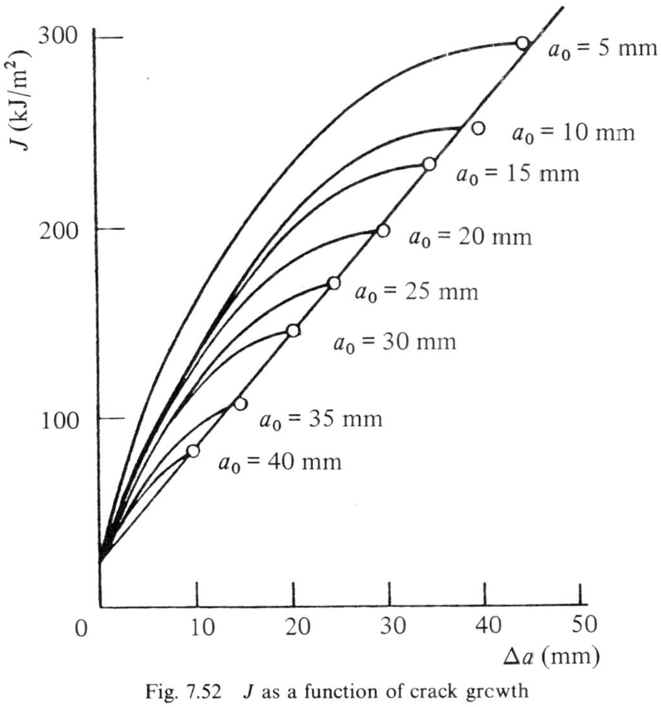

Fig. 7.52 J as a function of crack growth

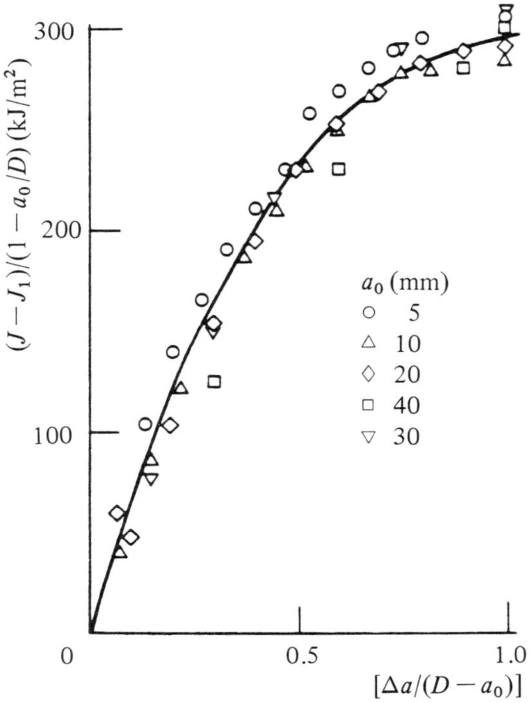

Fig. 7.53 J as a function of crack growth scaled over ligament length

For this case, $(1/P)(dP/da) = -1/(D-a)$, and we have:

$$G_{TOT} = G\left(\frac{2}{\eta} - 1\right) \tag{7.70}$$

where $\eta = (D-a)/C(dC/da)$. To determine η, the compliance C must be known and because this is a heavily distorted specimen, direct measurements were made by unloading a cracking specimen so that $\eta(a)$ could be found and thus G_{TOT} throughout each test. Fig. 7.54 shows these data and, while there is considerable scatter, an average of 24 kJ/m^2 is reasonable.

The crack growth prior to full yielding, where the load is increased, may be described in terms of the strain energy function W_0 in a manner analogous with the large strain elastic analysis, as in equation (2.32), i.e.

$$J = 2k\, W_0\, a \tag{7.71}$$

This is the representation given by Andrews [28] in his theory of generalised fracture mechanics. W_0 is found for a given load on the specimen by integrating the P–u curve for $a = 0$ and dividing by the specimen volume. For each load, J may be found for each crack length (using Fig. 7.51) so that $2k$ may be computed. Andrews suggests that k should be a function of W_0 because of non-

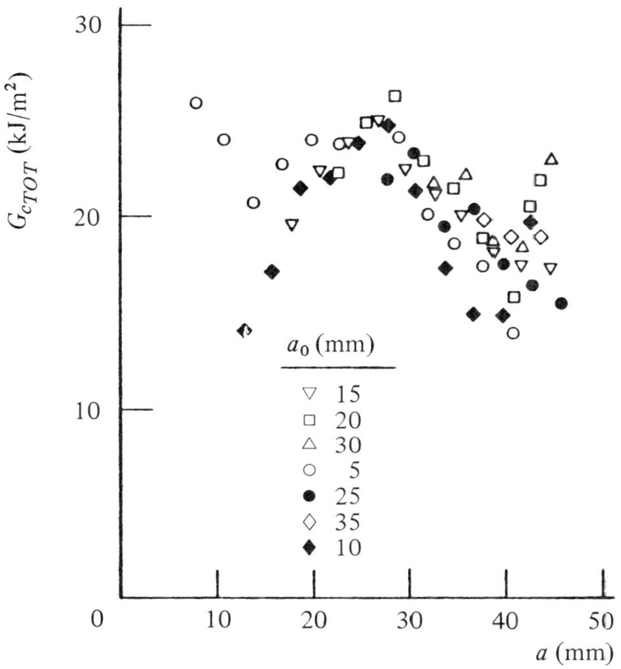

Fig. 7.54 Energy release rates for crack propagation (from reference [27])

linear effects and reports high k values for low W_0 which decrease as W_0 increases. When the data in Fig. 7.49 are analysed in this way, the origin of the high values is apparent when J/W_0 is plotted as a function of a, as shown in Fig. 7.55. There is a strong dependence on a and very little on load level, but there is clearly a finite value of J/W_0 as $a \to 0$, which would result in a rapid rise in apparent k. This effect probably arises from the plastic zone during propagation increasing the effective crack length so that equation (7.71) becomes:

$$\frac{J}{W_0} = 2k\, r_p + 2k\, a \qquad (7.72)$$

$2k$ may be found from Fig. 7.55 by determining r_p and since, as $a \to 0$, $2k \to 8$ (for the single edge notch geometry), then from the intercept in Fig. 7.55 we have that $r_p \simeq 10$ mm. This compares reasonably well with the plastic zone correction, $r_p = 1/2\pi\,(E\,J)/p_Y^2$, using $J_1 = 24$ kJ/m², $p_Y = 8$ MPa, and $E = 220$ MPa, which gives $r_p \simeq 13$ mm. Using $r_p = 10$ mm in conjunction with equation (7.72) yields $2k$ and this is shown as a function of a/D in Fig. 7.56. This may be compared with the linear solutions, since:

$$J \equiv G = Y^2\, p^2\, a$$

$$\frac{J}{W_0} = 2Y^2\, a \text{ where } W_0 = \frac{p^2}{2E}$$

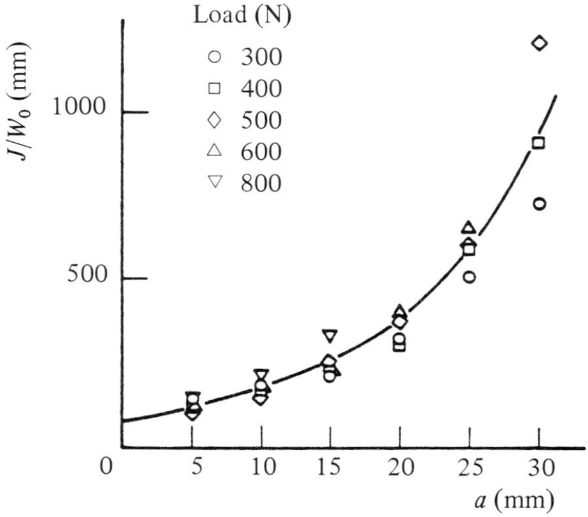

Fig. 7.55 J plotted using the energy per unit volume W_0

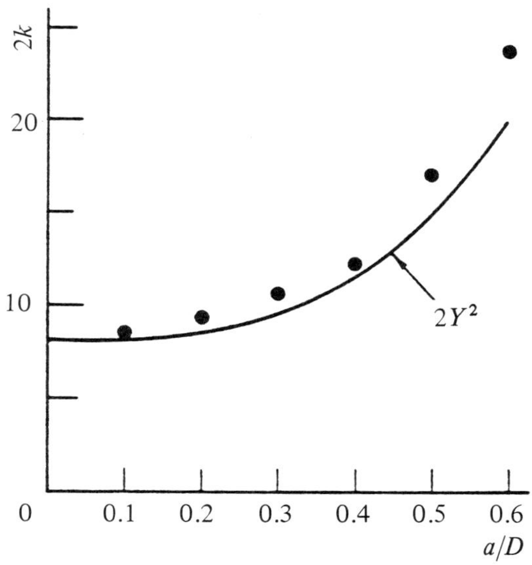

Fig. 7.56 $2k$ from Fig. 7.55 using $r_p = 10$ mm compared with $2Y^2$

and $2Y^2$ for the single edge notch geometry with a constant stress (no rotation) as used in this test is also shown in Fig. 7.56. The agreement is good, indicating that the correction factor is not significantly affected by non-linearity. The analysis is, of course, only of value up to the fully plastic condition, since beyond this the process is deflection controlled.

The Andrews analysis goes on to describe J in the form $J=(2\gamma_0)\,\Phi$, where Φ is a plastic work factor deduced from detailed strain measurements and γ_0 is the true surface work. The determination of Φ is not easy and there is considerable variation in γ_0, but there is some evidence that the method does isolate a crack tip specific value rather as discussed here.

7.6.2 *Ductile instability*

The stability of a fully plastic specimen has received some consideration [26, 29, 30], though the interpretation in terms of energy is still a topic of conjecture. The most simple derivation regards the deforming specimen as being in series with an elastic compliance consisting of the specimen and the machine, and postulates that the system will be unstable when the change of elastic energy will equal the change of plastic energy for an increment of crack growth. This may be written as:

$$P\left(\frac{du_e}{da}+\frac{du_p}{da}\right)>0 \tag{7.73}$$

for instability. This is really equivalent to the $G=G_c$ condition of normal elastic analysis but the dissipation term includes plastic work in this case. Since for fully plastic fracture, we have:

$$J=p_Y\,u_p$$

then:

$$\frac{du_p}{da}=\frac{1}{p_Y}\frac{dJ}{da}$$

The elastic deflection may be deduced from the compliance:

$$u_e=C\,P \text{ and } \frac{du_e}{da}=\frac{d}{da}(C\,P)$$

For the fully plastic case, $P\propto p_Y$ and also $C\propto E^{-1}$, so it is convenient to express equation (7.73) in the form:

$$\frac{d}{da}\left[(E\,C)\left(\frac{P}{p_Y}\right)\right]+\frac{E}{p_Y{}^2}\frac{dJ}{da}>0$$

The term $(E/p_Y{}^2)\,(dJ/da)$ has been called the tearing modulus T [29, 30] and given the subscript m to denote a material property, while $-(d/da)\,[(E\,C)\,(P/p_Y)]$ is the applied tearing modulus T_a, i.e.

$$T_a>T_m \tag{7.74}$$

where:

$$T_a=-\frac{d}{da}\left[(E\,C)\left(\frac{P}{p_Y}\right)\right] \text{ and } T_m=\frac{E}{p_Y{}^2}\frac{dJ}{da}$$

As an example, consider the single edge notch case and, in the first instance, ignore any compliance change due to a. We then have:

$$\left(\frac{P}{p_Y}\right) = B\,(D-a) \text{ and } (E\,C) = \frac{L}{B\,D}$$

giving a final condition of:

$$\frac{L}{D} > T_m \tag{7.75}$$

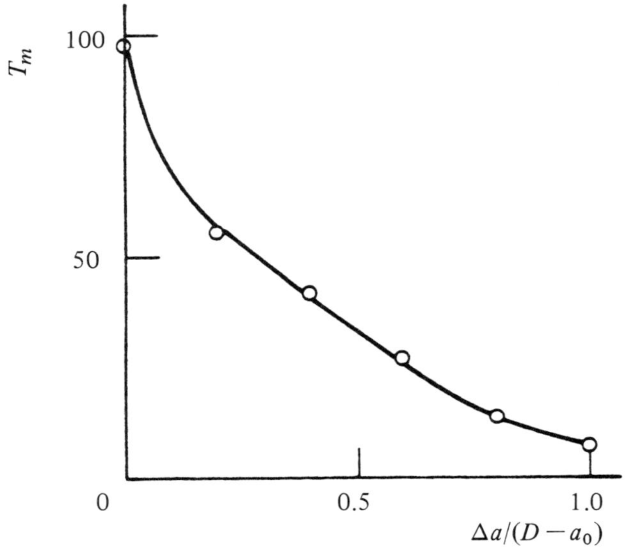

Fig. 7.57 Tearing modulus for polyethylene – from Fig. 7.53

Fig. 7.57 shows T_m as a function of $\Delta a/(D-a_0)$ taken from Fig. 7.53 and it can be seen to vary from about 100 at initiation down to 7 at final failure. For the specimens used, $T_a = 3$ and, as observed, were always stable. Long strips would be expected to show instability. A spring of stiffness k_s in series with the specimen would give:

$$E\,C = \frac{L}{B\,D} + \frac{E}{k_s}$$

and:

$$T_a = \frac{L}{D} + \frac{B\,E}{k_s}$$

Many solutions for other geometries have been worked out [29, 30] and there is some evidence [31] that they work reasonably well for polymers.

References

[1] Williams, J. G. & Marshall, G. P., *Proc. R. Soc. Lond.*, **A342** (1975) 55–77.

[2] Marshall, G. P., *Plastics & Rubber: Processing & Applications*, **2** (1982) 169–182.
[3] Greensmith, H. W. & Thomas, A. G., *J. Pol. Sci*, **18** (1955) 189–200.
[4] Leevers, P. S., *J. Mat. Sci.*, **17** (1982) 2469–2480.
[5] Stadler, B. & Kausch, H. H., *J. Mat. Sci.*, **17** (1982) 2481–2485.
[6] Marshall, G. P., Coutts, L. H. & Williams, J. G., *J. Mat. Sci.*, **9** (1974) 1409–1419.
[7] Morgan, G. P. & Ward, I. M., *Polymer*, **18** (1977) 87–91.
[8] Williams, J. G., *Adv. Pol. Sci.*, **27** (1978) 67–120, Springer-Verlag, Berlin.
[9] Marshall, G. P., Culver, L. E. & Williams, J. G., *Proc. R. Soc. Lond.*, **A319** (1970) 165–187.
[10] Graham, I. D., Williams, J. G. & Zichy, E. L., *Polymer*, **17** (1976) 439–442.
[11] Kinloch, A. J., *J. Mat. Sci.*, **15** (1980) 2141–2166.
[12] Kramer, E. J., Chapter 3 in *Developments in Polymer Fracture*, **1**, (E. H. Andrews, ed.), Applied Science Publishers (1979).
[13] Shanahan, M. E. R. & Schultz, J., *J. Pol. Sci., Pol. Phys. Ed.*, **18** (1980) 1747–1752.
[14] Williams, J. G., Marshall, G. P., Graham, I. D. & Zichy, E. L., *J. Pure & Appl. Chem.*, **39** (1974) 275–285.
[15] Kambour, R. P. & Yee, A. F., *Pol. Eng. Sci.*, **21**(4) (1981) 218–222.
[16] Chan, M. K. V. & Williams, J. G., *Polymer*, **24** (1983) 234–244.
[17] Shanahan, M. E. R. & Schultz, J., *J. Pol. Sci., Pol. Phys. Ed.*, **16** (1978) 803–812.
[18] Shanahan, M. E. R. & Schultz, J., *J. Pol. Sci., Pol. Phys. Ed.*, **17** (1979) 705–710.
[19] Hertzberg, R. W. & Manson, J. A., *Fatigue of Engineering Plastics*, Academic Press (1980).
[20] Constable, I., Williams, J. G. & Burns, D. J., *J. Mech. Eng. Sci.*, **12** (1970) 20–29.
[21] Williams, J. G., *J. Mat. Sci.*, **12** (1977) 2525–2533.
[22] Williams, J. G. & Osorio, A. M. B. A., in *Fatigue Thresholds, II*, (J. Bäcklund, A. F. Blom & C. J. Beevers, eds.), EMAS (1981).
[23] Williams, J. G., *Met. Sci.*, August–September (1980) 344–350.
[24] Schirrer, R., Schinter, M. G., Könczöl, L. & Döll, W., *Colloid & Pol. Sci.*, **259** (1981) 812–817.
[25] Mai, Y. W. & Williams, J. G., *J. Mat. Sci.*, **14** (1979) 1933–1940.
[26] Turner, C. E., Chapter 2 in *Post-Yield Fracture Mechanics*, (D. G. H. Latzko, ed.), Applied Science Publishers (1979).
[27] Hodgkinson, J. M. & Williams, J. G., *J. Mat. Sci.*, **16** (1981) 50–56.
[28] Andrews, E. H., *Makromol. Chem. Suppl.*, **2** (1979) 189–206.
[29] Paris, P. C., Tada, H., Zukhov, A. & Ernst, H., *ASTM STP 668* (1979) 5–36.
[30] Zukhov, A. & Paris, P. C., *Trans. ASME*, **103** (1981) 46–54.
[31] Donovan, J. A. & Prabhat, K., Unpublished work.

8

Impact testing and dynamic effects

8.1 Introduction

Impact testing is widely used in materials evaluation since it represents an attempt to model the most severe abuse to which a material can be subjected. The combination of some sort of notch or stress concentration, together with high rates, gives a severe test of a material's propensity to brittle failure. This may be understood in terms of the effect of these parameters on yield stress since they tend to increase p_Y, the first by increasing constraint and the second by a simple rate-dependence. The use of a bending rest configuration also promotes brittle behaviour, as discussed in Chapter 6, and this is the most common in impact tests. Historically, the tests arose from good intuitive insight rather than a very deep understanding and are embodied in the Charpy and Izod tests. The former uses a three-point bend test, while the latter employs a clamped cantilever bar notched close to the clamp. In both, the bars are struck with a pendulum and the energy to cause fracture is measured by the decrease in swing height. In essence, this is nothing more than a high speed bending test but the problems of measuring loads are considerable because of dynamic effects and the use of energy is a pragmatic solution to the difficulty. Impact strength is usually quoted as energy per unit of ligament area of the specimen and is thus related to the R and J concepts discussed here.

Impact tests have acquired a rather poor reputation in recent years which has arisen from their lack of reproducibility and variation between test methods. Part of the problem has been that they have been extended to model test conditions on finished parts and the connection between these results and the notched bar tests is often complex. In addition, the pendulum tests have been performed on rather poor equipment leading to errors and variability. Above all, however, the tests have not been subjected to rigorous analysis and the simple U/A interpretation is not adequate in deducing the true material parameters, G_c or J_c. Some care is also required in understanding the dynamic nature of the test and if this is not taken, errors occur even in the energy measurement method.

This chapter sets out the fracture mechanics of impact testing and describes how dynamic corrections may be made and G_c or J_c found. Testing under such conditions, while difficult, does give data which it is not possible to determine otherwise. It will be demonstrated that the tests are not unique, as is sometimes assumed, but are a coherent part of fracture mechanics as a whole.

While the impact test requires due attention to be paid to dynamic effects, it is even more important when circumstances involving high-speed crack propagation are concerned. There are cases of practical importance, such as pressurised pipelines, where the behaviour of a fast running crack, were it to occur, is a major concern. Here, kinetic energy effects and stress waves become the dominant factor and present considerable difficulties in both testing and analysis. There has been limited progress in understanding such cases and much remains to be done, but some basic analysis is available which will provide a starting point for future endeavours, and this will be outlined here.

8.2. Dynamic effects in impact testing

The nature of the dynamic character of the impact test can be understood by considering the spring mass model shown in Fig. 8.1. k_2 $(=1/C)$ is the stiffness of

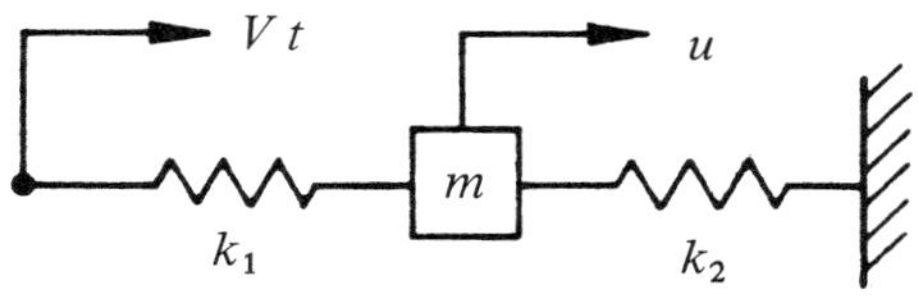

Fig. 8.1 Spring mass model of impact test

the sample and m is its mass. k_1 is the elastic contact stiffness of the striker–specimen interface. For the impact of a large striker, we may assume that it does not slow down appreciably during impact so that the end of the contact spring moves with velocity V. Since the striker and specimen are not attached, k_1 may only compress so that the contact force P, given by:

$$P=k_1\,(V\,t-u) \tag{8.1}$$

where u is the specimen displacement, must always be positive. When $P=0$, there will be loss of contact and a bounce will occur. The equation of motion for the system is:

$$m\,\ddot{u}+(k_1+k_2)\,u=k_1\,V\,t \tag{8.2}$$

with the boundary conditions $u=\dot{u}=0$ at $t=0$. The solution is:

$$u=\frac{V}{\omega}\left(\frac{k_1}{k_1+k_2}\right)(\omega t-\sin\,\omega t) \tag{8.3}$$

where $\omega^2 = (k_1 + k_2)/m$, the natural frequency. Substitution into equation (8.1) gives:

$$\left(\frac{1 + k_1/k_2}{k_1/k_2}\right)\frac{\omega\, P}{k_2\, V} = \left[\omega t + \left(\frac{k_1}{k_2}\right)\sin \omega t\right] \tag{8.4}$$

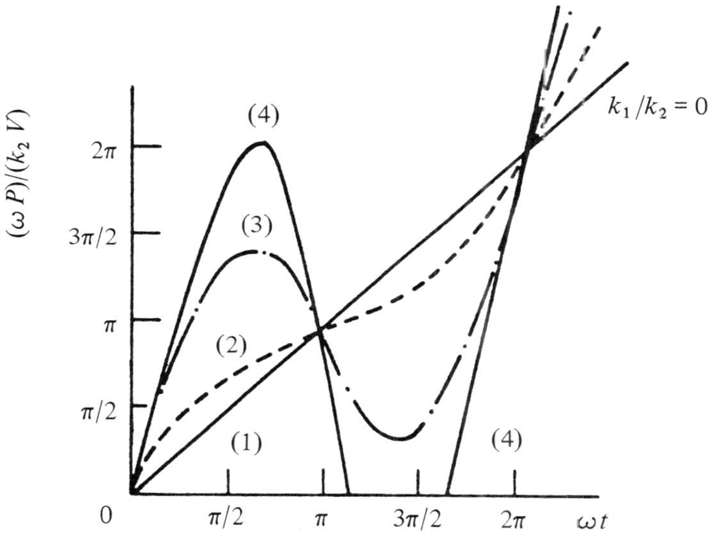

Fig. 8.2 Load–time diagrams

and Fig. 8.2 shows a schematic diagram of the form of this result. For $k_1/k_2 = 0$, we have the linear increase which would be desirable in a test but as k_1/k_2 increases, so does the amplitude of the load oscillations (curves (2) and (3)). The period of the oscillations is given by:

$$\bar{t} = \frac{2\pi}{\omega} = 2\pi\sqrt{m/k_2}\,\frac{1}{\sqrt{1 + k_1/k_2}} \tag{8.5}$$

and decreases as k_1/k_2 increases.

As k_1/k_2 increases, eventually P will go to zero at some value of $\omega t > \pi$, as shown on line (4) in Fig. 8.2, and we have a loss of contact, after which m undergoes a free oscillation governed by the equation:

$$m\,\ddot{u} + k_2\, u = 0$$

with the boundary conditions:

$$u = \frac{V}{\omega}\left(\frac{k_1}{k_1 + k_2}\right)(\hat{\omega t} - \sin \hat{\omega t})$$

and:

$$\dot{u} = V\left(\frac{k_1}{k_1+k_2}\right)(1-\cos\hat{\omega t}) \text{ at } \omega t = \hat{\omega t}$$

This will re-establish contact with the striker after some time when $u = V\,t$ and the load will rise again as shown in Fig. 8.2.

The limiting condition for bouncing to occur is when $k_1/k_2 > 4.62$ and all the bounces happen within the time-span $\pi < \hat{\omega t} < 4.5$. For high values of k_1/k_2, we may obtain a useful approximate solution to the loss of contact condition $(P=0)$:

$$\hat{\omega t} + \left(\frac{k_1}{k_2}\right)\sin\hat{\omega t} = 0$$

or:

$$\hat{\omega t} \simeq \pi\left(1+\frac{k_2}{k_1}\right) \tag{8.6}$$

It is clear from this analysis that, since contact stiffnesses are high compared with those of specimens, considerable load oscillations are likely over short time-periods. If the load rises to a large value (a tough specimen), then the time-scale is longer and problems are less likely, but at low loads they could be significant. This is an inherent problem in simply measuring the loads in impact tests and then applying fracture mechanics. Some efforts have been made [e.g. 1] to measure load-time traces and then find ω and k_1/k_2 in order to correct for the oscillation, but the corrections are large and full of uncertainties; particularly how well the model represents the contact behaviour.

This leads naturally to a consideration of energy and that lost by the striker may be computed from:

$$U = V\int_0^t P\,dt \tag{8.7}$$

which, on substitution from equation (8.4) and integrating, gives:

$$\frac{U}{m\,V^2} = \frac{k_1/k_2}{(1+k_1/k_2)^2}\left[\frac{(\omega t)^2}{2} + \left(\frac{k_1}{k_2}\right)(1-\cos\omega t)\right] \tag{8.8}$$

Again, there is an oscillating character to the response and with bouncing can give discrete values of energy. For the first bounce, we may use the result for high stiffness, equation (8.6), to give:

$$\frac{\hat{U}}{m\,V^2} \simeq 2 + \frac{k_2}{k_1} + \ldots \tag{8.9}$$

The result for $k_2/k_1 \to 0$ is, of course, that for rigid body impact and for more flexible contacts the energy absorbed actually increases. This model of contact behaviour assumes perfectly elastic deformation but, in reality, there are always some losses. It is convenient to model this effect by considering the impact of a large mass M (the striker) and the small mass m (the specimen). If u is now taken as the relative motion of the two bodies, we may write the equation of motion as:

$$m\,\ddot{u}+\left(1+\frac{m}{M}\right)f_e(u)+\left(1+\frac{m}{M}\right)f_d(\dot{u})=0 \tag{8.10}$$

where $f_e(u)$ is the elastic contact stiffness function, and $f_d(\dot{u})$ is a dissipation term. If linear behaviour is assumed, then $f_e=k\,u$, but for point contacts there is non-linearity, e.g. $\propto u^{3/2}$ for impacting spheres [2]. Equation (8.10) may be integrated after first multiplying through by $\dot{u}$ to give:

$$\frac{1}{2}m\,\dot{u}^2+\left(1+\frac{m}{M}\right)\int f_e(u)\,du+\left(1+\frac{m}{M}\right)\int f_d(\dot{u})\,\dot{u}\,dt=A$$

where A is a constant. At $t=0$, $\dot{u}=V$, $u=0$ and both integrals are zero, so that $A=\frac{1}{2}m\,V^2$, giving:

$$\dot{u}=\pm V\sqrt{1-\left(1+\frac{m}{M}\right)\int\frac{f_e(u)\,du}{\frac{1}{2}m\,V^2}-\left(1+\frac{m}{M}\right)\int\frac{f_d(\dot{u})\,\dot{u}\,dt}{\frac{1}{2}m\,V^2}}$$

For the subsequent loss of contact, $u=0$ again and thus so does $\int f_e(u)\,du$, but that for $\int f_d(\dot{u})\,\dot{u}\,dt$ will be finite over the contact time, $\bar{t}$. The relative velocity at loss of contact is thus:

$$V_1=-V\sqrt{1-\left(1+\frac{m}{M}\right)\int_0^{\bar{t}}\frac{f_d(\dot{u})\,\dot{u}\,dt}{\frac{1}{2}m\,V^2}} \tag{8.11}$$

This is true for any form of contact behaviour and reduces to the perfectly elastic case of $V_1=-V$ when there is no dissipation. It is convenient to express this relationship in terms of the Newtonian factor of a coefficient of restitution [3], e, so that:

$$e^2=1-\left(1+\frac{m}{M}\right)\int_0^{\bar{t}}\frac{f_d(\dot{u})\,\dot{u}\,dt}{\frac{1}{2}m\,V^2} \tag{8.12}$$

The energy lost by the striker is now:

$$\frac{U}{m\,V^2}=\left(\frac{1+e}{1+m/M}\right)\left[1-\frac{1}{2}\frac{m}{M}\left(\frac{1+e}{1+m/M}\right)\right] \tag{8.13}$$

and that acquired by the specimen is:

$$\frac{\check{U}}{m\,V^2}=\frac{1}{2}\left(\frac{1+e}{1+m/M}\right)^2 \tag{8.14}$$

in the form of kinetic energy. That lost in the impact is $\frac{1}{2}(1-e^2)/(1+m/M)$. (This analysis is restricted to linear motion and should include a rotational term [see equation (8.44)]. The simplification appears to be adequate for practical purposes).

e may be found for a given test situation by performing impact tests on unrestrained specimens and measuring the energy absorbed, $\hat{U}$. In most impact tests, $m/M \ll 1$ so that equation (8.13) becomes:

$$\hat{U}/(m\,V^2) \simeq 1 + e$$

and Fig. 8.3 shows data on a range of polymers for specimens of different masses and span to depth (L/D) ratios, where $\hat{U}$ is plotted versus $m\,V^2$. There is remarkably consistent behaviour with $e \simeq 0.5$. This number would vary with impact zone conditions and can be slightly higher than 0.5 for glassy polymers. The introduction of energy absorbent materials at the contact point as thin layers can greatly effect e. Elastic foam gave $e \simeq 0.4$, corrugated cardboard $\simeq 0.3$, and plasticine $\simeq 0.1$. This latter seems to be the lowest value achievable in practice.

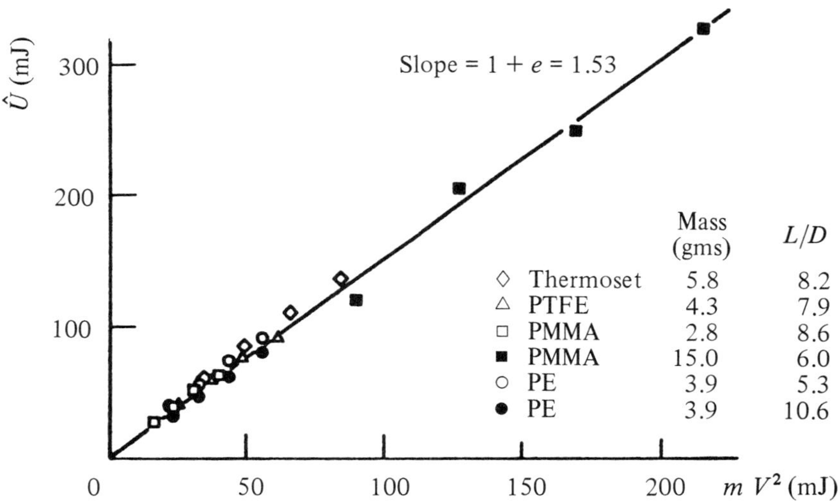

Fig. 8.3 Determination of e

The impact test may be quite accurately modelled by taking the impact between the striker and specimen as that of rigid bodies with a particular value of e. On first impact, the striker continues to move at approximately V (for small m/M), while the specimen moves off at $(1+e)\,V$. This kinetic energy is converted to strain energy as the specimen bends and slows down. At some instant, a second impact will occur when the striker catches up with the specimen and a further transfer of kinetic energy will occur. This is demonstrated in Fig. 8.4 where the specimen displacement is shown as a function of time for three values of e. In the fully elastic case, $e = 1$, the specimen is actually moving backwards at the second impact so that the relative velocity increases, giving a larger energy transfer, and thus, as further impacts result, ever increasing values of energy. For

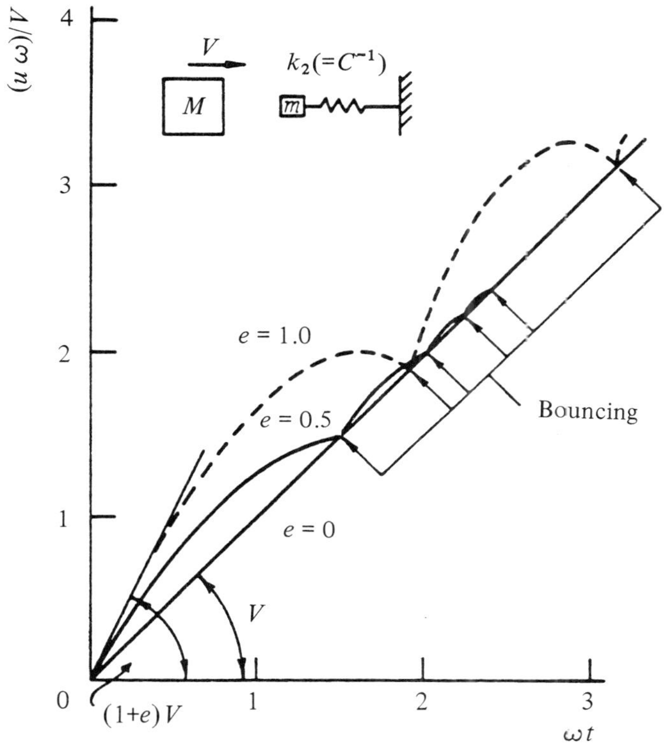

Fig. 8.4 Variation of specimen displacement with time, $\omega^2 = k_2/m$

lower values of e, the relative velocity decreases, thus giving decreasing energy transfer and, as shown for $e = 0.5$ in Fig. 8.4, the bounces will cease altogether, giving continuous contact between striker and pendulum. The limiting value between continuous bouncing and the dying-out condition is $e = \pi/2 - 1 = 0.57$. For $e = 0$, there is no bouncing at all with the striker picking up the specimen at zero time, as shown.

In an impact test, we are particularly concerned with the strain energy within the specimen, $\check{U}$, since it is this which effects fracture. The test measurements are those for that lost by the striker, $\hat{U}$, and it is instructive to consider the relationship between $\hat{U}/(m\,V^2)$ and $\check{U}/(m\,V^2)$ for the same three values of e, as shown in Fig. 8.5. For $e = 1$, the first impact gives a value of $\hat{U}/(m\,V^2)$ of 2 and, when this is transformed into strain energy, $\check{U}/(m\,V^2)$ increases to 2 but then decreases slightly as the specimen moves backwards to the second impact where a further parcel of about 3.2 is imparted, thus demonstrating the diverging nature of this process. For $e = 0$, there is no bouncing, giving a single line of slope unity and an intercept of unity ($\frac{1}{2}\,m\,V^2$ as kinetic energy, and $\frac{1}{2}\,m\,V^2$ dissipated). For $e = 0.5$, we see the repeated impact of converging energy amplitude giving

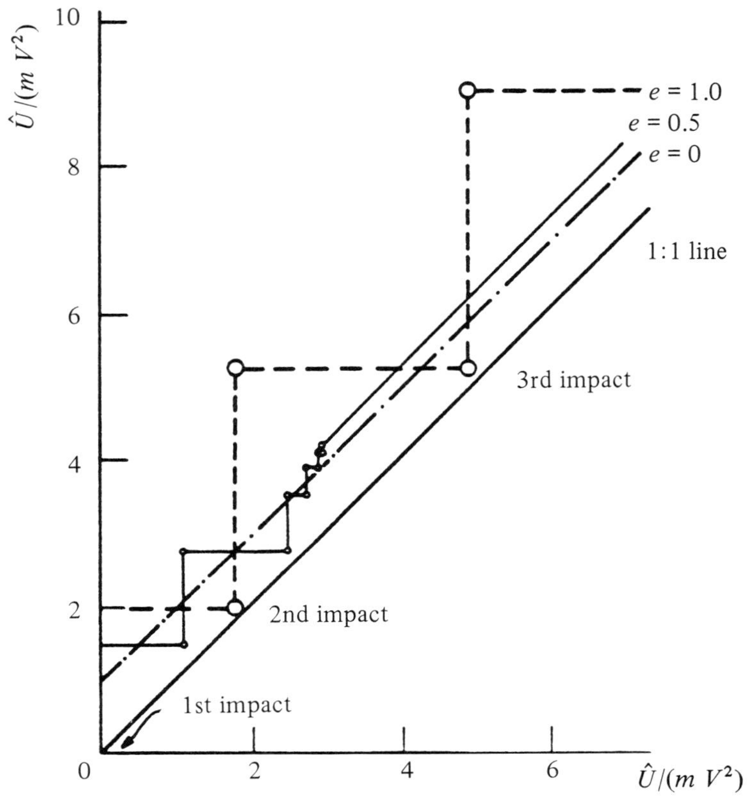

Fig. 8.5 Variation of striker energy loss $\hat{U}$ with strain energy in specimen $\check{U}$

steps of decreasing height until final permanent contact. The crucial fact here is that for $e \neq 0$, the recorded energy $\check{U}$ will be in discrete steps or parcels, at least for values of two or three times $(1+e)\,m\,V^2$ and this represents an inherent limitation in the ability of the test to discriminate energy values in this range. Data from experiments would consist of rows of points along the horizontal portions of the steps (say, from varying the crack length) and, as can be seen from Fig. 8.5, a mean line drawn through these sections for $e \simeq 0.5$ gives a good estimate of the required 1:1 slope.

8.3 Application to small-scale yielding (Class 1)

We will now consider the analysis of impact fractures where the elastic analysis requires no modification and the fracture initiation condition is defined in terms of G_c. If we employ the energy measurement method, then we may use the energy calibration factor, ϕ, described in Section 4.4, since:

$$U = G\,B\,D\,\phi \text{ where } \phi = \frac{C}{dC/d(a/D)} \tag{8.15}$$

and ϕ is listed for several geometries in Table 4.2. An experiment to determine $G = G_c$ would consist of measuring U for various a/D values and then plotting U as a function of $B\,D\,\phi$ to obtain G_c from the slope of the line. Equating U here with that lost by the striker (assuming all extraneous errors are accounted for) assumes that the fracture proceeds to completion immediately on initiation so that the load-deflection diagram is triangular in form with a rapid decrease in load following initiation. This appears to happen for most of the geometries used when brittle failures occur and probably reflects the fact that the initiation G_c value is larger than that required to propagate the running crack. It is possible to conceive of other phenomena occurring, however, and for this it is useful to consider the stability of the system.

For the fracture to be completed without the addition of any external work we must have instability under constant displacement for which (from equations (2.20)) we have:

$$G = \frac{1}{2B}\left(\frac{u}{C}\right)^2 \frac{dC}{da}$$

and the instability condition, from equation (2.20), is:

$$\frac{dG}{da} > 0$$

for constant $R = G_c$ ($dR/dA = 0$) and assuming that kinetic energy changes with crack length may be ignored. These two equations give:

$$C\frac{d^2C}{da^2} > 2\left(\frac{dC}{da}\right)^2$$

which may be expressed in terms of ϕ as:

$$-\frac{d\phi}{d(a/D)} > 1 \tag{8.16}$$

ϕ and $-d\phi/d(a/D)$ for the Charpy test geometry for $L/D = 6$ are shown in Fig. 8.6 and predict that unstable behaviour occurs for $a/D < 0.36$. Since G_c is assumed to be constant, we may also write a total energy balance for the propagation phase since, for propagation to be complete:

$$G_c\,B\,D\,\phi > G_c\,B\,D\left(1 - \frac{a}{D}\right)$$

i.e.

$$\phi + \frac{a}{D} > 1 \tag{8.17}$$

(A more complete discussion of this condition is given in Section 8.7.1). ($\phi + a/D$)

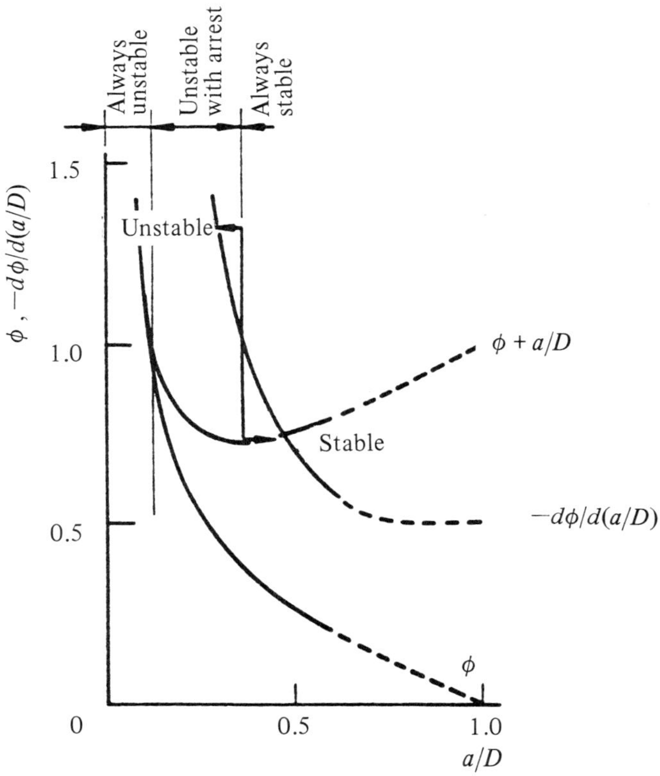

Fig. 8.6 Stability criterion for a Charpy specimen, $L/D = 6$

is also given in Fig. 8.6 and shows that for $a/D < 0.13$, the fracture is unstable and will proceed in this mode to complete failure. For $0.13 < a/D < 0.36$, the fracture will initiate in an unstable manner but will arrest (at $a/D = 0.36$) and complete in a stable manner. For $a/D > 0.36$, the fracture will always be stable. (Note that the condition $d\phi/d(a/D) = -1$ is equivalent to the minimum in $(\phi + a/D)$). Fig. 8.7 shows these various regions of behaviour as a/D plotted versus L/D for the Charpy test and demonstrates that high L/D values give the largest range of unstable behaviour.

In test specimens broken in the stable or stable and arrest conditions, there should be linearity on an U versus $(1 - a/D)$ basis since the total energy is determined by G_c. In the unstable mode, the linearity should be on the basis of U versus ϕ as previously discussed. The critical assumption here, of course, is that G_c in propagation, G_{c_p}, is the same as that for initiation, G_{c_i}, and it seems likely that with less than perfect notching, in reality $G_{c_i} > G_{c_p}$. The instability criterion

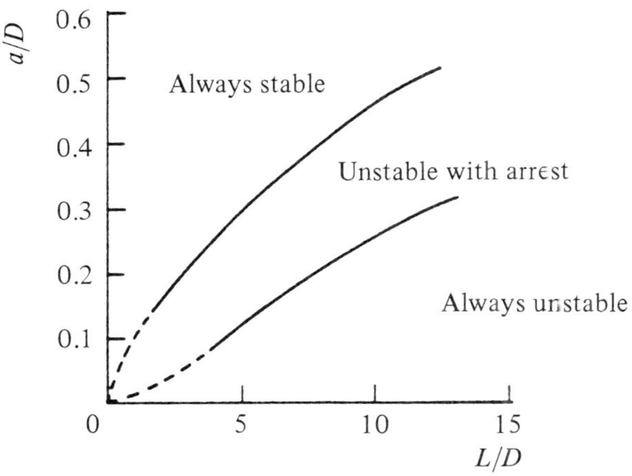

Fig. 8.7 Stability behaviour in the Charpy test for constant G_c

may now be written as:

$$G_{c_i}\, B\, D\, d\phi + G_{c_p}\, B\, da > 0$$

i.e.

$$-\frac{d\phi}{d(a/D)} > \frac{G_{c_p}}{G_{c_i}} \tag{8.18}$$

The data for $-d\phi/d(a/D)$ in Fig. 8.6 shows that for $G_{c_p}/G_{c_i} < \frac{1}{2}$, i.e. $G_{c_i} > 2G_{c_p}$, all tests will be unstable with no arrests.

An example of data without any significant complications is shown in Fig. 8.8, which is for a 12 mm thick HDPE sample tested at 20°C. Good linearity is apparent on the basis of U versus $B\,D\,\phi$ and the intercept is very close to the kinetic energy. A further demonstration of the power of the method is shown in Fig. 8.9, where data for an MDPE from the Charpy and Izod tests are plotted on the same basis (with their somewhat different kinetic energy values subtracted). There is very close agreement in these points, confirming that the analysis gives a geometry independent G_c. In both of these tests, the toughness was fairly high (5 to 8 kJ/m^2) so that, even for the shortest cracks, the energy was significantly greater than the kinetic energy. This may be usefully expressed in terms of a ratio of absorbed energy, $G_c\, B\, D\, \phi$, and the upper bound kinetic energy, $2m\, V^2$; i.e.

$$f = \frac{1}{2}\frac{G_c\, \phi}{L\, \rho\, V^2} \tag{8.19}$$

where ρ is the material density [4]. For the test used in Fig. 8.8, for example, $\phi > 0.4$, $L = 50$ mm (it is the total length here), $\rho \simeq 10^3$ kg/m^3, and $V = 3.36$ m/s, giving $f = 2$. In Fig. 8.10, we have data for a 90 mm long specimen and a slightly lower G_c for which $f \simeq 0.9$, and here we can see evidence of the cyclical variation in U which is expected from the dynamic analysis, and there is a good average line giving $G_c = 4.7$ kJ/m^2. Fig. 8.11 shows this effect exaggerated for a PP

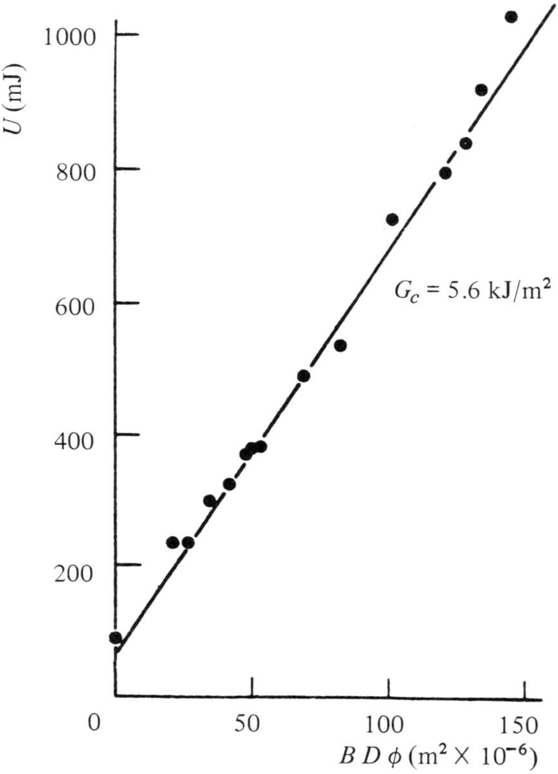

Fig. 8.8 Impact data on HDPE at 20°C ($L = 41$ mm, $B = 12$ mm, $D = 6$ mm), Charpy test

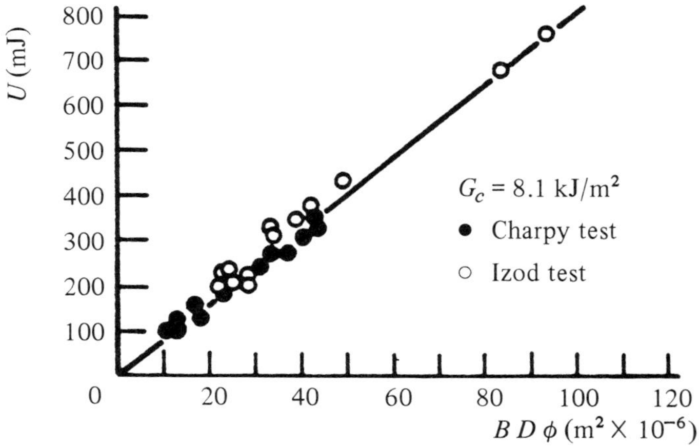

Fig. 8.9 Charpy and Izod data for a medium density polyethylene–kinetic energy subtracted (from reference [5])

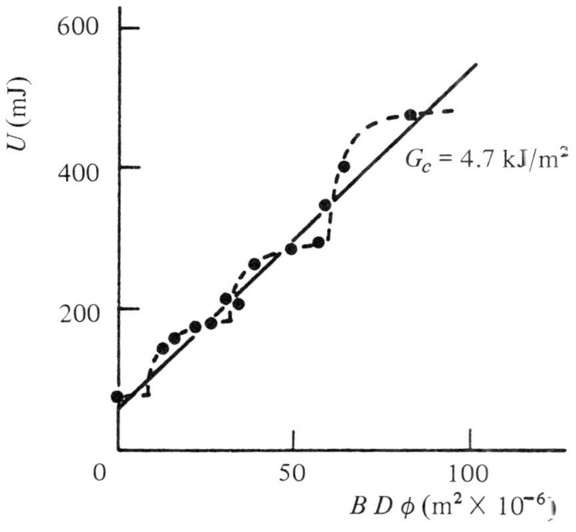

Fig. 8.10 Impact data for HDPE at 20°C ($L = 72$ mm, $B = D = 6$ mm), Charpy test

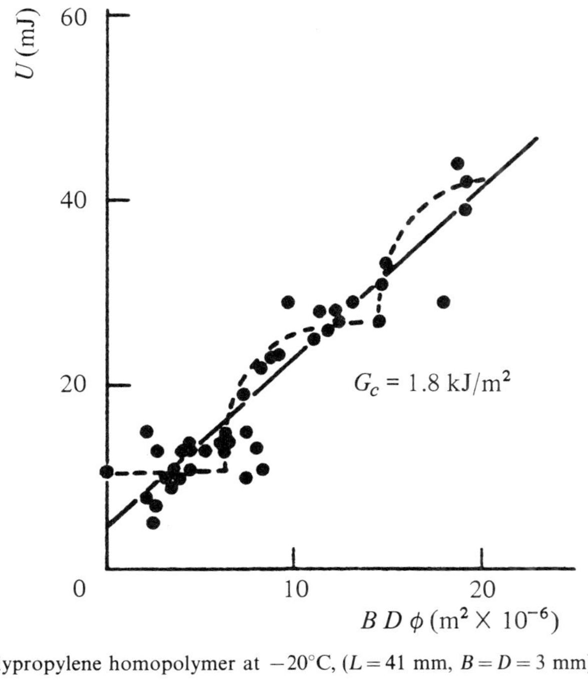

Fig. 8.11 Polypropylene homopolymer at −20°C, ($L = 41$ mm, $B = D = 3$ mm). Charpy test

homopolymer at $-20°C$ for which $G_c = 1.8$ kJ/m^2 and $f \simeq 0.6$. In both cases, however, G_c may be obtained with reasonable confidence, providing sufficient steps (say, >3) are available in the data which means that sufficiently small crack lengths are used. The pitfalls of not observing these rules are demonstrated in Fig. 8.12, which gives data for a low strength, rather dense, composite which, in the usual configuration, had $f \simeq 0.4$ and gave points only around the original plateau [4]. When f was increased to 0.8 by decreasing the span, the first rise is observed but only on slowing down the test so that $f = 2$ did sufficient information become available to determine G_c ($\simeq 0.83$ kJ/m^2) for this range of crack lengths. It might be supposed that the addition of, say, a plasticine layer which gives $e \simeq 0.1$ could be used to overcome the bouncing effect. This is not a very satisfactory solution since in fracture tests the higher loads result in rupture of the layer, very high intercept values and thus high corrections.

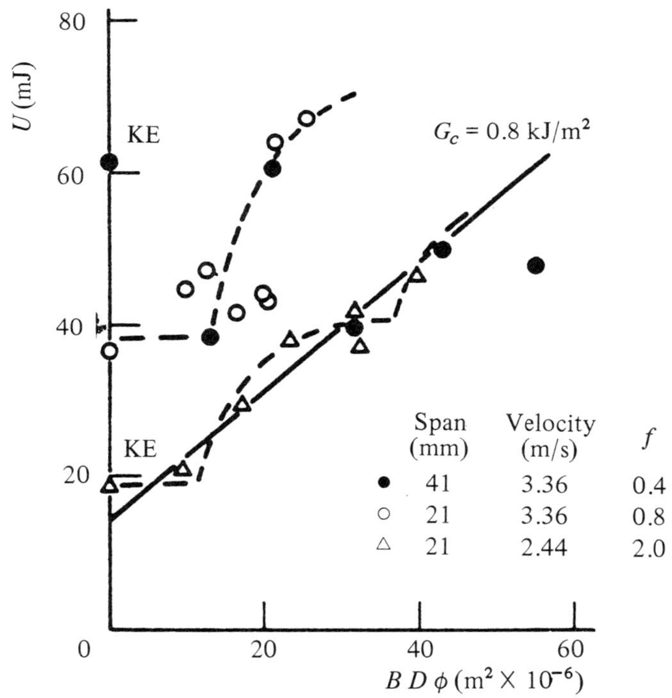

Fig. 8.12 Changes in impact data on brittle materials at low f values (data taken from reference [4])

Fig. 8.13 shows data for PMMA at 20°C where there is some evidence of one bounce at low ϕ values but there is a good line as ϕ increases ($G_c = 0.9$ kJ/m^2). There are also some high points in the low ϕ region. It is possible that these are for specimens which had stable, or perhaps partly stable, crack growth for which

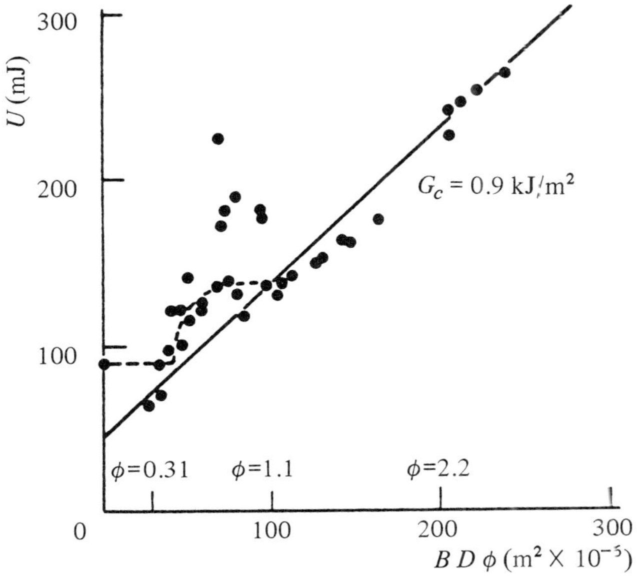

Fig. 8.13 Impact data on PMMA at 20°C ($L=41$ mm, $B=D=9.5$ mm), Charpy test

the correlation is really with $B\,D\,(1-a/D)$ which would give an apparent G_c of about twice that for the long crack specimens, as observed. The results will be discussed further in Section 8.5, when rate effects are considered.

8.3.1 *Temperature and thickness effects*

These two phenomena may be treated together conveniently since it seems likely that both arise from variations in the yield stress. Fig. 8.14 shows G_c values for several polymers over a range of temperatures [6] and there is clearly a very similar sigmoidal form to the curves. The most reasonable explanation of this is that the materials have essentially temperature independent plane stress and plane strain values of G_c, G_{c2} and G_{c1}, and that the measured toughness is an average of these governed by the plastic zone size and thus yield stress, as discussed in Section 5.3. For a layer model using the G_c values, we have:

$$G_c' = G_{c1} + \frac{E\,G_{c2}}{\pi\,B\,{p_Y}^2}\,(G_{c2} - G_{c1}) \tag{8.20}$$

(This is derived on the same basis as equation (5.11) but uses an average G_c rather than K_c). Since p_Y increases with decreasing temperature, then $G_c' \to G_{c1}$ at low temperatures and, similarly, thicker specimens will also show lower G_c' values. Estimating p_Y for use in an impact analysis presents some difficulty and in Fig. 8.15 values from a moderate rate tensile test are used to plot G_c' for three

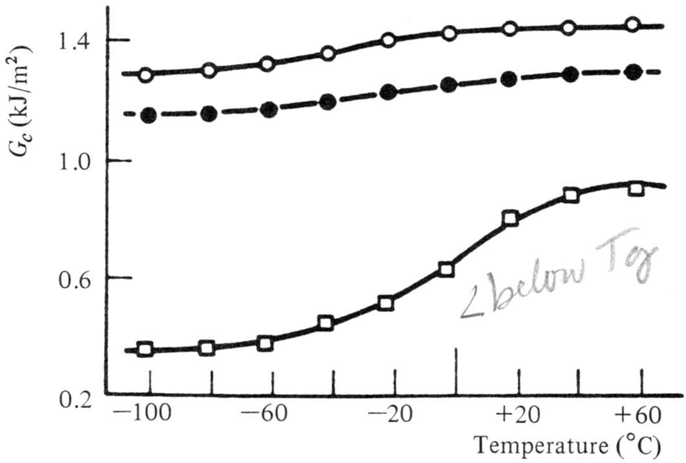

(a) Low impact strength materials; ○ PVC, ● PMMA, □ PS

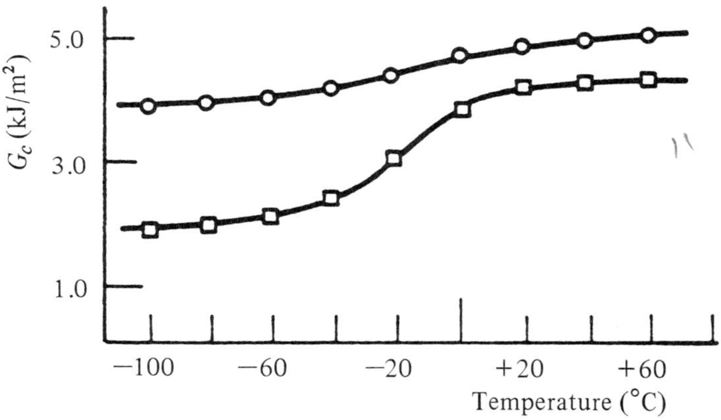

(b) Medium impact strength materials; ○ polycarbonate, □ nylon-66 (dry)

thicknesses versus ${p_Y}^2$ taken at the appropriate temperatures for a MDPE. The absolute values of p_Y may be in error but the trend is likely to be correct and the linearity with extrapolation to the same G_{c1} value confirms this assumption. In Fig. 8.16, the slopes of these lines, $dG_c'/d{p_Y}^{-2}$, are plotted versus inverse thickness for both the MDPE and for similar data on PC, again confirming the relationship. G_{c1} may be estimated directly from G_c' versus B^{-1} plots, as demonstrated in Fig. 8.17 which shows data for an HDPE at 20°C. The two sets of points are for the different rates (see Section 8.5) and illustrate the small variation in G_{c1} with rate over this range.

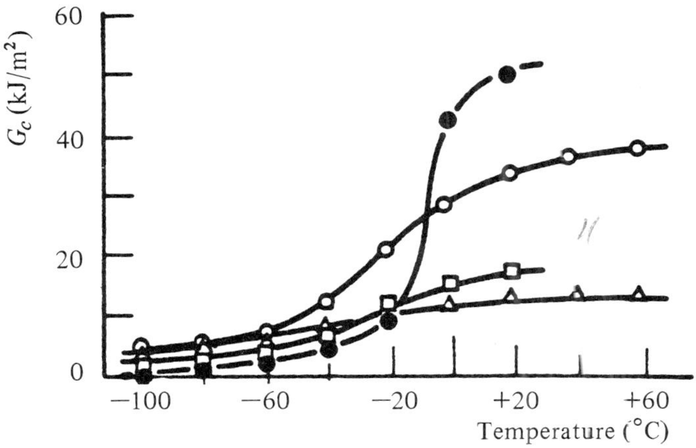

(c) High impact strength materials; ○ LDPE, △ MDPE, □ HIPS, ● ABS

Fig. 8.14 Variations of G_c in impact with temperature (from reference [6])

The ductile–brittle transition can also be couched in these terms since it may be taken as the condition when $G_c' = G_{c2}$, i.e.

$$r_p = \frac{1}{2\pi}\frac{E\,G_{c2}}{p_Y^{\,2}} = \frac{B}{2}$$

If p_Y is taken as approximately linear and going to zero at T_s with a reference value of p_0 at T_0, then we have:

$$\bar{T} = T_s - (T_s - T_0)\sqrt{2r_0/B} \tag{8.21}$$

where $\bar{T}$ is the ductile–brittle transition temperature, and $r_0 = (1/2\pi)\,(E\,G_{c2})/p_0^{\,2}$. T_s will often be T_g, but it should be noted that $\bar{T} \simeq T_g$ only for large B and, in general, $\bar{T} < T_g$.

Table 8.1 lists some estimates of G_{c1} and G_{c2} for several polymers. There is only an approximate correlation with the low rate initiation J_{c1} and J_{c2} values given in Chapter 5. These differences are important and give a genuine indication of the degree of rate dependence of the energy absorbing mechanisms involved. Of particular note is PP, where the J_{c1} at low rates of 10 kJ/m^2 reduces to about 0.5 kJ/m^2 in impact, a reflection of the absence of microvoiding in impact.

8.3.2 *Blunt cracks and loss peak effects*

Many practical impact tests are carried out with notches of a finite tip radius. This is frequently rationalised on the basis that they are more easily reproduced than sharp cracks and that, although less severe, they are a more reasonable representation of practical conditions. In fracture mechanics terms, they present

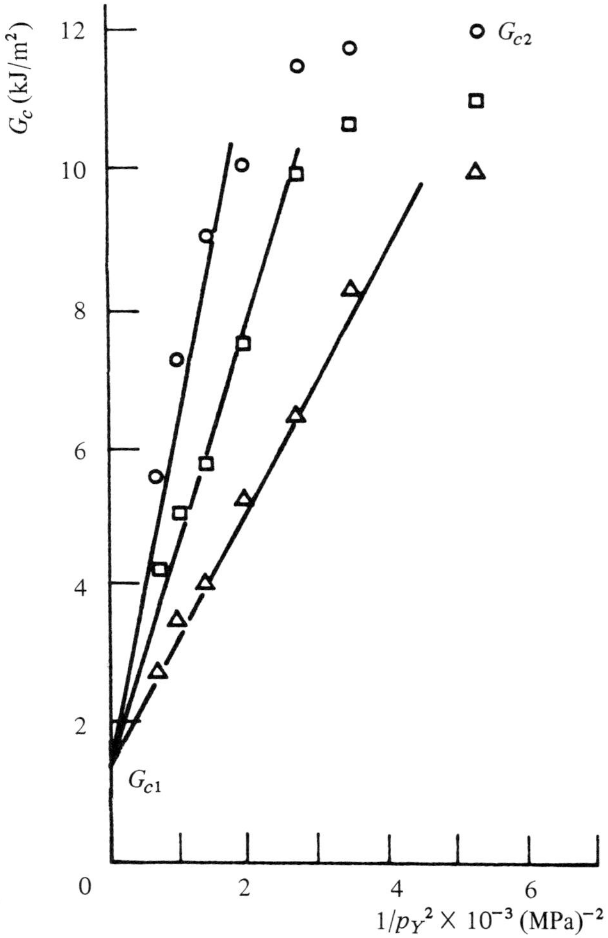

Fig. 8.15 Variation of impact toughness with yield stress for MDPE for three specimen thicknesses; ○ $B = 3$ mm, □ $B = 6$ mm, △ $B = 11$ mm

some difficulty in having a non-singular stress field but by using the concept of a critical stress, p_c, at a critical distance, l_0, an equivalent fracture parameter for a blunt notch may be deduced so that their effects may be encompassed within the usual analysis. The fracture criteria may be taken from equation (4.9):

$$p\sqrt{\pi a} = K_I \frac{(1+\rho/2l_0)^{3/2}}{(1+\rho/l_0)} \tag{8.22}$$

from which a blunt crack G_c value, G_B, may be deduced:

$$\frac{p^2 \pi a}{p_c{}^2 \, 2\pi l_0} = \frac{G_B}{G_c} = \frac{(1+\rho/2l_0)^3}{(1+\rho/l_0)^2} \tag{8.23}$$

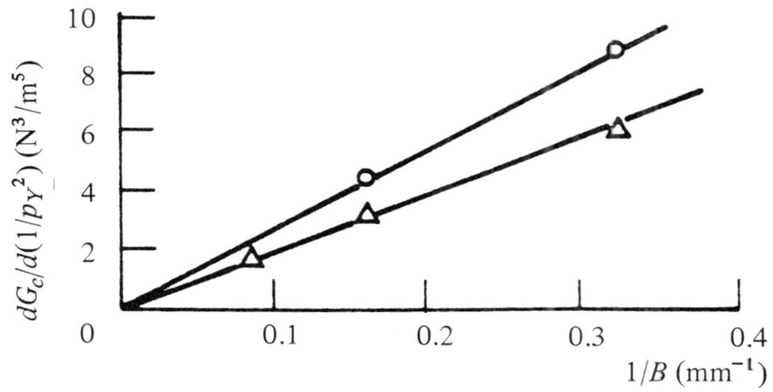

Fig. 8.16 Variation of slope of G_c versus $1/p_Y{}^2$ lines with with specimen thickness; ○polycarbonate, △polyethylene (medium density) (from reference [6])

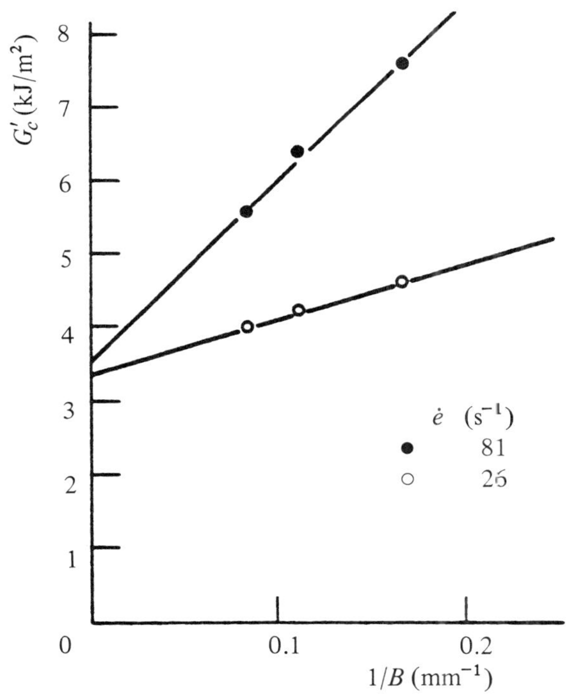

Fig. 8.17 Apparent impact toughness G_c' as a function of inverse thickness for HDPE at 20°C

For $\rho \gg l_0$, this may be approximated to:

$$\frac{G_B}{G_c} = \frac{1}{2} + \frac{\rho}{8l_0} \tag{8.24}$$

so that if G_B is plotted as a function of ρ, a straight line should result with a slope

Table 8.1

Plane stress, plane strain G_c and critical distance values

Material	G_{c1} (kJ/m^2)	G_{c2} (kJ/m^2)	l_0 at 20°C (μm)
PMMA	1.06	1.28	88
PS	0.35	0.90*	—
uPVC	1.23	1.44	28
PVC (modified)	—	10.0†	—
PC	3.5	5.02	14
Nylon 66 (dry)	0.25	4.15	—
HDPE	0.7	6.0	89
MDPE	1.3	11.9	17
LDPE	5.0	35.0*	—
HIPS	1.0	15.0	—
ABS	1.6‡	27.0*	—
PP	0.5	—	—

*Estimated values, no yield stress data available
†Room temperature value, approximation
‡Low temperature value, approximation

of $(8l_0)^{-1}$ and an intercept of $G_c/2$. To obtain a G_B value in the usual way, we must have an unstable brittle fracture and blunt notches do tend to induce ductile failure because of the higher G_B values. The method is thus most suitable for rather brittle materials and Fig. 8.18 shows some data on an unmodified

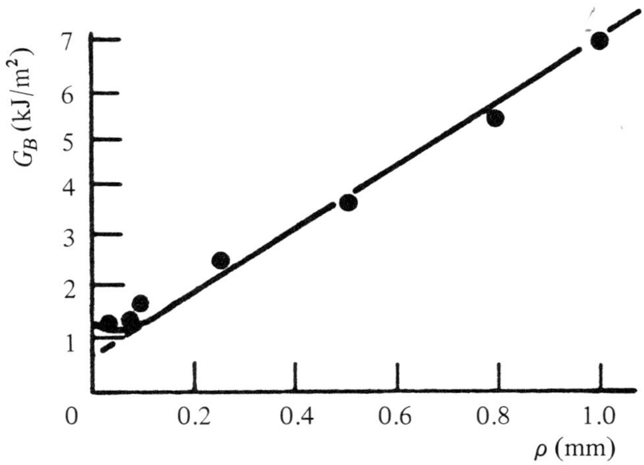

Fig. 8.18 Variation of G_B with notch root radius for PVC (from reference [5])

PVC which is of this type and gives the expected form of relationship. Table 8.1 includes a column of l_0 values measured at room temperature which are in the range 10 to 100 μm. Equation (8.24) shows that low values of l_0 will ensure a rapid increase in G_B with blunt notches which, in turn, will give a high impact toughness under these conditions. It says nothing, of course, about the sharp notch case. The data for PC and PMMA are of particular interest here and it is also worth noting that notch preparation can greatly affect l_0. For PMMA, for example, the 88 μm values are for smooth milled notches but this increased to 180 μm when rather rougher drilled notches were used. In the latter, crazing on the notch root could be seen and it seems likely that this figure represents the craze size.

Rather interesting results are obtained by performing blunt notched tests over a temperature range. Fig. 8.19 shows G_B as a function of temperature for PC

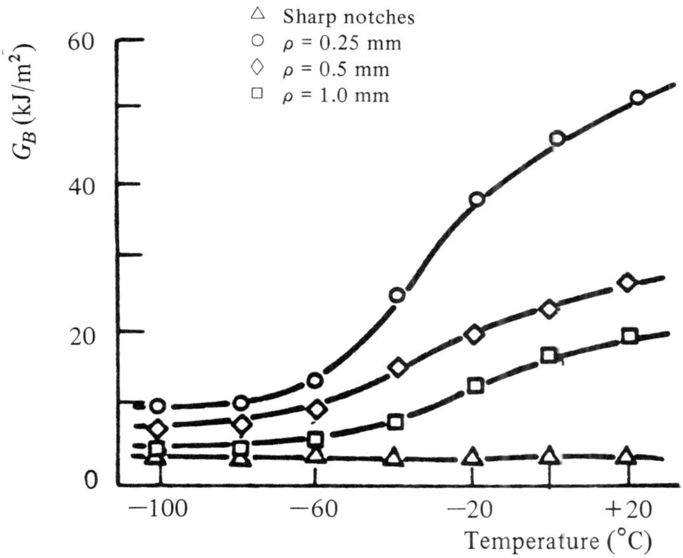

Fig. 8.19 Blunt notch data for polycarbonate (from reference [6])

where it can be seen there is a pronounced amplification in the variation with temperature as ρ increases. This is more dramatically illustrated in Fig. 8.20 for PMMA where a peak appears at about −60°C which is greatly amplified as ρ increases. These data may be used to deduce l_0 as a function of temperature and some results for PMMA, PC and uPVC are shown in Fig. 8.21. It would be expected that l_0 would stay sensibly constant with an apparent drop at higher temperatures where the onset of more ductile failures would give an increase in G_B not due solely to the l_0 effect. PVC shows this pattern with a superimposed small dip at about −20°C. In PMMA, the dip is more distinct and at about −55°C, while in PC it is broad (−80°C to 20°C) and obscures the general trend.

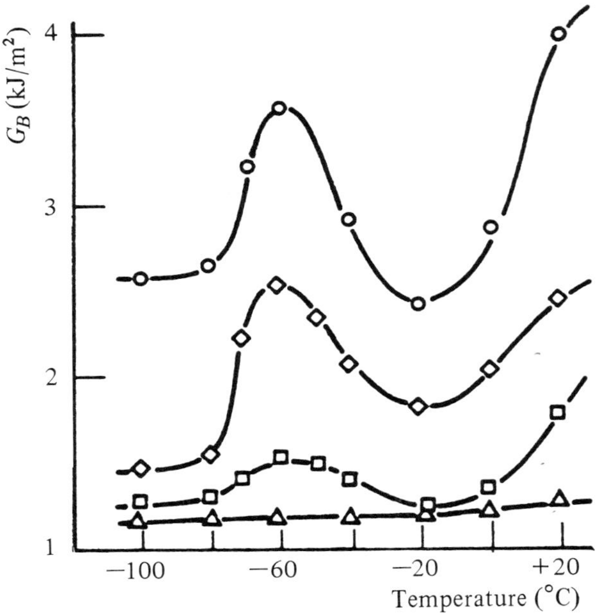

Fig. 8.20 Blunt notch data for PMMA; △sharp notches, □$\rho=0.25$ mm, ◇ $\rho=0.5$ mm, ○ $\rho=1.0$ mm (from reference [6])

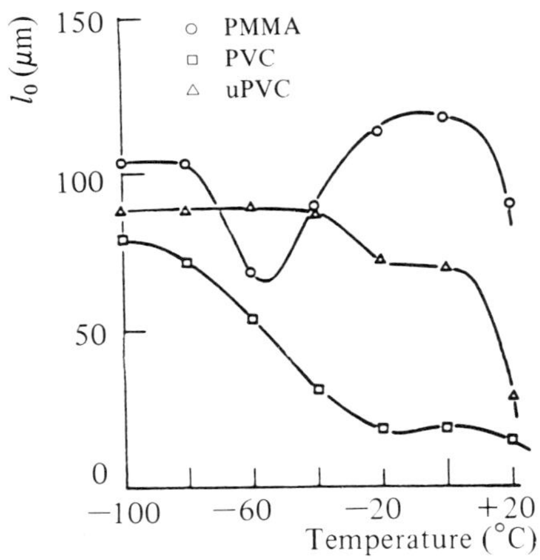

Fig. 8.21 Variation of critical distance l_0 with temperature in impact

In all three cases, the presence of the dips corresponds to and has the characteristics of a β peak in tan δ.

In some materials, there are peaks in sharp notch data which also correlate quite well with loss peaks and Figs. 8.22 and 8.23 illustrate this for PTFE with

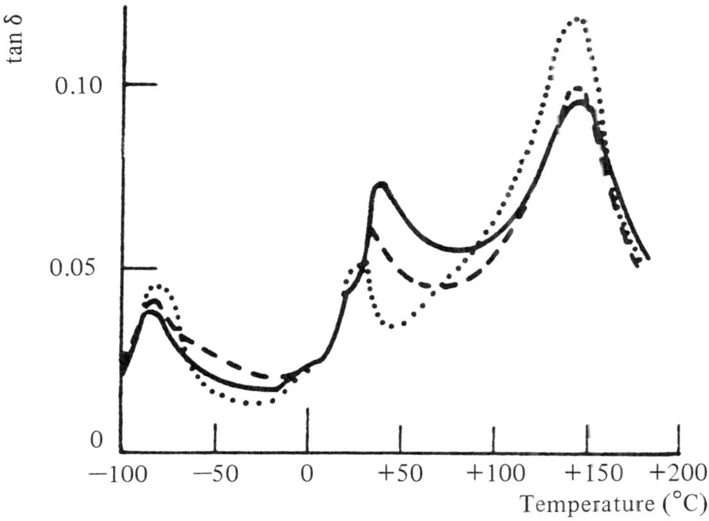

Fig. 8.22 Loss factor versus temperature for the commercial and heat treated PTFE at 11 Hz. — Commercial, - - - slow cooled, ··· quenched (from reference [7])

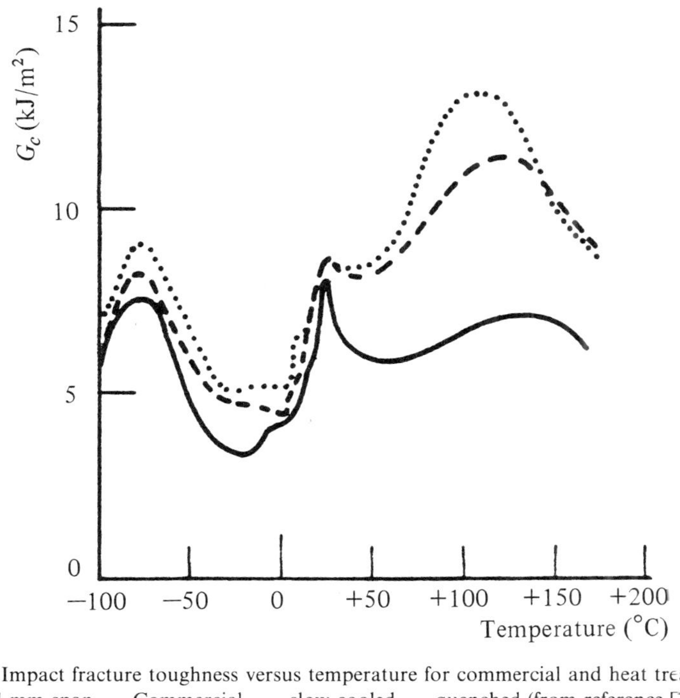

Fig. 8.23 Impact fracture toughness versus temperature for commercial and heat treated PTFE, 41 mm span. — Commercial, - - - slow cooled, ··· quenched (from reference [7])

three types of heat treatment [7]. There are four peaks within the temperature range $-100°C$ to $+160°C$ and the G_c data on sharp notches reflect all of them. There has been much debate on the meaning of such correlations and it seems likely that it does not arise directly from the energy absorbing process which gives the tan δ peak, but that the presence of the peak promotes an additional process, e.g. crazing or microvoiding. Such processes are stabilised by the presence of visco-elasticity and it seems likely that an increase will give an increase in energy absorption and thus in G_c (or G_B). PTFE has a particulate type of structure giving rather diffuse energy absorption, even with sharp notches, while for other polymers, such as PMMA, PC and uPVC, it is necessary to promote more wide-scale absorption, as with blunt notches, before the tan δ effect becomes apparent. The decreases in l_0 noted in Fig. 8.21 are thus really reflections of increase in G_B when the additional process becomes active.

8.4 Application to contained yielding (Class 2)

For tougher materials tested in the conventional sizes, the presence of the plastic zone can influence the results and corrections are sometimes necessary. If we use the concept of adding r_p to the crack length where (see Section 5.4):

$$r_p = \frac{1}{2\pi} \frac{K_c^2}{p_Y^2}$$

then the corrected G_c values (J_c) may be found. Errors from larger r_p sizes can be detected when the U versus $B\,D\,\phi$ graphs are plotted. This is illustrated in Fig. 8.24, where $U/(B\,D\,G_c)$ is plotted versus apparent ϕ, where the a/D values have been reduced for various fixed r_p/D values for $L/D=6$, i.e. ϕ has been computed on each line for $(a-r_p)/D$ to give the apparent value. (Kinetic energy is not included). Thus, data containing this type of error do not follow the straight line but fall below with curvature. The broken line is an estimate of the fully plastic condition when $a=r_p$. (A more precise form may be obtained by integrating the Y_p^2 expressions from Section 4.4).

An example of data showing this effect is given in Fig. 8.25 for MDPE. The original data are shown, together with those for the same points with 0.5 and 1 mm added to a ($D=6$ mm). It is clear that $r_p=0.5$ mm provides a suitable correction and that 1 mm is too great. In general, it is difficult to calculate r_p in impact tests since p_Y is not known. A procedure of least squares fitting in which a best straight line is deduced by increasing r_p until the standard deviation is a minimum can be employed. This can be done numerically and, providing r_p/D is not too large (<0.2), good results are achieved. For small r_p/D, a useful approximation is to draw a tangent line through the kinetic energy to the uncorrected points. It should be noted that data of this type do not give linearity on the basis of ligament area, as is shown in Fig. 8.26 where the same points are plotted in this way.

In some cases, the presence of a substantial plastic zone will result in stable

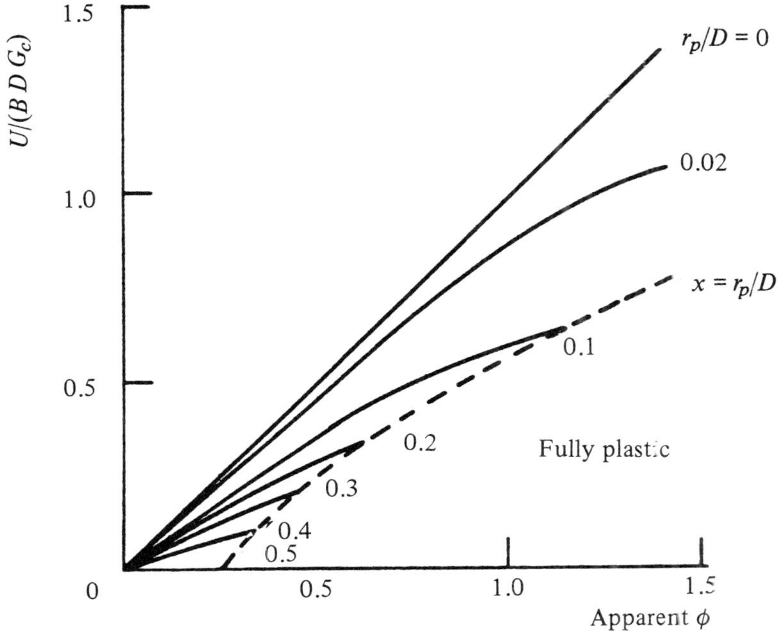

Fig. 8.24 Plastic zone size effects in impact data

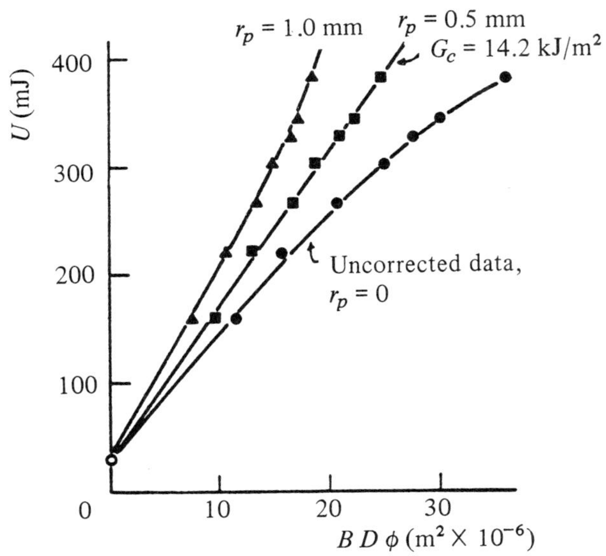

Fig. 8.25 Plastic zone correction applied to MDPE data (from reference [4])

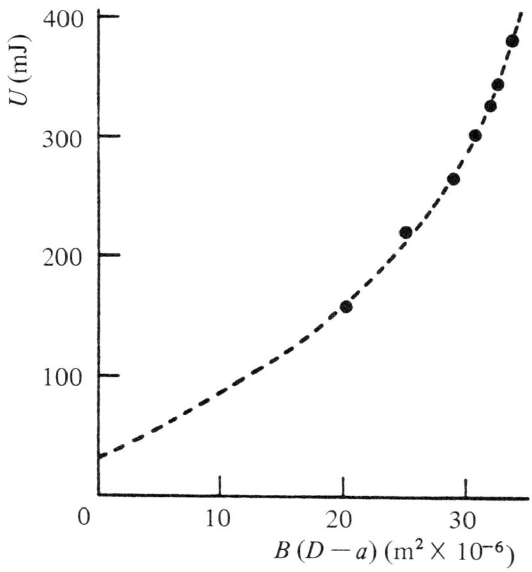

Fig. 8.26 Data from Fig. 8.25 plotted on a ligament area basis (from reference [4])

crack growth so that the correlation becomes linear on a ligament area basis. Such a process seems to require a very robust toughening mechanism and has been observed in toughened polymers. Fig. 8.27 shows data for ABS at 60°C for two spans in the Charpy test [8] and there is very good linearity with ligament area, except for the very smallest crack lengths where there is some divergence. The intercept is negative here and it is possible that this arises from the development of the plastic zone, as illustrated in Fig. 8.28. In Fig. 8.28(a), we have a large ligament so that the zone develops to its full depth, r_0, for propagation and, if we suppose an energy per unit volume of ω_p, then the energy is:

$$U = B\,(D-a)\,r_0\,\omega_p - B\,r_0^{\,2}\,\omega_p\,k$$

where k is a factor determined by the zone shape, and we may deduce G_c from the slope, i.e.

$$G_c = -\frac{dU}{B\,da} = r_0\,\omega_p$$

giving a final result:

$$U = B\,(D-a)\,G_c - B\,r_0\,G_c\,k \qquad (8.25)$$

For small ligament area, the zone is not fully developed, as shown in Fig. 8.28(b), so that:

$$U = (1-k)\,B\,(D-a)^2\,\omega_p = (1-k)\,\frac{B}{r_0}\,(D-a)^2\,G_c \qquad (8.26)$$

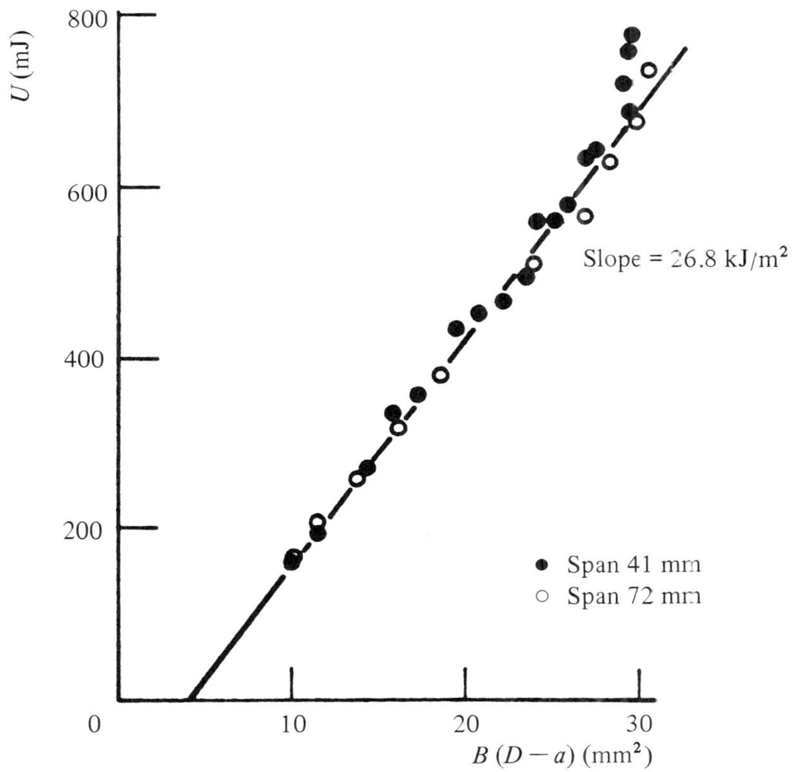

Fig. 8.27 Energy as a function of ligament area – ABS at +60°C (from reference [8])

The expected form of the energy versus ligament area graph is shown in Fig. 8.28(c) and from Fig. 8.27 we have $G_c = 26.8$ kJ/m^2 and, from the intercept, $k\ r_0 = 1$ mm. Direct measurement of r may be made in ABS since stress whitening occurs and $r/2$ versus $(D-a)$ at two temperatures is shown in Fig. 8.29. The zone shape assumption is that:

$$r\,(D-a) = (D-a)^2\,(1-k)$$

i.e.

$$\frac{r}{2} = \left(\frac{1-k}{2}\right)(D-a)$$

and the linear form in Fig. 8.29 is as predicted with $k = 0.83$ at $+20$°C and 0.75 at $+60$°C (a circular zone would give $k = \pi/4 = 0.78$). Using this latter figure, we have $r_0 = 1.3$ mm at $+60$°C and this is also shown in Fig. 8.29. There is a minimum value of zone size of 0.1 mm which suggests, as in the slow rate tests, the existence of a crack tip specific energy. Since r_0 is known, we may find ω_p

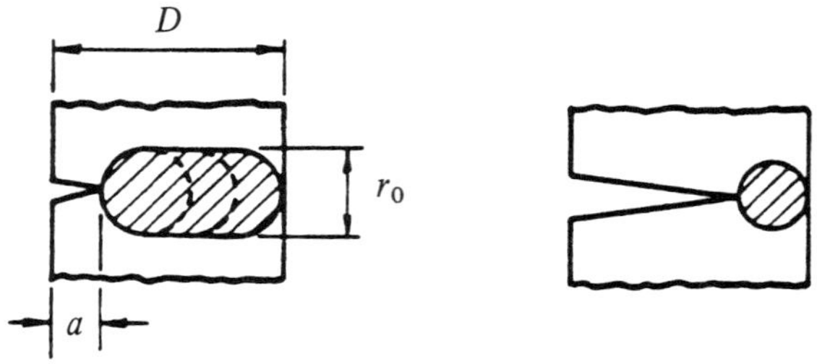

(a) Large ligament area, $(D-a)>r_0$ (b) Small ligament area, $(D-a)<r_0$

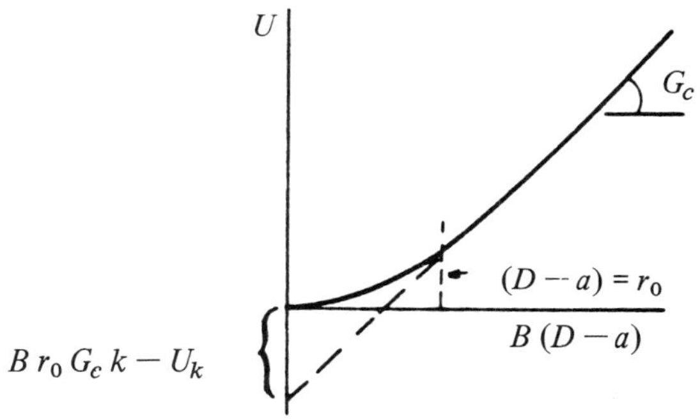

(c) Energy–ligament area graph

Fig. 8.28 Stable crack propagation with a large plastic zone

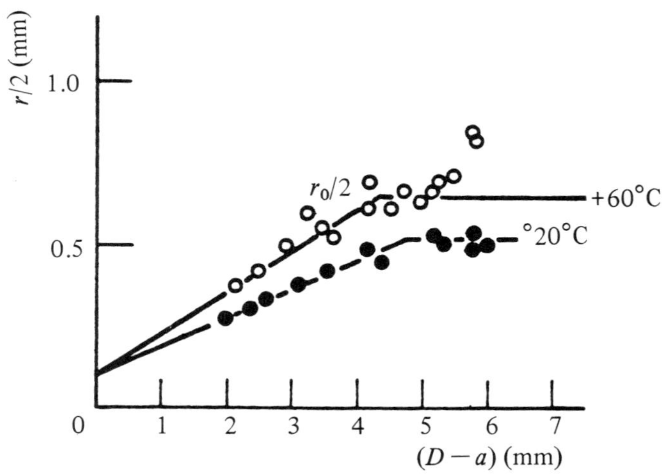

Fig. 8.29 Craze zone half width as a function of ligament length

since:

$$\omega_p = \frac{G_c}{r_0} = 21 \text{ MJ/m}^3$$

and thus the crack specific G_c, $G_{c1} = 2.1 \text{ kJ/m}^2$. These figures are further confirmed in Fig. 8.30 where energy is plotted as a function of crazed material volume measured directly from the specimens for a range of temperatures. It can be seen that $\omega_p = 21 \text{ MJ/m}^3$ is constant over the temperature range with an intercept 50 mJ, the kinetic energy.

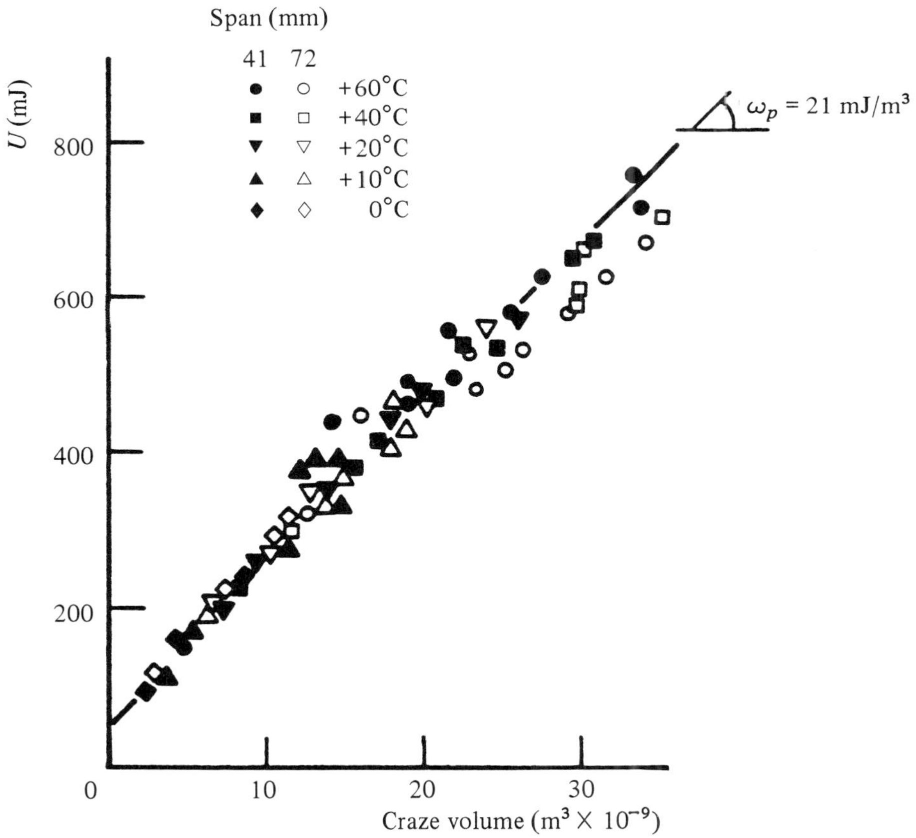

Fig. 8.30 Energy to failure in ABS as a function of volume of crazed material

At temperatures below about 0°C, ABS shows a transition from stable to unstable behaviour. Specimens with long crack lengths have a fully whitened ligament as at higher temperatures but for shorter cracks, there is a banded surface where the whitened crack tip is followed by an uncrazed dark surface.

Fig. 8.31 shows data for $-10°C$ plotted as energy versus ligament area and the sharp transition is apparent. The unstable data do not give linearity on a $B\,D\,\phi$ basis unless the energy is reduced by that used in the stable growth given by $G_c\,B\,\Delta a$, where G_c is obtained from the stable data and Δa is the zone length. In addition, ϕ must be calculated by adding a correction to the original crack length and about $\frac{3}{4}\Delta a$ gives good linearity. When this correction is made, the unstable points from Fig. 8.31 give the data shown in Fig. 8.32, where an unstable G_c value of 7.8 kJ/m^2 is obtained which compares with 11.3 kJ/m^2 in stable growth.

Below $-20°C$, all the failures are unstable in this geometry, giving linearity on a $B\,D\,\phi$ basis with no evidence of whitening. The resulting G_c values as a function of temperature are shown in Fig. 8.33 where the three zones of behaviour are indicated. A possible interpretation of these data is that the higher values, which involve the whitening mechanism, give rise to high initiation values which are sustained during propagation at high temperatures. The lower values, without whitening, represent a low energy propagation so that, at $-10°C$ for example, we have a ratio of $G_{c_p}/G_{c_i} \simeq 0.7$ which gives an unstable–stable a/D ratio of 0.5 using equation (8.18) and Fig. 8.6. Fig. 8.31 gives a transition at $a_0/D \simeq 0.42$ which is in quite close agreement with the prediction. The slight upturn in the data at $+60°C$ seen in Fig. 8.27 may indicate some

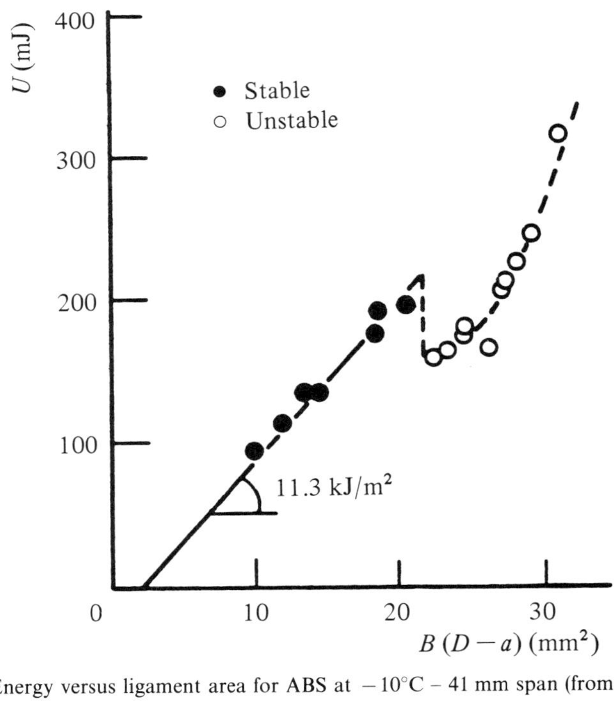

Fig. 8.31 Energy versus ligament area for ABS at $-10°C$ – 41 mm span (from reference [8])

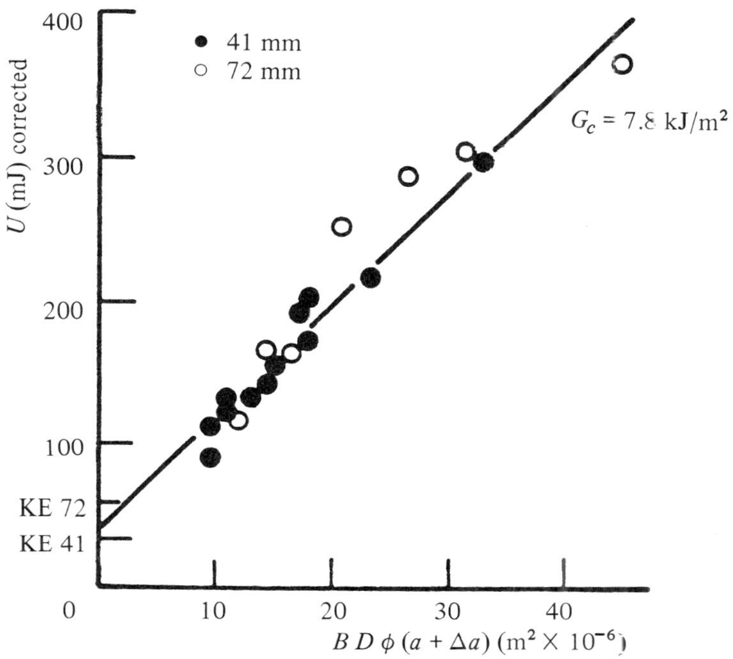

Fig. 8.32 Unstable data for ABS at −10°C plotted in corrected form (from reference [8])

unstable behaviour for very short crack lengths, even at this high temperature, but the effect is too slight to allow any measurements and suggests that $G_{c_p} \simeq G_{c_i}$.

Similar effects have been observed in HIPS, although the toughness values are rather lower [9]. Fig. 8.34 shows data for a 2.2% rubber content material plotted on a ligament area basis and, as in Fig. 8.31, there is linearity at low areas but a considerable divergence at higher values. On a $B\,D\,\phi$ basis the data show pronounced curvature. In this case, there is no banding on the surface and the whole surface is whitened, suggesting that the ligament is completely yielded in all cases and that this is then broken in either a stable or an unstable mode. In the latter case, we can write down a total energy of:

$$U = B\,(D-a)\,r_0\,\omega_p - B\,r_0^2\,\omega_p\,k + U_k + B\,D\,\phi\,G_c \tag{8.27}$$

Here, G_c is for unstable propagation and it does not appear that the stable propagation case of $G_c = r_0\,\omega_p$ occurs. From the slopes and intercepts in Fig. 8.34, we have $r_0\,\omega_p \simeq 2.7$ kJ/m^2 and $k\,r_0 \simeq 1.8$ mm. G_c may be found by rearranging equation (8.27) in the form:

$$\frac{[U + B\,r_0^2\,\omega_p\,k - U_k]}{B\,D\,\phi} = G_c + \left[\frac{1-a/D}{\phi}\right] r_0\,\omega_p$$

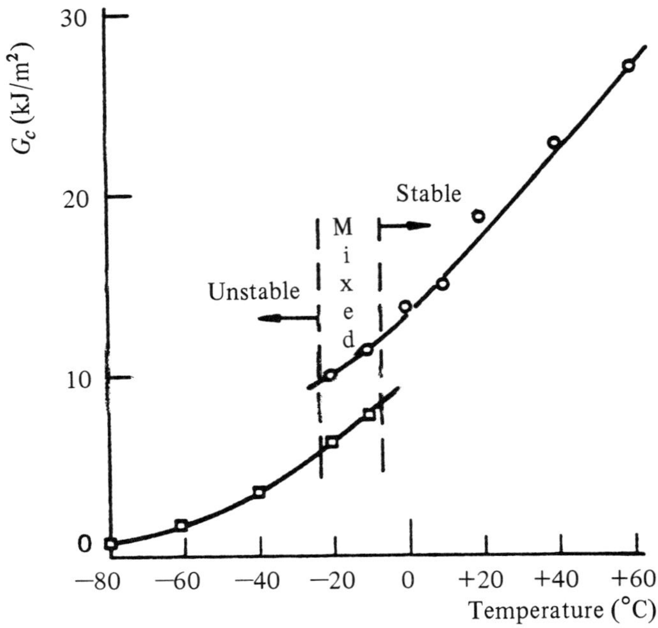

Fig. 8.33 Impact fracture of ABS

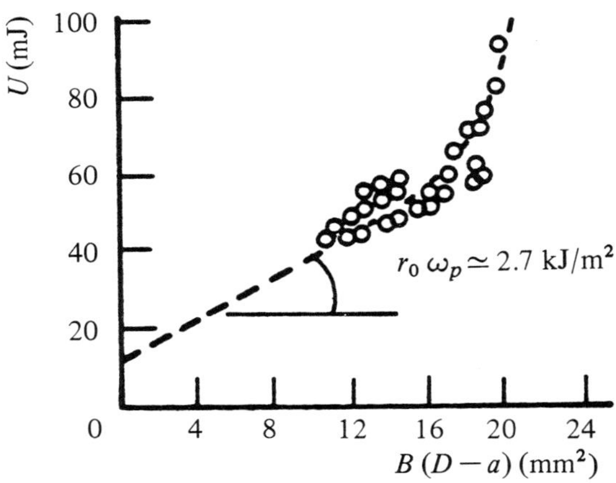

Fig. 8.34 Impact test data on HIPS at 20°C – Charpy ($L=41$ mm, $B=3.1$ mm, $D=6.3$ mm) (from reference [9])

which agrees with the separate values. The G_c value is very close to that expected and Fig. 8.35 shows the data from Fig. 8.34 plotted on this basis, giving $G_c=0.3$ kJ/m^2 and $r_0\,\omega_p=2.5$ kJ/m^2. It should be noted that for long crack lengths, $\phi\rightarrow\frac{1}{2}(1-a/D)$, so that the slope in Fig. 8.34 would tend to $(r_0\,\omega_p+(G_c/2))$

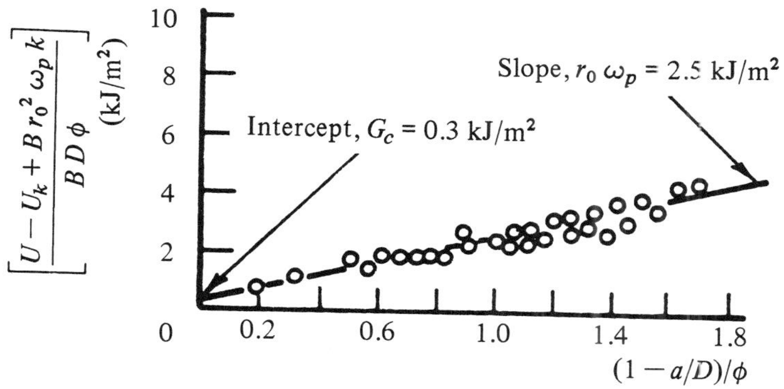

Fig. 8.35 Data from Fig. 8.34 plotted as a function of $(1-a/D)/\phi$ (from reference [9])

for unmodified PS and suggests that the whitening process does not affect the subsequent crack propagation. Since k is not known, ω_p may only be estimated but is about 1.4 MJ/m^3, i.e. considerably less than ABS, as expected. ω_p would also be expected to be a strong function of the rubber content, while G_c would not and this is confirmed in Fig. 8.36 which gives data for a series of materials, in which only the rubber content was varied.

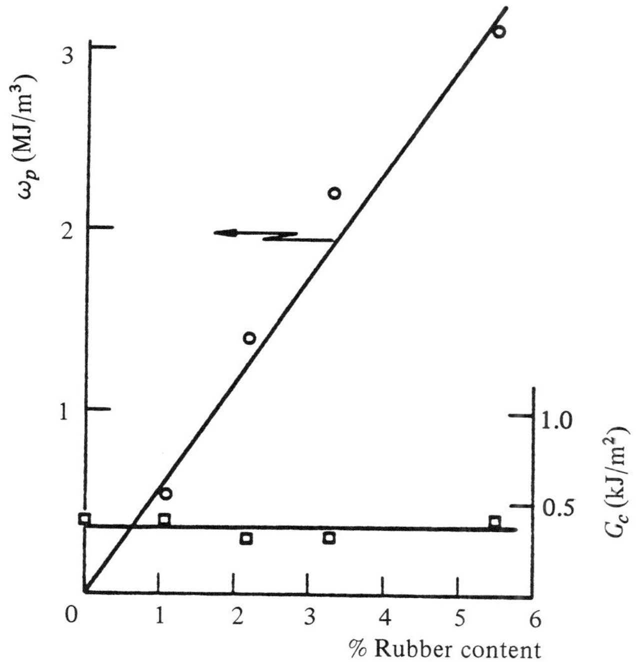

Fig. 8.36 Energy per unit volume and G_c as a function of rubber content in HIPS

It should be noted that the type of behaviour observed in HIPS could be described as Class 3 since there is a fully yielded ligament, but since the subsequent fracture is unstable its inclusion under Class 2 is considered justified. No truly Class 3 failures have been analysed since the slow tearing process does not result in complete failure in this form of impact test, but only in heavily distorted specimens.

8.5 Rate effects

Since a major objective of impact testing is to model behaviour at high rates, it is important to know if the chosen rate is appropriate. Although the loading rates are considerably higher than conventional tests, they are usually far below ballistic speeds which may be of interest in some applications. In many, the rate is sufficiently high to describe accidental blows and drops, and thus gives useful information on behaviour in practice. The question still remains, however, whether there is any variation in, say, G_c with rate at these values since this will affect the comparison of materials for a given application. The simplest basis for comparison is to consider the nominal strain rate (i.e. ignoring the effect of the crack) in the Charpy specimen given by:

$$\dot{e}=6\left(\frac{V}{D}\right)\left(\frac{D}{L}\right)^2 \tag{8.28}$$

The conventional geometry has $L\simeq 41$ mm, $D=6$ mm and $V\simeq 3.5$ m/s. The latter value is largely determined by the energy requirement for failure which governs the size of the pendulum. These values give $\dot{e}\simeq 75\ \mathrm{s}^{-1}$ which may be compared with that achieved in a conventional tension test with a gauge length of 10^{-1} m at an extension rate of 10^{-4} m/s (0.5 cm/min), i.e. $\dot{e}\simeq 10^{-3}\ \mathrm{s}^{-1}$. $\dot{e}$ may be varied by changing V but only modest changes are achievable because of practical limitations and the largest effect is produced by changing L. There are dangers in this method since other effects may inadvertently be introduced, but the use of the appropriate ϕ functions would be expected to give satisfactory results.

Quite significant variations are observed [10], as is demonstrated in Fig. 8.37, which gives data for PMMA at 20°C at $72\ \mathrm{s}^{-1}$ and $274\ \mathrm{s}^{-1}$. There are high points at low ϕ values (long crack lengths) as previously observed, but there is a distinct change in all the points giving a significant variation in G_c. This is further demonstrated in Fig. 8.38 which shows G_c as a function of $\dot{e}$, obtained by varying both L and V, and a line for $G_c \propto \sqrt{\dot{e}}$. This type of dependence is unlikely to arise from normal visco-elastic effects and suggests some other controlling mechanism. More light is shed when the same results are compared on the basis of loading time which may be computed for each specimen, since:

$$t=\frac{u}{V}=\frac{(2\,C\,U)^{1/2}}{V} \tag{8.29}$$

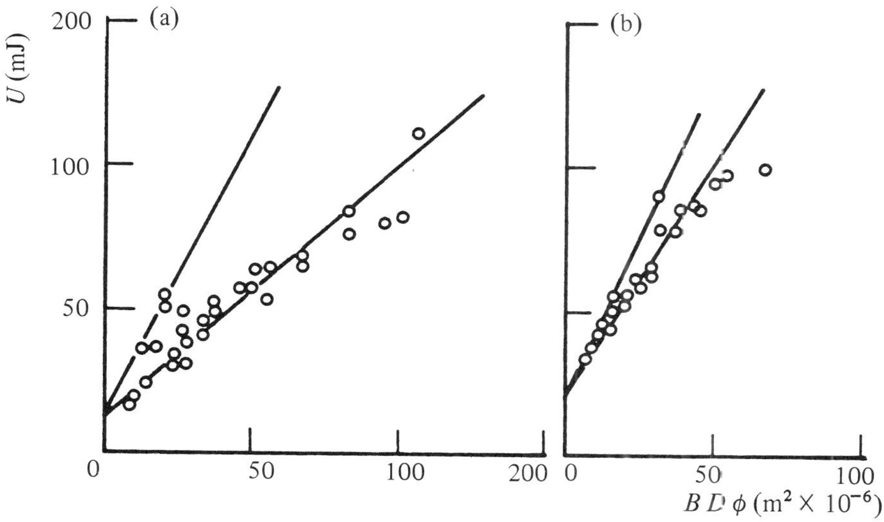

Fig. 8.37 Charpy impact results for PMMA at 20°C; (a) $\dot{e}=72\ s^{-1}$ ($L=41$ mm, $B=D=6$ mm, $V=3.36$ m/s); (b) $\dot{e}=274\ s^{-1}$ ($L=21$ mm, $B=D=6$ mm, $V=3.36$ m/s) (from reference [10])

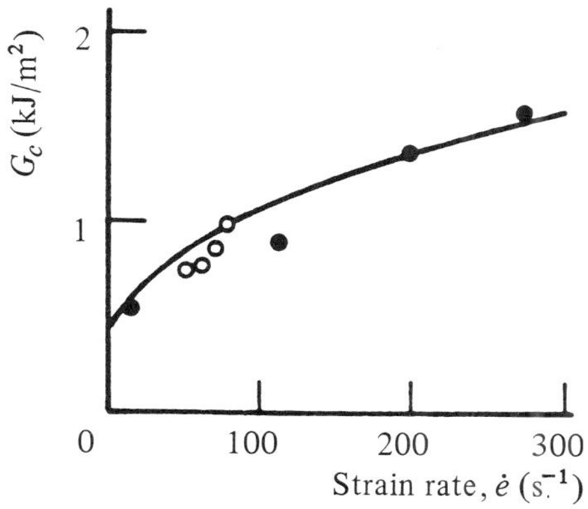

Fig. 8.38 Fracture toughness as a function of strain rate for PMMA at 20°C (● varying span, ○ varying pendulum velocity); line for $\dot{e}=0.4+0.07\sqrt{\dot{e}}$ (from reference [10])

This is only approximate, particularly for the short cracks because of the dynamic effects discussed previously, and this is reflected in the considerable scatter shown in Fig. 8.39 which gives G_c as a function of t for each specimen. The high points arising from crack arrest are quite clear on this basis (if halved, they fall in the general trend) but, even with this degree of scatter, a distinct trend is apparent with G_c remaining constant for times >0.3 ms and then showing a pronounced increase as t decreases. The phenomenon is more clearly demon-

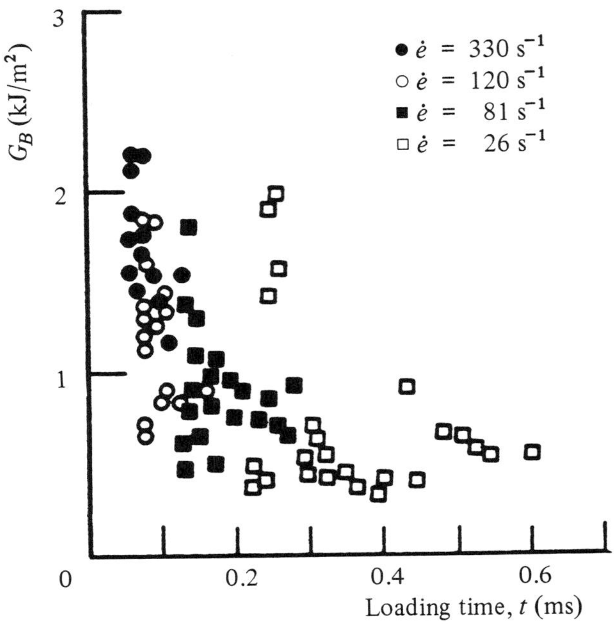

Fig. 8.39 Fracture toughness as a function of loading time for PMMA at 20°C (from reference [10])

strated for tougher materials where there is less scatter, and this is shown in Fig. 8.40 which gives data for HDPE at 0°C. The increase in G_c with rate is again apparent and when plotted against loading time in Fig. 8.41, we see more clearly the trend described for PMMA.

The rather strong time dependence of the form $t^{-1/2}$ is observed over a wide temperature range and for several polymers [10], and suggests something other than a basic material property. A clue is obtained by considering the isothermal–adiabatic transition time which may be computed from equation (7.12):

$$\Delta T = \frac{G_c}{\sqrt{\pi \rho C k}} \frac{1}{\sqrt{t}}$$

If we suppose that ΔT is the temperature rise required to increase the temperature from T_0 to a softening value, T_s (i.e. $\Delta T = T_s - T_0$), then we may compute the time to achieve this, t_1, for some typical values. Taking $G_c \simeq 1$ kJ/m^2, $\Delta T \simeq 100$°C and noting that $\sqrt{\pi \rho C k} \simeq 10^3$ J/m^2°C$\sqrt{\text{s}}$ for most polymers and temperatures, we have that $t_1 \simeq 0.1$ ms, which is similar to the times at which G_c increased rapidly in the impact data.

It has previously been argued (Section 7.2.3) that adiabatic heating resulted in

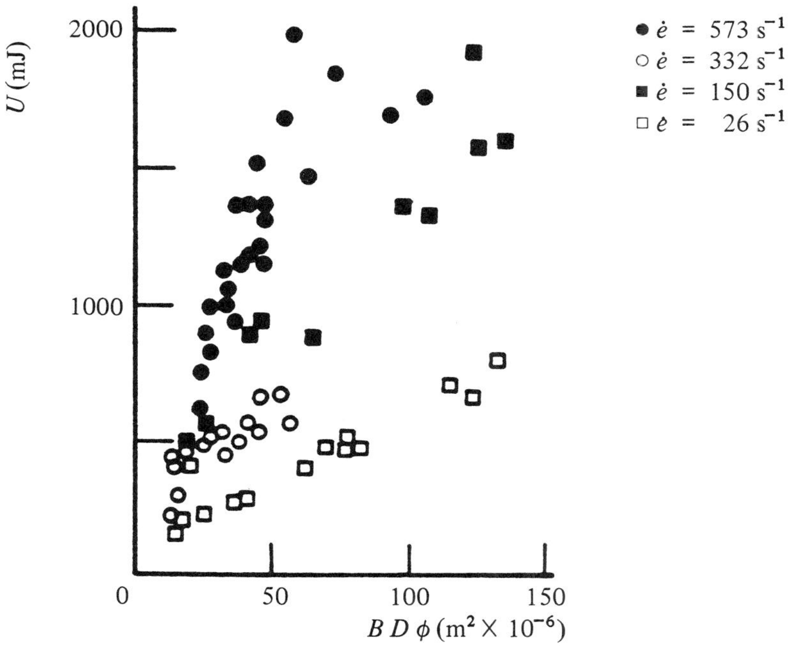

Fig. 8.40 Fracture energy against $B\,D\,\phi$ for HDPE at 0°C (from reference [10])

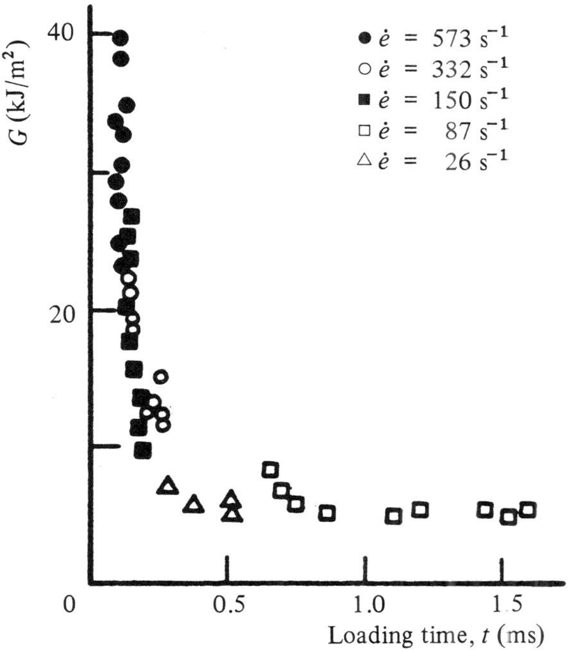

Fig. 8.41 Toughness as a function of loading time for HDPE at 0°C (from reference [10])

a decrease in G_c, but in that case the analysis was for a running crack. Here, we have a stationary crack at initiation, where the self-sharpening effect of propagation is absent, and we may argue that the increase in temperature results in a decrease in yield stress which, in turn, gives crack tip blunting and an increase in G_c. A useful model of this effect is obtained by using the blunt crack result discussed in Section 8.3 such that:

$$\frac{G_B}{G_c} \simeq \frac{1}{2} + \frac{\rho}{8l_0} \tag{8.30}$$

and then assuming that ρ may be modelled by the crack opening displacement such that:

$$\rho \simeq \frac{\delta}{2} \simeq \frac{2}{\pi} \frac{K_c^2}{E\, p_Y}$$

from equation (5.15), taking $M=1$. We may write $K_c^2 = 2\pi\, l_0\, p_c^2$ and also assume a constant yield strain, $e_y = p_Y/E$, giving a final result of:

$$\frac{\rho}{l_0} = 4e_y \left(\frac{p_c}{p_Y}\right)^2 \tag{8.31}$$

The variation of p_Y with T may be taken as approximately linear, such that:

$$\frac{p_Y}{p_0} = \frac{T_s - T}{T_s - T_0} = 1 - \sqrt{t_1/t} \tag{8.32}$$

with:
$$\sqrt{t_1} = \frac{G_c}{(T_s - T_0)\sqrt{\pi\, \rho\, C\, k}}$$

Combining equations (8.30), (8.31) and (8.32), we have:

$$\frac{G_B}{G_c} = \frac{1}{2} + \left[\frac{N}{1 - \sqrt{t_1/t}}\right]^2 \tag{8.33}$$

where:
$$N = \sqrt{e_y/2}\left(\frac{p_c}{p_Y}\right)$$

which may be re-written as:

$$\left[\frac{G_B}{G_c} - \frac{1}{2}\right]^{-1/2} = \frac{1}{N}\left[1 - \sqrt{t_1/t}\right] \tag{8.34}$$

N is a measure of the nature of the material since p_c/p_0 is the crack tip constraint factor and is in the range 3 to 5. For brittle materials, $e_y \simeq 0.05$, giving $N \simeq 0.5$, while for more ductile materials e_y is larger, giving higher N values. Fig. 8.42 shows equation (8.34) plotted as G_B/G_c versus t/t_1 and the form is clearly as described previously. Low N values give constant G_c values down to short times and then a very rapid rise as t_1 is approached. For the more ductile materials,

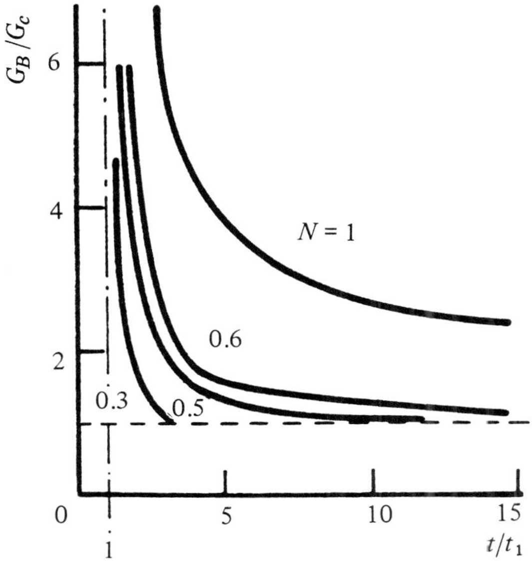

Fig. 8.42 The variation of apparent fracture toughness with loading time, from equation (8.34) (from reference [10])

the blunting effects spread over a larger time range and for $N > 1/\sqrt{2}$, for which $(p_c/p_0)^2\ e_y > 1$, the G_c value will always include a self-blunting effect.

For $t \gg t_1$, equation (8.34) may be approximated to:

$$\frac{G_B}{G_c} \simeq \frac{1}{2} + N^2\ (1 + 2\sqrt{t_1/t})$$

and since $t \simeq e_y/\dot{e}$, we have:

$$G_B \simeq (\tfrac{1}{2} + N^2)\ G_c + N^2\ G_c\ 2\sqrt{t_1/e_y}\ \dot{e}^{1/2} \tag{8.35}$$

which is the form observed for the PMMA data shown in Fig. 8.38, from which $(\frac{1}{2} + N^2)\ G_c \simeq 0.4$ kJ/m^2. More precise values are obtained by plotting the average times as a function of G_B/G_c in accordance with equation (8.34) and this is shown in Fig. 8.43 using $G_c \simeq 0.55$ kJ/m^2, from which we have $N \simeq 0.56$ and $t_1 = 0.033$ ms.

This type of behaviour has also been observed in HDPE and PTFE. For the latter, N decreases from 0.68 at 150°C to 0.40 at -100°C with t_1 in the range 0.3 to 0.1 ms. For HDPE, N was only just below 0.7 at -100°C and was as high as 0.9 at 20°C, suggesting that blunting effects are always present in this material since $N > 2^{-1/2}$. The results highlight the need for caution in the interpretation of data, particularly at the high strain rates. The thermal blunting may be regarded as an artefact of the test since it is presumably possible to obtain initiation at G_c without the effect, although this has not been seen in practice.

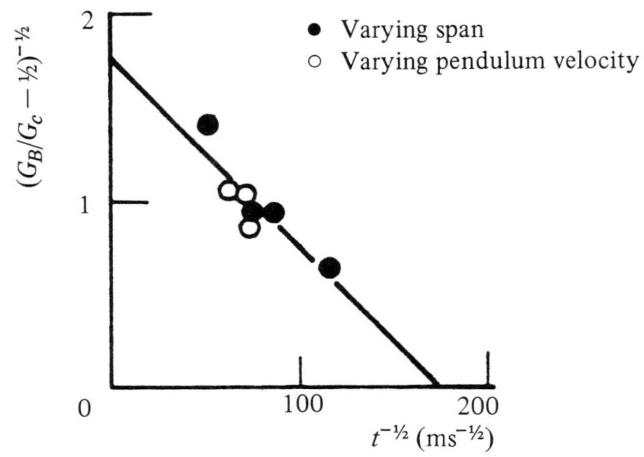

Fig. 8.43 The determination of N and t_1 for PMMA at 20°C (from reference [10])

The evidence so far available would suggest that the true G_c varies very little in the rate ranges accessible by this method.

8.6 Load measurements

Relatively few studies have been made comparing K_c values obtained from instrumented pendulum tests with those deduced from G_c measured from energy. The difficulty is, of course, the modulus, but such evidence as there is suggests good agreement. In one such study [11], the modulus was taken as the in-phase value at 11 Hz in a dynamic test and the loads measured using a load transducer mounted on the specimen supports. The materials used were a series of copolymers of PP and a homopolymer, and the data are shown in Fig. 8.44. The agreement is quite good with the instrumented values being about 10% higher, probably due to the inappropriate modulus which is most likely to be low.

Some load measurements have also been made on PMMA using laser–Doppler anemometry on a small falling weight machine [12] and Fig. 8.45 shows ${K_B}^{-1}$ plotted versus $t^{-1/2}$ for these data. Equation (8.33) may be recouched in terms of K to give:

$$\frac{K_B}{K_c} = \frac{N}{1-\sqrt{t_1/t}} \tag{8.36}$$

and the appropriate line is drawn, giving $t_1 = 0.1$ ms and $N\,K_c = 1.06$ MPa$\sqrt{\text{m}}$ and, using $N = 0.56$ from the G_c data in Section 8.5, we have $K_c = 1.9$ MPa$\sqrt{\text{m}}$, i.e. in reasonable agreement with the higher rate measurements discussed in Section 6.3. The G_c and K_c value so deduced give a modulus of 6.6 GPa.

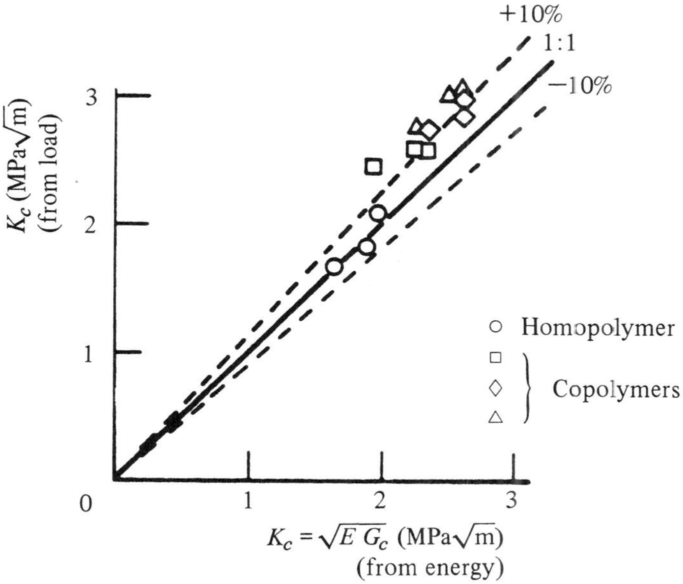

Fig. 8.44 Comparison of K_c values from load and energy measurements – PP (data from reference [11])

8.7 General dynamic analysis

Impact test analysis involved the consideration of dynamic effects since significant errors can arise if only statics are employed. This is because the rates are sufficiently high to give significant kinetic energy to the body and this must be included in any energy balance. There are several other cases of practical importance where this is true also and, in this section, we will consider the inclusion of dynamic effects in more detail. As in the static case, it is convenient to start with energy considerations and then look at local solutions.

8.7.1 *Energy analysis*

If we return to the basic energy balance discussed in Section 2.2, we may write:

$$\delta U_1 - \delta U_2 - \delta U_3 - \delta U_4 = 0 \tag{8.37}$$

(equation (2.1)), and note that:

$$\frac{dU_2}{dA} = R \text{ and } \frac{dU_1}{dA} - \frac{dU_3}{dA} = G$$

If we confine our attention to cracks in uniform sheets, we may rewrite this in a form convenient for dynamic analysis as:

$$\left[(G-R) - \frac{dU_4}{B\,da}\right] = 0 \tag{8.38}$$

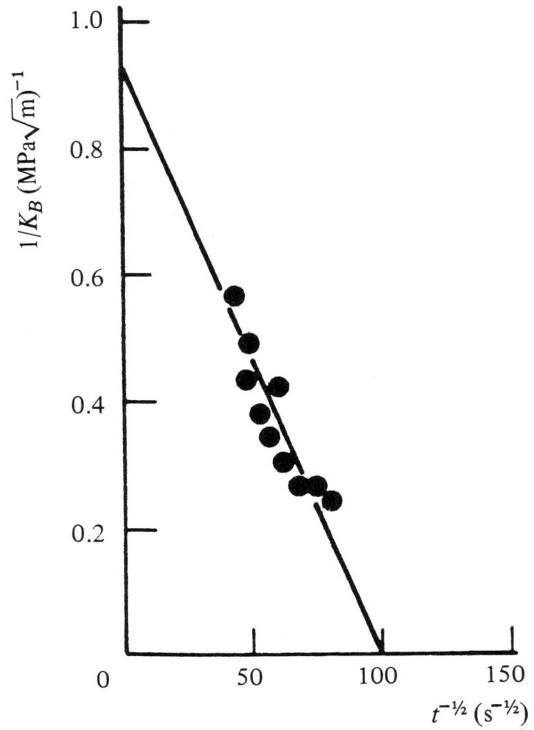

Fig. 8.45 $K_B{}^{-1}$ versus $t^{-1/2}$ for instrumented impact test

and the problem is concerned with the determination of U_4. For cartesian coordinates, we may express this as:

$$U_4 = B \int_A \tfrac{1}{2} \rho \, (\dot{u}^2 + \dot{v}^2) \, dx \, dy \tag{8.39}$$

where $\dot{u}$ and $\dot{v}$ are determined using the equilibrium equations containing inertia terms, i.e.

$$\left.\begin{aligned} \frac{\partial p_{xx}}{\partial x} + \frac{\partial p_{yx}}{\partial y} &= \rho \, \ddot{u} \\ \text{and:} \qquad \frac{\partial p_{yy}}{\partial y} + \frac{\partial p_{yx}}{\partial x} &= \rho \, \ddot{v} \end{aligned}\right\} \tag{8.40}$$

Substitution of Hooke's law and the displacement definition of strain gives the equations in terms of displacements which, by appropriate manipulation, may be rewritten in the form:

$$\left.\begin{aligned} \nabla^2\left(\frac{\partial u}{\partial x}+\frac{\partial v}{\partial y}\right)&=\frac{1}{C_d^2}\frac{\partial^2}{\partial t^2}\left(\frac{\partial u}{\partial x}+\frac{\partial v}{\partial y}\right) \\ \text{and:}\qquad \nabla^2\left(\frac{\partial u}{\partial y}-\frac{\partial v}{\partial x}\right)&=\frac{1}{C_s^2}\frac{\partial^2}{\partial t^2}\left(\frac{\partial u}{\partial y}-\frac{\partial v}{\partial x}\right) \end{aligned}\right\} \tag{8.41}$$

where $C_d=\sqrt{E/((1-v^2)\,\rho)}$ and $C_s=\sqrt{E/(2(1+v)\,\rho)}$ are the dilatation and shear wave velocities, respectively. The first of these equations describes the propagation of dilatation waves, while the second is that for the propagation of rotations. As the equations show, the two are separable, resulting in waves propagating at different speeds.

For a given problem, the solution for displacement may be written, for example, as:

$$u=u_1 f_1\left(\frac{a}{D},\frac{L}{D},\frac{C_{d,s}\,t}{D},\frac{x}{D},\frac{y}{D}\right)+u_0 \tag{8.42}$$

where u_1 is a characteristic displacement, and f_1 is a dimensionless function of specimen dimensions and time as determined by equation (8.41) and the boundary conditions. u_0 is the rigid body motion not determined by the state of stress. The velocity is thus:

$$\dot{u}=\left(\frac{u_1\dot{a}}{D}\right)\frac{\partial f_1}{\partial(a/D)}+\dot{u}_1 f_1+\left(\frac{u_1 C_{d,s}}{D}\right)\frac{\partial f_1}{\partial((C_{d,s}\,t)/D)}+\dot{u}_0 \tag{8.43}$$

and a similar expression for $\dot{v}$. The separate terms represent the different sources of motion; the first arises from the crack velocity, the second from imposed rates of loading, the third from the stress waves, and the last from the rigid body motion. This latter is the case used in the impact test analysis and the general expression for kinetic energy has the simple form:

$$U_4=\tfrac{1}{2}(M\,V^2+I\,\omega^2) \tag{8.44}$$

where M and I are the mass and moment of inertia, respectively, and V and ω are the translational and rotational velocities of the body. All other cases require f_1 (and f_2 for v) and solutions are very difficult to obtain. Some useful insights can be obtained, however, by restricting our attention to effects arising solely from high speed crack propagation and assuming that stress waves do not interfere significantly with the solutions. This type of analysis was discussed by Berry [13, 14]. In this case:

$$\dot{u}=\left(\frac{u_1\dot{a}}{D}\right)\left[\frac{df_1}{d(a/D)}+\frac{\dot{u}_1 D}{\dot{a}\,u_1}f_1\right] \text{ and } \dot{v}=\left(\frac{u_1\dot{a}}{D}\right)\left[\frac{df_2}{d(a/D)}+\frac{\dot{u}_1 D}{\dot{a}\,u_1}f_2\right]$$

and we have:

$$U_4 = \frac{B\rho}{2}\dot{a}^2\left(\frac{u_1}{D}\right)^2 \int_A \left(\frac{df_1}{d(a/D)} + \frac{\dot{u}_1 D}{\dot{a}\,u_1} f_1\right)^2 + \left(\frac{df_2}{d(a/D)} + \frac{\dot{u}_1 D}{\dot{a}\,u_1} f_2\right)^2 dx\,dy \tag{8.45}$$

The stored strain energy can also be written in terms of the characteristic displacement via the compliance function (see Section 2.4.1), so that:

$$U_3 = \frac{1}{2}\frac{u_1^2}{C(a)}$$

and we may write:

$$U_4 = U_3\left(\frac{\dot{a}}{c}\right)^2 \beta \tag{8.46}$$

where:

$$c^2 = \frac{E}{\rho}$$

and:

$$\beta = \frac{BE}{D^2} C \int_A \left(\frac{df_1}{d(a/D)} + \frac{\dot{u}_1 D}{\dot{a}\,u_1} f_1\right)^2 + \left(\frac{df_2}{d(a/D)} + \frac{\dot{u}_1 D}{\dot{a}\,u_1} f_2\right)^2 dx\,dy$$

The fact that $U_4 \propto U_3$ arises because the displacement rates are a consequence of elastic deformations.

The equations of motion of a crack may now be defined using $R = G_c(\dot{a})$ and from equation (8.38), we have:

$$[G - G_c(\dot{a})] = \frac{d}{da}\left[\frac{U_3}{B}\left(\frac{\dot{a}}{c}\right)^2 \beta\right]$$

or, in integral form: (8.47)

$$\left(\frac{\dot{a}}{c}\right)^2 = \left[\frac{B}{U_3}\right]\frac{1}{\beta}\int_{a_0}^{a} [G - G_c(\dot{a})]\,da$$

for $\dot{a} = 0$ at $a = a_0$.

A dynamic energy release rate, G_D, may be defined for a constant crack speed using:

$$G_D = -\frac{dU_4}{B\,da}$$

and using equation (8.45), and noting that $da = -dx$, we have:

$$G_D = \frac{\rho}{2}\dot{a}^2\left(\frac{u_1^2}{D}\right) \int_L \left(\frac{df_1}{d(a/D)} + \frac{\dot{u}_1 D}{\dot{a}\,u_1} f_1\right)^2 + \left(\frac{df_2}{d(a/D)} + \frac{\dot{u}_1 D}{\dot{a}\,u_1} f_2\right)^2 dy \tag{8.48}$$

where the integral is evaluated over y in this case, as in G determinations.
Some examples of the determination of β will now be given.

(i) *Double cantilever beam (DCB)*
This is the geometry discussed in Section 2.4.3 and shown in Fig. 2.6(i). The static analysis gives:

$$C=\frac{u_1}{P}=\frac{8}{E\,B}\left(\frac{a}{D}\right)^3$$

and an expression may be obtained for the displacement profile in terms of x, the distance from the load point, such that:

$$v=u_1 f_2, f_2=\frac{1}{2}\left[\frac{1}{2}\left(\frac{D}{a}\right)^3 \xi^3-\frac{3}{2}\left(\frac{D}{a}\right)\xi+1\right]$$

where $\xi=x/D$. Here, $u=0$ so that $f_1=0$ and the expression for β becomes:

$$\beta=8\left(\frac{a}{D}\right)^3\int_0^{a/D}\left(\frac{df_2}{d(a/D)}+\left(\frac{\dot{u}_1 D}{\dot{a}\,u_1}\right)f_2\right)^2 d\xi$$

$\dot{u}_1$ is determined by the loading conditions and we shall consider two special cases. For fixed displacement, $\dot{u}_1=0$ and:

$$\beta_1=18\left(\frac{D}{a}\right)\int_0^{a/D}\xi^2\left(1-\left(\frac{D}{a}\right)^2\xi^2\right)^2 d\xi=\left(\frac{64}{105}\right)\left(\frac{a}{D}\right)^2 \qquad (8.49)$$

with $U_3=(u_1^2/16)\;E\;B\;(D/a)^3$. For fixed load:

$$\dot{u}_1=\frac{24}{E\;B}\frac{P}{D^3}\,a^2\,\dot{a}\text{ and }\frac{\dot{u}_1 D}{\dot{a}\,u_1}=3\left(\frac{D}{a}\right)$$

giving:

$$\beta_2=6\left(\frac{a}{D}\right)^2$$

and:

$$U_3=\frac{4P^2}{E\,B}\left(\frac{a}{D}\right)^3 \qquad (8.50)$$

The equations of motion may now be derived if we assume a constant G_c value for propagation. For the fixed displacement case:

$$G=\frac{3}{16}\,u_1^2\,\frac{E\;D^3}{a^4}$$

and if $a=a_0$ at initiation, we may write:

$$G=n^2\left(\frac{a_0}{a}\right)^4 G_c$$

where $n = u_1/u_0$, the ratio of the applied displacement to the initiation value. Since:

$$\frac{U_3}{B} = \frac{u_1^2}{16} E \left(\frac{D}{a}\right)^3 = \frac{n^2}{3} \frac{a_0{}^4}{a^3} G_c \text{ and } \beta_1 = \frac{64}{105}\left(\frac{a}{a_0}\right)^2 \left(\frac{a_0}{D}\right)^2$$

we may write equation (8.47) as:

$$\left(\frac{\dot{a}}{c}\right)^2 = \frac{105}{64}\left(\frac{D}{a_0}\right)^2 \left(\frac{a}{a_0}\right) \frac{3}{n^2} \int_1^{a/a_0} \left[n^2 \left(\frac{a_0}{a}\right)^4 - 1\right] d\left(\frac{a}{a_0}\right)$$

i.e.
$$\left(\frac{a_0}{D}\right)^2 \left(\frac{\dot{a}}{c}\right)^2 = \frac{105}{64}\left\{\left[\frac{3}{n^2}\left(1-\frac{a}{a_0}\right)+1\right]\frac{a}{a_0} - \left(\frac{a_0}{a}\right)^2\right\} \tag{8.51}$$

Similarly, for a constant load, we have:

$$\left(\frac{a_0}{D}\right)^2 \left(\frac{\dot{a}}{c}\right)^2 = \frac{1}{6}\left(\frac{a_0}{a}\right)^2 \left\{1 - \left[1 - \frac{3}{n^2}\left(1-\frac{a}{a_0}\right)\right]\left(\frac{a_0}{a}\right)^3\right\} \tag{8.52}$$

where $n = P/P_0$. These two expressions are shown in Fig. 8.46(a) and (b), respectively. For constant displacement, there is no solution for $n = 1$ since G decreases below G_c immediately but for $n > 1$ there is an increasing crack speed to a maximum, after which the speed decreases to final arrest. For this loading, G decreases and the broken line in Fig. 8.46(a) is the condition for $G = G_c$. For much of the growth, the static energy release rate is less than that required to propagate the crack and the additional energy comes from that available as kinetic energy, as discussed in Section 2.3. This behaviour also provides a good illustration of the dynamic stability criterion given in equation (2.10). Here, we may take unstable to be when $\ddot{a} > 0$, i.e. when the speed is increasing and this is:

$$\frac{d^2}{da^2}[\beta\, U_3] > 0$$

the case for the initial stages of growth. For constant load (Fig. 8.46(b)), the speed increases very rapidly initially and is followed by a rapid decrease but not to arrest. There is very little load dependence in this case.

The limiting condition discussed in Section 8.3 concerning total energy requirements at fixed displacement may be written for this case as:

$$n^2 G_c B D \phi > G_c B (a - a_0)$$

i.e.
$$\frac{a}{a_0} < \frac{n^2}{3} + 1$$

These limits are shown in Fig. 8.46 and are close to, and, of course, greater than, actual arrest.

(ii) *Infinite plates*

Because of the unbounded nature of this case, it is preferable to express

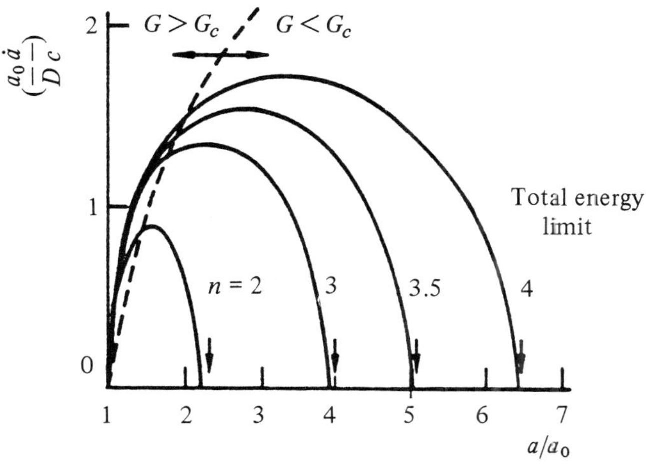

(a) Constant displacement, $n = u_1/u_0$

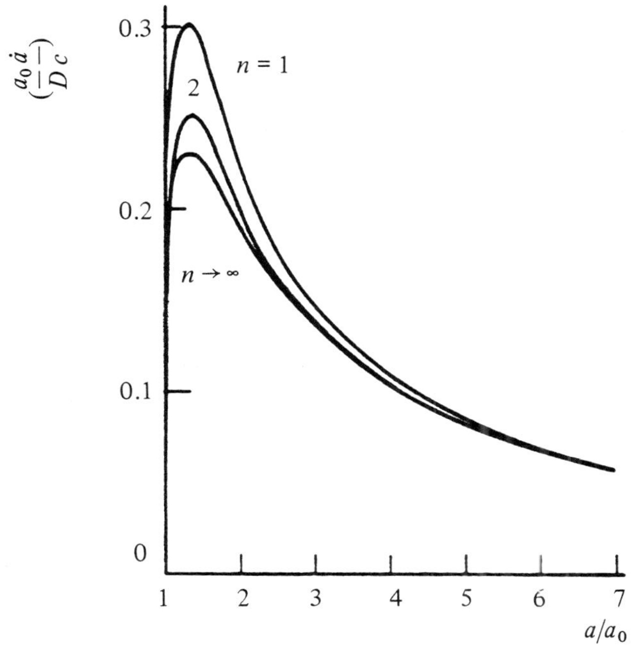

(b) Constant load, $n = P/P_0$

Fig. 8.46 Velocity as a function of crack length for the DCB specimen for two loading conditions; (a) constant displacement, and (b) constant load (computations kindly performed by M. J. Williams) equation (8.45) in terms of the strain energy density function, W_0, as used in Section 2.4.3(ix). It is also more appropriate to express the velocity in terms of a characteristic remote strain, e, so that we may write:

$$\dot{u}=e\,\dot{a}\frac{df}{da}(x,y,a)$$

and:

$$U_4=\frac{1}{2}\rho\,e^2\,\dot{a}^2\int_A\left[\frac{df}{da}(x,y,a)\right]^2 dx\,dy$$

This may only be non-dimensionalised on a basis of a, so that:

$$U_4=\left(\frac{e^2\,E}{2}\right)\left(\frac{\dot{a}}{c}\right)^2\beta_0\,a^2$$

where:

$$\beta_0=\int_A\left[\frac{df}{da}\left(\frac{x}{a},\frac{y}{a}\right)\right]^2 d\left(\frac{x}{a}\right)d\left(\frac{y}{a}\right)=\text{a constant}$$

in this case. Now, $(e^2\,E)/2=W_0$ and we may write:

$$U_4=\beta_0\,W_0\,a^2\left(\frac{\dot{a}}{c}\right)^2$$

For the constant stress case, $W_0=p^2/2E$ with $G=(\pi\,p^2\,a)/E$ and equation (8.47) becomes:

$$\int_{a_0}^{a}\left(\frac{\pi\,p^2\,a}{E}-G_c\right)da=\beta_0\frac{p^2}{2E}a^2\left(\frac{\dot{a}}{c}\right)^2$$

and for constant G_c:

$$\left(\frac{\dot{a}}{c}\right)^2=\left[\frac{\pi}{\beta_0}\right]\left(1-\frac{a_0}{a}\right)\left[1+\left(1-\frac{2}{n^2}\right)\frac{a_0}{a}\right] \tag{8.53}$$

where $n=p/p_0$ and p_0 is the initiation stress for $a=a_0$, and, for $a\gg a_0$, $\dot{a}\rightarrow c\sqrt{\pi/\beta_0}$, a limiting speed. This form of solution was first derived by Mott [15] and an estimate of the constant factor was made by Roberts & Wells [16] using an approximate analysis which gave $\sqrt{\pi/\beta_0}=0.38$, i.e. $\beta_0\simeq 22$.

(iii) *Parallel strip (Section 2.4.3(vi)):*
This is a further example of only one finite dimension, the depth in this case, giving a constant β_0. In this case, we may non-dimensionalise x on a and y on D, giving:

$$U_4=\left(\frac{e^2\,E}{2}\right)\left(\frac{\dot{a}}{c}\right)^2\beta_0\,a\,D$$

where:

$$\beta_0=2\int_0^1\int_0^1\left[f\left(\frac{x}{a},\frac{y}{D}\right)\right]^2 d\left(\frac{x}{a}\right)d\left(\frac{y}{D}\right)=\text{constant}$$

Here, $G=(p^2/2E)\,2D$ and $W_0=p^2/2E$, so that:

$$\int_{a_0}^{a}\left(\frac{p^2}{E}D-G_c\right)du=\beta_0\frac{p^2}{E}\,a\,D\left(\frac{\dot{a}}{c}\right)^2$$

giving:

$$\left(\frac{\dot{a}}{c}\right)^2=\frac{1}{\beta_0}\left(1-\frac{1}{n^2}\right)\left(1-\frac{a_0}{a}\right) \tag{8.54}$$

Again, there is a limiting velocity which is load-dependent in this case and tends to $c/\sqrt{\beta_0}$ for large loads. β_0 is likely to be different from the infinite plate case.

8.7.2 *Local crack tip stresses*

A considerable insight into the dynamic case may be obtained by considering the crack tip stress field set up by a running crack. We now start with the dynamic equilibrium equations (equations (8.40)) which may be written in terms of displacement using Hooke's law. For dynamic solutions, it is convenient to express the displacements in terms of two potential functions, ϕ and ψ, as given by Kolsky [17] when analysing stress wave problems, i.e.

$$u=\frac{\partial\phi}{\partial x}+\frac{\partial\psi}{\partial y}\quad\text{and}\quad v=\frac{\partial\phi}{\partial y}-\frac{\partial\psi}{\partial x}$$

It should be noted that the dilatation is given by $\nabla^2\phi$ and the rotation by $\nabla^2\psi$. Equations (8.40) now become:

$$\frac{\partial\nabla^2\phi}{\partial x}+\left(\frac{1-\nu}{2}\right)\frac{\partial\nabla^2\psi}{\partial y}=\frac{1}{C_d^2}\left[\frac{\partial}{\partial x}\left(\frac{\partial^2\phi}{\partial t^2}\right)+\frac{\partial}{\partial y}\left(\frac{\partial^2\psi}{\partial t^2}\right)\right]$$

$$\frac{\partial\nabla^2\phi}{\partial y}-\left(\frac{1-\nu}{2}\right)\frac{\partial\nabla^2\psi}{\partial x}=\frac{1}{C_s^2}\left[\frac{\partial}{\partial y}\left(\frac{\partial^2\phi}{\partial t^2}\right)-\frac{\partial}{\partial x}\left(\frac{\partial^2\psi}{\partial t^2}\right)\right]$$

which are satisfied if:

$$\nabla^2\phi=\frac{1}{C_d^2}\frac{\partial^2\phi}{\partial t^2}\quad\text{and}\quad\nabla^2\psi=\frac{1}{C_s^2}\frac{\partial^2\psi}{\partial t^2} \tag{8.55}$$

If there are no time effects, i.e. a static solution, then $\nabla^2\phi=\nabla^2\psi=0$ which yields the trivial case of zero effects. This form does not, therefore, give the static solution which requires the solution of a biharmonic equation ($\nabla^4=0$) but only the dynamic part which may be added to the static case in a manner analogous to complementary functions and particular integrals in ordinary differential equations. Following the analysis of Yoffe [18], as developed by Rice [19], we may now express the crack problem in terms of moving coordinates as shown in Fig. 8.47. If we confine the analysis to rate effects arising solely from the crack motion, then we have:

$$\frac{\partial^2}{\partial t^2}=\dot{a}^2\frac{\partial^2}{\partial x^2}$$

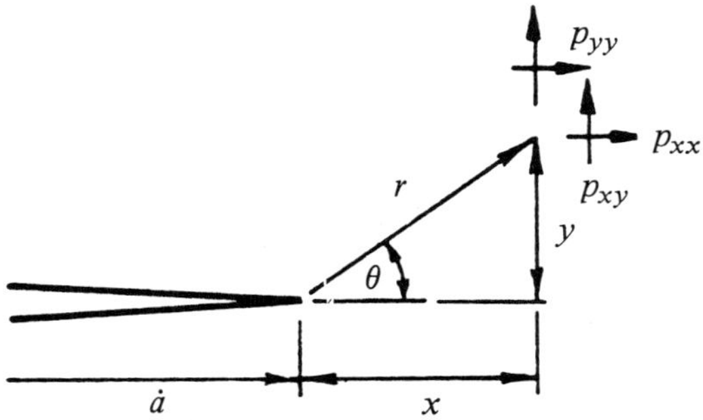

Fig. 8.47 Coordinates of the moving crack tip

for constant crack speed and $\dot{a} = -\dot{x}$ so that equations (8.55) become:

$$\left.\begin{aligned} \left[1-\left(\frac{\dot{a}}{C_d}\right)^2\right]\frac{\partial^2\phi}{\partial x^2}+\frac{\partial^2\phi}{\partial y^2}=0 \\ \left[1-\left(\frac{\dot{a}}{C_s}\right)^2\right]\frac{\partial^2\psi}{\partial x^2}+\frac{\partial^2\psi}{\partial y^2}=0 \end{aligned}\right\} \quad (8.56)$$

These may be expressed as Laplace equations, $\nabla^2(\quad)=0$ in terms of transformed variables x and $y_{d,s}=\alpha_{d,s}y$, where:

$$\alpha_{d,s}{}^2 = 1-\left(\frac{\dot{a}}{C_{d,s}}\right)^2 \quad (8.57)$$

If we confine the analysis to mode I, then we may expect symmetrical harmonic functions of the form:

$$\phi = A\, r_d{}^n \cos n\,\theta_d \text{ and } \psi = -B\, r_s{}^n \sin n\,\theta_s \quad (8.58)$$

where:

$$\left.\begin{aligned} r_{d,s}{}^2 &= r^2(\cos^2\theta+\alpha_{d,s}{}^2\sin^2\theta) \\ \tan\theta_{d,s} &= \alpha_{d,s}\tan\theta \end{aligned}\right\} \quad (8.59)$$

and A, B and n are constants.

The stresses may now be written in terms of the potential functions and the transformed variables as, for example:

$$p_{yy} = \frac{E}{1-\nu^2}\left[\alpha_d{}^2\frac{\partial^2\phi}{\partial y_d{}^2}+\nu\frac{\partial^2\phi}{\partial x^2}-(1-\nu)\,\alpha_s\frac{\partial^2\psi}{\partial x\,\partial y_s}\right]$$

and performing the appropriate differentiation, we have:

$$\left.\begin{aligned}
p_{xx} &= \frac{n(n-1)E}{1-\nu^2}\left[(1-\nu\alpha_d^2)\,A\,r_d^{\,n-2}\cos(n-2)\,\theta_d\right.\\
&\qquad\left.-(1-\nu)\,\alpha_s\,B\,r_s^{\,n-2}\cos(n-2)\,\theta_s\right]\\
p_{yy} &= \frac{n(n-1)E}{1-\nu^2}\left[(\nu-\alpha_d^2)\,A\,r_d^{\,n-2}\cos(n-2)\,\theta_d\right.\\
&\qquad\left.+(1-\nu)\,\alpha_s\,B\,r_s^{\,n-2}\cos(n-2)\,\theta_s\right]\\
p_{xy} &= \frac{n(n-1)E}{2(1+\nu)}\left[-2\alpha_d\,A\,r_d^{\,n-2}\sin(n-2)\,\theta_d\right.\\
&\qquad\left.+(1+\alpha_s^2)\,B\,r_s^{\,n-2}\sin(n-2)\,\theta_s\right]
\end{aligned}\right\}\qquad(8.60)$$

The boundary conditions for stress-free crack faces are $p_{yy}=p_{xy}=0$ at $\theta=\pm\pi$ when $r_{d,s}=r$ and $\theta_{d,s}=\pm\pi$, giving the eigen conditions:

$$[(\nu-\alpha_d^2)\,A+(1-\nu)\,B\,\alpha_s]\cos n\,\pi=0$$

$$[-2\alpha_d\,A+(1+\alpha_s^2)\,B]\sin n\,\pi=0$$

from which we have a singular term with $n=3/2$ and $B=2\alpha_d/(1+\alpha_s^2)\,A$. Substituting in equations (8.59) and noting that $(1-\nu)/2=(1-\alpha_d^2)/(1-\alpha_s^2)$, we have:

$$\left.\begin{aligned}
p_{xx} &= \frac{3\,E\,A}{\sqrt{r}\,8(1+\nu)}\left[(1+2\alpha_d^2-\alpha_s^2)\cos\frac{\theta_d}{2}\sqrt{r/r_d}\right.\\
&\qquad\left.-\frac{4\alpha_s\alpha_d}{1+\alpha_s^2}\cos\frac{\theta_s}{2}\sqrt{r/r_s}\right]\\
p_{yy} &= \frac{3\,E\,A}{\sqrt{r}\,8(1+\nu)}\left[-(1+\alpha_s^2)\cos\frac{\theta_d}{2}\sqrt{r/r_d}\right.\\
&\qquad\left.+\frac{4\alpha_s\alpha_d}{1+\alpha_s^2}\cos\frac{\theta_s}{2}\sqrt{r/r_s}\right]\\
p_{xy} &= \frac{3\,E\,A}{\sqrt{r}\,4(1+\nu)}\,\alpha_d\left[\sin\frac{\theta_d}{2}\sqrt{r/r_d}-\sin\frac{\theta_s}{2}\sqrt{r/r_s}\right]
\end{aligned}\right\}\qquad(8.61)$$

with the associated displacements:

$$\left.\begin{aligned} u &= 3A\sqrt{r}\left[\frac{1}{2}\cos\frac{\theta_d}{2}\sqrt{r_d/r} - \frac{\alpha_s\,\alpha_d}{1+\alpha_s^{\,2}}\cos\frac{\theta_s}{2}\sqrt{r_s/r}\right] \\ v &= 3A\sqrt{r}\left[-\frac{\alpha_d}{2}\sin\frac{\theta_d}{2}\sqrt{r_d/r} + \frac{\alpha_d}{1+\alpha_s^{\,2}}\sin\frac{\theta_s}{2}\sqrt{r_s/r}\right] \end{aligned}\right\} \quad (8.62)$$

As in the static case, there is a factor, A in this case, determined by the boundary conditions and this can be expressed in terms of a dynamic stress intensity factor, K_D, i.e.

$$K_D = p_{yy}\sqrt{2\pi\, r}_{\theta=0} = \frac{3\;E\,A}{8\,(1+v)}\sqrt{2\pi}\left[\frac{4\,\alpha_s\,\alpha_d}{1+\alpha_s^{\,2}} - (1+\alpha_s^{\,2})\right] \quad (8.63)$$

The function of α here is the stress wave effect excluded from the discussions previously and, as expected, is zero at zero speed ($\alpha_d = \alpha_s = 1$). There is a second zero condition when:

$$4\,\alpha_s\,\alpha_d = (1+\alpha_s^{\,2})^2$$

which gives the speed of propagation along a stress-free surface and is the Rayleigh speed, $C_r \simeq 0.93 C_s$ for $v = \frac{1}{2}$. K_D is positive (for positive A) for $0 < \dot{a} < C_r$ and negative for $\dot{C}_r < \dot{a} < C_s$. The form of dependence on $\dot{a}/C_s$ is shown in Fig. 8.48.

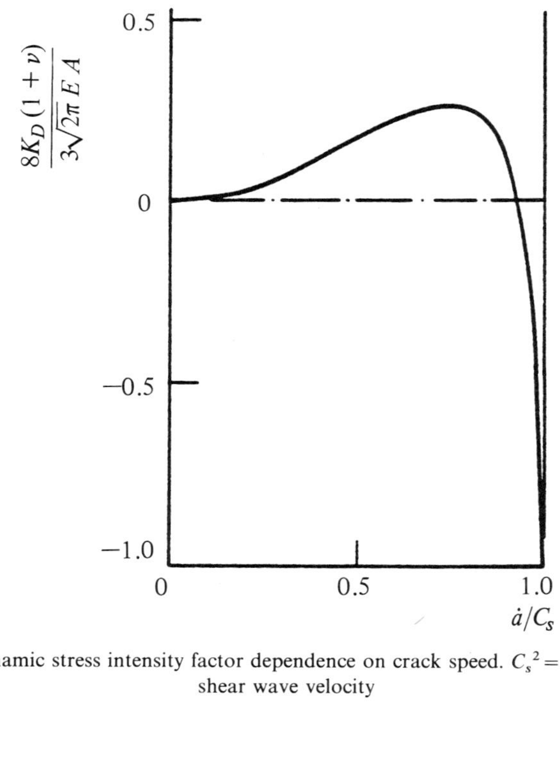

Fig. 8.48 Dynamic stress intensity factor dependence on crack speed. $C_s^{\,2} = E/(2(1+v)\rho)$, the shear wave velocity

The displacement expressions in equations (8.62) are the local form of equation (8.42), where the x and y dependency is now defined in terms of r and θ, and the wave speed effects are embodied in α_d and α_s. $(3A)$ may be written as:

$$(3A) = \frac{u_1}{\sqrt{D}}\, g\left(\frac{a}{D}, \frac{L}{D}\right)$$

and we may deduce G_D, the dynamic energy release rate, by the method given in Section 3.3 using p_{yy} at $\theta=0$ and $2v$ at $\theta=\pi$. The result is:

$$G_D = \frac{K_D{}^2}{E}\left(\frac{1+\nu}{2}\right)\frac{\alpha_d\,(1-\alpha_s{}^2)}{(1+\alpha_s{}^2)} \tag{8.64}$$

which may be written in terms of u_1 and $\dot{a}$ to give:

$$G_D = \frac{\rho}{2}\,\dot{a}\,\frac{u_1^2}{D}\left\{\frac{\pi}{16}\,g^2\,\frac{\alpha_d}{1+\alpha_s{}^2}\left[\frac{4\,\alpha_s\,\alpha_d}{1+\alpha_s{}^2}-(1+\alpha_s{}^2)\right]^2\right\} \tag{8.65}$$

This may be compared with equation (8.48) to give the stress wave dependence of f used in the energy analysis.

This form of solution contains only dynamic effects and when static loads are present, it must be combined with the usual static result. The effective stress intensity factor in mode I is thus:

$$K_T = K_I + K_D$$

and the combined energy release rate is:

$$G_T = \frac{1}{E}\left[K_I + K_D\left(\frac{1+\nu}{2}\right)\frac{\alpha_d\,(1-\alpha_s{}^2)}{(1+\alpha_s{}^2)}\right][K_I + K_D] \tag{8.66}$$

For an infinite plate, we may deduce the form of the crack speed dependence of $G\,(=K_I{}^2/E)$ for a constant G_c as:

$$\begin{aligned} G_c = G&\left\{1+\beta_0\left(\frac{1+\nu}{2}\right)\frac{\alpha_d\,(1-\alpha_s{}^2)}{(1+\alpha_s{}^2)}\left[\frac{4\,\alpha_s\,\alpha_d}{1+\alpha_s{}^2}-(1+\alpha_s{}^2)\right]\right\} \\ &\times\left\{1+\beta_0\left[\frac{4\,\alpha_s\,\alpha_d}{1+\alpha_s{}^2}-(1+\alpha_s{}^2)\right]\right\} \end{aligned} \tag{8.67}$$

where β_0 is a constant, and $G=G_c$ at C_r, as expected. The second term is the dominant one and gives a limiting speed between C_r and C_s for $\beta_0>1$.

This analysis has also been used [18] to provide a possible explanation of crack bifurcation. If we consider the hoop stress, $p_{\theta\theta}$, acting at angles other than zero, then, by analogy with the mixed mode analysis discussed in Section 4.6, we might expect the crack to move from its own line if the maximum in $p_{\theta\theta}$ occurred other than at $\theta=0$. We may deduce $p_{\theta\theta}$ from equations (8.61) using:

$$p_{\theta\theta} = p_{xx}\sin^2\theta + p_{yy}\cos^2\theta - 2p_{xy}\sin\theta\cos\theta$$

and in Fig. 8.49 we have $p_{\theta\theta}/p_0$ versus θ for various $\dot{a}/C_s$ values, where $p_0=(3\,E\,A)/(8(1+\nu))$. For $\dot{a}/C_s$ up to 0.6, the maximum is at $\theta=0$, but soon afterwards it moves away from the crack line and gives a pronounced peak by $\dot{a}/C_s=0.8$. Fig. 8.50 shows the angle of this maximum as a function of $\dot{a}/C_s$ and it can be seen to rise very rapidly for $\dot{a}/C_s>2/3$ up to about 30° and then more slowly thereafter.

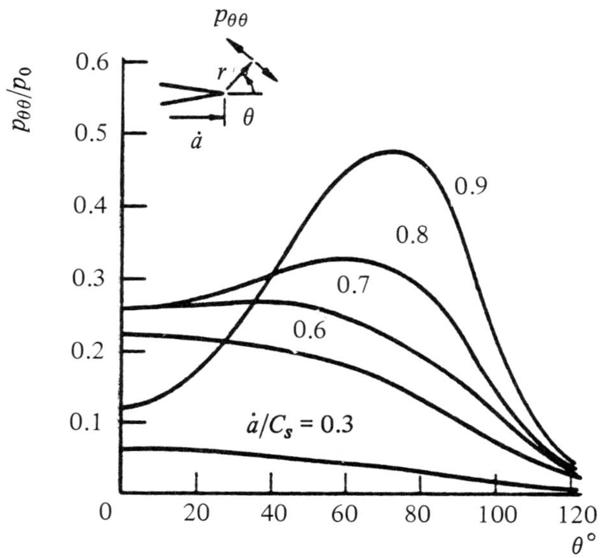

Fig. 8.49 Variation of hoop stress with angle at the tip of a running crack

8.7.3 *Experimental results*

Experimental results for high-speed cracks are scarce. This is principally because it is very difficult to measure loads and crack lengths for cracks moving at these high speeds over the short distances of conventional specimen sizes. Even when they are measured, the analysis of the data to yield $G_c(\dot{a})$ requires the ability to perform the necessary dynamic analysis to remove the inertia terms. The analysis described in preceding sections gives some idea of how this may be carried out, but there is usually considerable uncertainty present. Most results suggest a constant G_c value up to close to some limiting crack speed, after which there is a rapid rise. This may be genuine, but it would also occur if there were any shortcomings in the analysis. An additional problem is that bifurcation does occur at quite low speeds and at values as low as $0.3C_s$ surface roughening has been observed [20] which increases with $\dot{a}$. This may be regarded as small-scale bifurcation but it raises considerable problems about interpreting measured G_c values. A popular method of finding G_c is to measure the local stress field by using stress optical methods [21] and rather similar results are obtained. Since

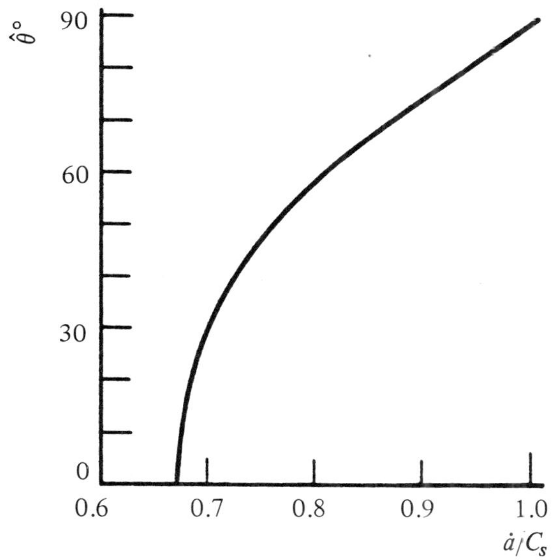

Fig. 8.50 Variation of hoop stress maximum angle with crack speed

the same analytical problems exist in interpreting the results, these do not provide much extra support.

The most reliable data are those for crack velocities and the results of Cotterell [20] are of particular interest. He reports crack speeds in PMMA ranging between 500 m/s and 900 m/s. There was a gradual increase in general surface roughness as the crack speed increased, as mentioned previously, but there was a distinct transition to bifurcations at 670 m/s with angles of about 20°. Since we may estimate C_d and C_s for most polymers using $E \simeq 3$ GPa, $\rho = 10^3$ kg/m^3 and $\nu = 0.35$, we have that $C_d \simeq 1700$ m/s and $C_s \simeq 1000$ m/s. The occurrence of bifurcation at 670 m/s is thus in good accord with the dynamic analysis given in Section 8.7.2, i.e. at $2/3C_s$. The higher speeds were obtained using grooved specimens to inhibit bifurcation and approach the Rayleigh wave speed of $0.93C_s$, again as discussed previously. The 20° bifurcation angle is also a reasonable expectation from Fig. 8.50. The limiting speed of $0.38c$, from [16], gives 650 m/s which is not a limiting speed in this case but does happen to correspond to the bifurcation condition.

Cotterell noted the onset of surface roughening at speeds as low as 200 m/s and the rise in K_c for epoxy [21] occurs at about 340 m/s, i.e. in the range 0.2 to $0.3C_s$. This condition probably corresponds to the onset of fractures initiating ahead of the main crack at higher K_c values. Finally, it should be noted that average speeds of about 340 m/s have been observed for axial cracks in PVC pipe [22] with peak values of up to 2000 m/s in a sinusoidal propagating path. The pressure used was about 1.5 times the initiation value, so the average speed is consistent with equation (8.54) ($n = 1.5$) and $\beta_0 \simeq 20$ as in the infinite plate case. The behaviour is rather complex, however, and illustrates the need for much further work in this area.

References for Chapter 8

[1] Glover, A. P., Johnson, F. A., Radon, J. C. & Turner, C. E., Paper 8, Conference on Dynamic Fracture Toughness, Welding Institute, London (1976).
[2] Timoshenko, S. P. & Goodier, J. N., *Theory of Elasticity*, 3rd edn, McGraw-Hill (1970) 420–422.
[3] Lamb, H., *Dynamics*, Cambridge University Press (1951) 123–127.
[4] Williams, J. G. & Birch, M. W., Proceedings of 4th International Conference on Fracture (ICF4), University of Waterloo, Canada, **1** (June 1977) 501–528.
[5] Plati, E. & Williams, J. G., *Polym. Eng. Sci.*, **15** (1975) 470–477.
[6] Plati, E. & Williams, J. G., *Polymer*, **16** (1975) 915–920.
[7] Kisbenyi, M., Birch, M. W., Hodgkinson, J. M. & Williams, J. G., *Polymer*, **20** (1979) 1289–1297.
[8] Newmann, L. V. & Williams, J. G., *Polym. Eng. Sci.*, **18** (1978) 893–899.
[9] Nikpur, K. & Williams, J. G., *Plastics & Rubber: Materials & Applications*, (November 1978) 163–166.
[10] Williams, J. G. & Hodgkinson, J. M., *Proc. R. Soc. Lond.*, **A375** (1981) 231–248.
[11] Hodgkinson, J. M., Savadori, A. & Williams, J. G., *J. Mat. Sci.*, **18** (1983) 2319–2336.
[12] Hodgkinson, J. M., Vlachos, N. S., Whitelaw, J. H. & Williams, J. G., *Proc. R. Soc. Lond.*, **A379** (1982) 133–144.
[13] Berry, J. P., *J. Mech. Phys. Solids*, **8** (1960) 194–206.
[14] Berry, J. P., *J. Mech. Phys. Solids*, **8** (1960) 207–216.
[15] Mott, N. F., *Engineering*, **165** (1948) 16–18.
[16] Roberts, P. K. & Wells, A. A., *Engineering*, **178** (1954) 820–821.
[17] Kolsky, H., *Stress Waves in Solids*, Clarendon Press, Oxford (1953).
[18] Yoffe, E. H., *Phil. Mag.*, **42** (1951) 739–750.
[19] Rice, J. R., in *Fracture: Volume II* (ed. H. Liebowitz), Academic Press (1968) 235–238.
[20] Cotterell, B., *Appl. Mat. Res.*, (October 1965) 227–232.
[21] Metcalf, J. T. & Kobayashi, T., in *Crack Arrest Methodology and Applications* (eds. G. T. Hahn & M. F. Kanninen), *ASTM STP 711* (1980) 128–145.
[22] Shannon, R. W. E. & Wells, A. A., *Int. J. Fracture*, **10** (1974) 471–486.

Index of Authors

Index